Supercritical Carbon Dioxide

ACS SYMPOSIUM SERIES **860**

Supercritical Carbon Dioxide

Separations and Processes

Aravamudan S. Gopalan, EDITOR
New Mexico State University

Chien M. Wai, EDITOR
University of Idaho

Hollie K. Jacobs, EDITOR
New Mexico State University

Sponsored by the
ACS Division of Industrial and Engineering Chemistry, Inc.

American Chemical Society, Washington, DC

Library of Congress Cataloging-in-Publication Data

Supercritical carbon dioxide : separations and Processes / Aravamudan S. Gopalan, editor, Chien M. Wai, editor, Hollie K. Jacobs,editor ; sponsored by the ACS Division of Industrial and Engineering Chemistry, Inc.

p. cm.—(ACS symposium series ; 860)

Includes bibliographical references and index.

ISBN 0–8412–3836–7

1. Liquid carbon dioxide. 2. Supercritical fluid extraction.

I. Gopalan, Aravamudan S., 1954- II. Wai, Chien M. III. Jacobs, Hollie K., 1962- IV. American Chemical Society. Division of Physical Chemistry. IV. American Chemical Society. Division of Industrial and Engineering Chemistry, Inc. V. Series.

TP244.C1S76   2003
660'.284248—dc21   2003052007

The paper used in this publication meets the minimum requirements of American National Standard for Information Sciences—Permanence of Paper for Printed Library Materials, ANSI Z39.48–1984.

PRINTED IN THE UNITED STATES OF AMERICA

Foreword

The ACS Symposium Series was first published in 1974 to provide a mechanism for publishing symposia quickly in book form. The purpose of the series is to publish timely, comprehensive books developed from ACS sponsored symposia based on current scientific research. Occasionally, books are developed from symposia sponsored by other organizations when the topic is of keen interest to the chemistry audience.

Before agreeing to publish a book, the proposed table of contents is reviewed for appropriate and comprehensive coverage and for interest to the audience. Some papers may be excluded to better focus the book; others may be added to provide comprehensiveness. When appropriate, overview or introductory chapters are added. Drafts of chapters are peer-reviewed prior to final acceptance or rejection, and manuscripts are prepared in camera-ready format.

As a rule, only original research papers and original review papers are included in the volumes. Verbatim reproductions of previously published papers are not accepted.

ACS Books Department

Contents

Organics

Analytical and Modeling Methods

Novel Materials

Reactions in Supercritical Carbon Dioxide

Indexes

Preface

Supercritical carbon dioxide has drawn increasing attention as an environmentally friendly alternative to traditional organic solvents. In addition, carbon dioxide is inexpensive, noncombustible, and relatively nontoxic. It is a potential alternative to organic solvents that are used in large-scale industrial processes. Supercritical carbon dioxide is a unique solvent/reaction media for chemical processes because small variations in pressure and/or temperature lead to differential solubility, reactivity, and selectivity.

The past decade has witnessed a tremendous growth in the study and applications of supercritical carbon dioxide. The studies range from theoretical modeling to a myriad of practical applications including environmental remediation. In addition, the use of microemulsions and surfactants has further enhanced the utility of supercritical carbon dioxide in reactions and separations, and has increased its attractiveness as a media for the synthesis of new materials.

This book is based on the results presented at a recent, very successful symposium titled *Separations and Processes Using Supercritical Carbon Dioxide* that was held at the 223rd American Chemical Society (ACS) National Meeting in Orlando, Florida, April 7–11, 2002. Although a number of books and reviews on supercritical carbon dioxide have been published, this book attempts to capture the recent excitement in this area. The reader will be impressed by the diversity of research and potential applications of supercritical carbon dioxide. This book will be useful for scientists and engineers with a wide variety of interests including separation technologies, materials science, and environmental remediation. Topics covered in this book include extraction/separation of transition metals, actinide ions, and organics. Other areas covered in the book include development of novel materials including nanoparticles, monitoring methods, molecular modeling, reactivity and catalysis, and formation/use of microemulsions and surfactants in supercritical carbon dioxide.

xi

Acknowledgements

We gratefully acknowledge the financial support of the Donors of The Petroleum Research Fund, administered by the ACS, and the Separation Science and Technology Subdivision of the ACS Division of Industrial and Engineering Chemistry, Inc. The ACS Books Department is thanked for their encouragement and support in the publication of this book. We thank all of the researchers, scientists, engineers, and students for their active participation in the symposia. Finally we thank the authors for their contributions and the reviewers for their time and valuable comments.

Aravamudan (Amudhu) S. Gopalan
Department of Chemistry and Biochemistry
New Mexico State University
Las Cruces, NM 88003
505–646–2589 (telephone)
505–646–2649 (fax)
agopalan@nmsu.edu

Chien M. Wai
Department of Chemistry
University of Idaho
Moscow, ID 83844–2343
208–885–6552 (telephone)
208–885–6173 (fax)
cwai@uidaho.edu

Hollie K. Jacobs
Department of Chemistry and Biochemistry
New Mexico State University
Las Cruces, NM 88003
505–646–5245 (telephone)
505–646–2649 (fax)
hjacobs@nmsu.edu

Supercritical Carbon Dioxide

Overview

1

Chapter 1

An Introduction to Separations and Processes Using Supercritical Carbon Dioxide

Chien M. Wai[1], Aravamudan S. Gopalan[2], and Hollie K. Jacobs[2]

[1]Department of Chemistry, University of Idaho, Moscow, ID 83844–2343
[2]Department of Chemistry and Biochemistry, New Mexico State University, Las Cruces, NM 88003

Introduction

Supercritical fluid carbon dioxide (SF-CO_2) with its moderate critical constants, nonflammable nature and low cost provides an attractive alternative for replacing organic solvents traditionally used in chemical manufacturing processes. Minimizing liquid waste generation, easy separation of solutes and fast reaction rates are some of the advantages of the supercritical fluid extraction (SFE) technology over conventional solvent extraction methods. The solvation strength of supercritical CO_2 can be tuned by changing the density of the fluid phase, thus selective extraction and dissolution of compounds may be achieved in SFE processes. Because CO_2 is not regulated by the US EPA as a volatile organic compound, CO_2-based SFE processes are environmentally acceptable. The CO_2 used in SFE is produced by other industrial processes (e.g. fermentation processes), therefore the gas vented into the atmosphere from the SFE processes does not contribute to the "green house" effect.

Liquid and supercritical CO_2 are effective solvents for dissolving non-polar and slightly polar organic compounds. Because CO_2 is a linear molecule with no dipole moment, SF-CO_2 is actually a poor solvent for dissolving polar compounds and ionic species. In the past decade, several new techniques have been developed for dissolving polar or ionic substances in supercritical CO_2. These new techniques have significantly expanded potential applications of supercritical fluid technology. One important technique developed in the past decade is the *in situ* chelation method for dissolving metal species as CO_2-soluble metal chelates in supercritical CO_2 (1,2). Another significant development is the use of water-in-CO_2 microemulsions for dispersing and transporting highly polar compounds and ionic species in supercritical CO_2 (3-5). These and other new SF-CO_2 techniques not only have found applications in environmental remediation and metal separation processes but also have led us to new frontiers of materials research. For example, metal chelates dissolved in supercritical CO_2 can be chemically reduced to the elemental state for metal deposition in the fluid phase, a process of considerable interest to the electronics industry (6,7). The unique properties of SF-CO_2 should enable the deposition of materials in small structures that are essential for developing minielectronic devices. Other examples such as synthesizing nanomaterials and utilizing metal nanoparticles for catalysis in SF-CO_2 may lead to advanced materials development and new chemical processes (8).

Using CO_2 as a solvent or a starting material for chemical synthesis and reactions has been a subject of extensive research since the early 90s (9,10). Some of the developments in this area have already led to new processes in chemical separations and manufacturing. Carbon dioxide-based dry cleaning techniques and synthesis of fluoropolymers in SF-CO_2 are examples of industrial applications of these new supercritical fluid-based techniques. Demonstrations for the remediation of toxic metals in solid wastes and for reprocessing of spent nuclear fuel in supercritical CO_2 have also been initiated recently. It is likely that research and development in CO_2-based technologies for chemical separations and materials processing will continue to expand in this decade. This chapter presents a review of some of the topics presented at the 223rd ACS National Meeting symposium entitled "Separations and Processes using Supercritical Carbon Dioxide" held in Orlando, Florida in April 2002.

Dissolution of Metal Species

Extraction of transition metal ions using a fluorinated dithiocarbamate chelating agent was first reported by Laintz et al. in 1992 (1). The extraction was based on an observation in the previous year that the fluorinated metal dithiocarbamates showed extremely high solubilities in supercritical CO_2 (by 2 to 3 orders of magnitude) relative to the non-fluorinated analogues (1). After the initial reports by Laintz et al., this *in situ* chelation/metal extraction technique quickly expanded to other metal ions particularly to lanthanides and actinides because of its potential applications in nuclear waste management.

Extraction of the f-block elements using different chelating agents in SF-CO$_2$ including fluorinated β-diketones and organophosphorus reagents were reported by different research groups (1). Soon, it was found that tri-n-butylphosphate (TBP) forms CO$_2$-soluble complexes with trivalent lanthanides and uranyl ions in the presence of nitrate ions or β-diketones. Particularly, the uranyl nitrate-TBP complex, UO$_2$(NO$_3$)$_2$(TBP)$_2$, was found to have a high solubility in supercritical CO$_2$, reaching a concentration of 0.45 mol/L in CO$_2$ at 40 $^{\circ}$C and 200 atm (1). This uranium complex has important implications in the nuclear industry because it plays an important role in the conventional PUREX (Plutonium Uranium Extraction) process for recovering uranium from spent nuclear fuel. Very recently, it was demonstrated that solid uranium dioxide could be dissolved directly in supercritical CO$_2$ using a CO$_2$-soluble TBP-nitric acid complex of the form TBP(HNO$_3$)$_x$(H$_2$O)$_y$ leading to the formation of UO$_2$(NO$_3$)$_2$(TBP)$_2$ (11). A demonstration of this technology for reprocessing spent nuclear fuels (the Super-DIREX process) is currently in progress in Japan based on the method described by Enokida et al. in Chapter 2. Other approaches of extracting uranium and plutonium using TBP and fluorinated β-diketones also appear promising as described by Lin and Smart (Chapter 3) and by Fox and Mincher (Chapter 4). Wenclawiak et al. investigated the structure and solubility of metal dithiocarbamate chelates in SF-CO$_2$ (Chapter 5). For selective removal of metals, Glennon and coworkers reported selective extraction of gold in SF-CO$_2$ using fluorinated molecular baskets and thiourea ligands (Chapter 6). In another metal related SFE technology demonstration, Kersch et al. reported a pilot-scale plant in Netherlands for ligand assisted extraction of toxic heavy metals from sewage fly ashes using Cyanex 302, a trimethylphenyl substituted monothiophosphinic acid, as an extractant (Chapter 7). Supercritical fluid technology for metal removal may find a number of industrial applications in the future including nuclear waste treatment and cleaning of semiconductor devices in metal deposition processes. These high-value product applications should provide economic incentives for developing supercritical fluid-based technologies by various industrial sectors.

Extraction of Organic Compounds

Selective extraction of organic compounds using green solvents is attractive from economic and environmental points of view. Carbon dioxide and water are two of the cheapest and most environmentally acceptable solvents on the earth. As explained in the introduction section, liquid and supercritical CO$_2$ are able to dissolve non-polar or slightly polar organic compounds. Water is an excellent solvent for ionic compounds because of its high dielectric constant (ε =78.5 at 25 $^{\circ}$C) which decreases with increasing temperature due to the weakening of hydrogen bonds between water molecules. Therefore, water at high temperatures also becomes effective for dissolving organic compounds. Utilizing the two green solvents (CO_2 and H_2O) in sequence may allow us to extract polar and non-polar organic compounds separately from plants or environmental samples without involving harmful solvents. Coupled processing methods using supercritical CO_2 and water or with a modifier for tailored isolation of specific organic compounds are new approaches for separation and processing of agricultural and natural products as described by King in Chapter 8. Selective extraction of natural products using supercritical CO_2 is illustrated by Mannila et al. for isolation of the active compound hyperforin from St. John's wort, an herb widely used for treatment of depression (Chapter 9). Using hot water for extraction of active ingredients from rosemary is described by Gan and Yang in Chapter 10. In environmental applications, Yu and Chiu described the removal of dispersed hydrophobic dye stuff from textile waste water using supercritical CO_2 (Chapter 11). The feasibility of dechlorination of polychlorinated biphenyls with a zero valent metal or bimetallic mixture in a supercritical CO_2 stream at elevated temperatures was demonstrated by Marshall and coworkers in Chapter 12. These chapters are just a few examples of green extraction techniques that may have important applications for environmental remediation and processing agricultural and natural products in the future.

Analytical and Modeling Methods

In the analytical area, a time resolved laser induced fluorescence (TRLIF) technique for on-line monitoring of uranium complexes in supercritical CO_2 was described by Addleman (Chapter 13). Park et al. reported solubility measurements in liquid and SF-CO_2 using a quartz crystal microbalance under sonication (Chapter 14). Recently, modeling supercritical fluid metal extraction processes using molecular dynamic simulations were initiated by Wipff and coworkers (Chapter 15). The molecular dynamic simulations provide valuable information on the migration of various relevant species involved in the supercritical fluid extraction process that can not be obtained by experimental methods. Modeling the solubility of β-diketones using the Peng-Robinson equation of state was discussed by Lubbers et al. (Chapter 16). Clifford et al. reported a thermodynamic modeling of the Baylis-Hillman reaction in SF-CO_2 (Chapter 17). Modeling supercritical processes in terms of thermodynamics, kinetics and molecular dynamics are important for understanding molecular interactions and mechanisms of solute-solvent interactions in the fluid phase.

Novel Materials Preparation

A historical perspective of solubility of materials in supercritical CO_2 was given by Eric Beckman at the symposium. Our current knowledge about the solubility of materials in supercritical CO_2 is apparently insufficient to precisely predict and design CO_2-philic compounds. Generally speaking, compounds containing fluorine, silicon, and phosphorus are considered CO_2-philic. However, a recent report by Wallen and coworkers showed that the acetylation of hydroxy groups in carbohydrates resulted in materials that were nearly CO_2-miscible (12). An understanding of the interactions of CO_2 with carbonyl compounds is given in Chapter 18 of this book.

Reverse micelles and microemulsions formed in supercritical CO_2 allow highly polar compounds and electrolytes to be dispersed in the non-polar fluid phase. Searching for CO_2-soluble surfactants that would form stable water-in-CO_2 microemulsions started a decade ago. A review of the design and performance of surfactants for making stable water-in-CO_2 microemulsions with all relevant references is presented by Eastoe in Chapter 19.

The rapid expansion of supercritical fluid (RESS) process was reported more than a decade ago by Smith and coworkers as a method of making fine particles (13). It is a physical process of expanding a high-pressure supercritical fluid solution rapidly through a restrictor or a nozzle turning solutes into aerosols in an evacuated container. The aerosol particles can be collected using a filter or deposited on a substrate. Sun and coworkers modified the RESS process by trapping silver nitrate particles in a solution containing the reducing agent sodium borohydride to make silver nanoparticles (Chapter 20). Sievers and coworkers developed a new process utilizing carbon dioxide assisted nebulization with a bubble dryer to micronize solutes to fine particles (Chapter 21). Potential applications of this process include fine powder generation and drug delivery. Stabilization of monodispersive nanoparticles in SF-CO_2 utilizing perfluorodecanethiol ligands is described by Shah et al (Chapter 22). Münüklü et al. have used the PGSS (particle from gas-saturated solution) process for the preparation of fat particles of different shapes (Chapter 23). Supercritical CO_2 was also used as a medium for preparing siloxane-based self-assembled monolayers inside nanoporous ceramic supports by Fryxell and coworkers at PNNL (Chapter 24). In the polymer synthesis area, Cooper et al. showed the use of SF-CO_2 and liquid 1,1,1,2-tetrafluoroethane as alternative solvents for the synthesis of crosslinked polymer materials with control over structural features (Chapter 25). Using SF-CO_2 as a medium for synthesizing novel materials particularly in nanomaterials preparation is likely to attract increasing research support from various industries and from government agencies in the future.

Chemical Reactions in Supercritical CO_2

Catalytic hydrogenations utilizing metal particles stabilized in supercritical CO_2 offer several advantages over conventional metal catalyzed hydrogenation conducted in solvents. In supercritical CO_2 systems, high concentrations of hydrogen can be easily introduced into the fluid phase because H_2 is miscible with CO_2. Faster reaction rates and easy separation of products are other advantages of conducting catalytic hydrogenations in supercritical CO_2. Using rhodium catalysts supported on fluoroacrylate copolymers, Flores et al. showed rapid hydrogenation of olefins in supercritical CO_2 (Chapter 26). Ohde et al. stabilized palladium and rhodium nanoparticles in a water-in-CO_2 microemulsion for catalytic hydrogenation of olefins and arenes in SF-CO_2 (Chapter 27). In these reactions, the microemulsion was formed using a mixture of perfluoropolyether phosphate (PFPE-PO_4) surfactant and a conventional sodium bis(2-ethylhexyl)sulfosuccinate (AOT) surfactant. Very rapid hydrogenation rates were observed probably due to dispersing of nanometer-sized metal catalyst by the CO_2 microemulsion in SF-CO_2. Dong and Erkey used a sodium salt of a fluorinated AOT surfactant to support a water-in-CO_2 microemulsion and studied hydroformylation of a number of olefins catalyzed by water soluble complexes (Chapter 28). These reports demonstrated that water-in-CO_2 microemulsions could be used as nanoreactors for catalysis in SF-CO_2. Reactions in SF-CO_2 will continue to be a main research topic in the foreseeable future because of the advantages this unique solvent system offers for developing green chemical processes.

Summary

The new techniques described in this introoudction chapter offer many opportunities for developing faster and more efficient chemical separations and processes using supercritical CO_2 that are environmentally sustainable. Nanomaterials synthesis, catalysis, and chemical manufacturing are some forseeable applications of the supercritical technology for the chemical industry. Remediation of organic and metal pollutants particularly nuclear waste management using supercritical CO_2 as a medium are potentially important for solving many environmental and energy related problems. Selective isolation of agricultural and natural products using supercritical CO_2 may greatly improve methods of processing agricultural products and benefit alternative medicine and health related industries. The chapters presented in this book only represent a few examples in these potential application areas. In summary, supercritical fluid CO_2 with its many unique properties is likely to play an important role in

8

leading us to challenge and develop new technologies that are needed for our societies in the future.

References

1. Wai, C.M. Metal Processing in Supercritical Carbon Dioxide in Supercritical Fluid Technology in Materials Science and Engineering, Sun, Y.P. ed., Marcel Dekker, **2002**, p. 351-386.
2. Darr, J.: Poliakoff, M. Chem. Rev. **1999**, 99, 495-541.
3. Johnston, K.P.; Harrison, K.L.; Clarke, M.J.; Howdle, S.M.; Heitz, M.P.; Bright, F.V.; Carlier, C.; Randolph, T.W. Science **1996**, 271, 624.
4. Yates, M.Z.; Apodaca, D.L.; Campbell, M.L.; Birnbaum, E.R.; McCleskey, T.M. Chem. Commun. **2001**, 25-26.
5. Ji, M.; Chen, X.: Wai, C.M.; Fulton, J.L. J. Amer. Chem. Soc. **1999**, 121, 2631-2632.
6. Watkins, J.J.; Blackburn, J.M.; McCarthy, T.J. Chem. Mater. **1999**, 11, 213-215.
7. Ye, X.; Wai, C.M.; Zhang, D.; Kranov, Y.; McIlory, D.N.; Lin, Y.; Engelhard, M. Chem. Mater. **2003**, 15, 83-91.
8. Ohde, H.; Wai, C.M.; Kim, H.; Kim, J.; Ohde, M. J. Amer. Chem. Soc. **2002**, 124, 4540-4541.
9. Jessup, P.G.; Ikariya, T.; Noyori, R. Nature **1994**, 368, 231.
10. Desimone, J.M.; Guan, Z.; Elshernd, C.S. Science **1992**, 257, 945.
11. Enokida, Y.; El-Fatah, S.A.; Wai, C.M. Ind. Eng. Chem. Res. **2002**, 41, 2282-2286.
12. Raveendran, P.; Wallen, S.L. J. Amer. Chem. Soc. **2002**, 124, 7274-7275.
13. Matson, D.W.; Fulton, J.L.; Petersen, R.C.; Smith, R.D. Ind. Eng. Chem. Res. **1987**, 26, 2298.

Metals

Chapter 2

Extraction of Uranium and Lanthanides from Their Oxides with a High-Pressure Mixture of TBP–HNO₃– H₂O–CO₂

Youichi Enokida[1], Ichiro Yamamoto[2], and Chien M. Wai[3]

[1]Research Center for Nuclear Materials Recycle and [2]Department of Nuclear Engineering, Nagoya University, Nagoya 4648603, Japan [3]Department of Chemistry, University of Idaho, Moscow, ID 83844–2343

The solubility of tri-n-butylphosphate (TBP) complex with nitric acid in supercritical CO_2 (SF-CO_2) is very high and $(HNO_3)_{1.8}(H_2O)_{0.6}$TBP is completely miscible in SF-CO_2 if the mixture is pressurized more than 14 MPa at 323 K. A mixture of 0.5 mol·dm^{-3} $UO_2(NO_3)_2$(TBP)$_2$ and SF-CO_2 is also miscible if the mixture is pressurized more than 13 MPa at 323 K. Uranium was extractable from the solid matrices of UO_2 with the high pressure solution of TBP-HNO₃-H₂O-CO₂ at 323 K and 13 MPa. Molecular ratios of HNO_3 to TBP in the high pressure mixture affected extraction rates and the higher ratio gave the higher extraction rates. Currently an application of supercritical fluid extraction to the nuclear fuel reprocessing is under investigation in Japan. Another promising field of application is treatment of uranium wastes. Decontamination factor of UO_2 or U_3O_8 was determined to be higher than 3×10^2 for the decontamination using the simulated uranium waste. The solid waste after the supercritical fluid leaching treatment was dry and contained low concentrations of TBP and HNO₃.

Uranium is a widely and dispersedly distributed element on the earth. Apart from its long-standing though small-scale use for coloring glass and ceramics, its only significant use is as a nuclear fuel (1). The nuclear fuel has a distinct feature that a spent fuel can be recycled if it is reprocessed properly (2-4). Therefore separation of uranium from the other elements is very important in the processes related to the nuclear fuel cycle. For this reason, separation of uranium as well as plutonium is indispensable. The conventional reprocessing process is known as the PUREX process, which involves a solvent extraction using tri-n-butylphosphate (TBP). The PUREX process is used commercially without almost any exception, because we can expect economical recovery of the products with a high purity. However, recently several alternative processes have been investigated in pursuit of a more economical process. The economical decline of the PUREX process was brought about mainly by increasing cost of waste management. An innovative uranium separation method is also expected for treatment of uranium wastes which arise from nuclear fuel fabrication. In these fields, avoidance of generating the secondary contaminated wastes is important point of view to develop new processes.

Lanthanide elements are used for miscellaneous purposes (1) and recently they are used as additives in fluorescence light tubes to improve their spectra. Additionally, recycling of fluorescence light tubes has just started because mercury is contained in the tubes. A new method of recovering lanthanides is required to improve economical aspect of the recycle.

Extraction of uranium and lanthanides from solid matrices is required in the industrial processes previously mentioned. Conventionally, leaching with strong acids such as nitric acid is widely used in these processes, but the treatment and recycling of waste acids are costly and their potential impact on the environment is a major concern. Supercritical fluid extraction (SFE) techniques have been reported for recovering uranium and lanthanides from their oxides and from environmental samples using fluorinated β-diketones and TBP as extractants (5-7). The reasons for developing SFE technologies are mostly due to the changing environmental regulations and increasing costs for disposal of conventional liquid solvents.

Supercritical fluids exhibit gas-like mass transfer rates and yet have liquid-like solvating capability. The high diffusivity and low viscosity of supercritical fluids enable them to penetrate and transport solutes from porous solid matrices. From this point of view, SFE is an ideal method to extract uranium and lanthanides from solid wastes. Carbon dioxide (CO_2) is most frequently used in SFE because of its moderate critical pressure (P_c) and temperature (T_c), inertness, low cost, and availability in pure form. Figure 1 illustrates moderate values of P_c and T_c compared with those of water.

This new extraction technology appears promising for effective processing with marked reduction in waste generation because no aqueous solutions and organic solvents are involved and phase-separation can be easily achieved by depressurization. Recently, the authors found a CO_2 soluble TBP complex with nitric acid was very effective for dissolution of UO_2 and U_3O_8 and extracted as $UO_2(NO_3)_2(TBP)_2$ in supercritical CO_2 (SF-CO_2)(8-11).

In this chapter, applicability of SFE of uranium and lanthanides using TBP complex with nitric acid in SF-CO_2 is explained and discussed.

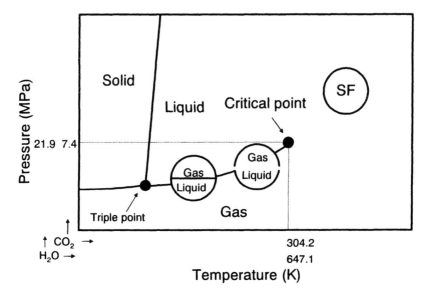

Figure 1. Schematic phase diagram of CO₂ and H₂O

Introduction of Acid into Supercritical CO₂

Extraction of metal compounds by SF-CO_2 is feasible when they are complexed with organic ligands (7). For example, nitrates of uranium and lanthanides are easily complexable with TBP and solubilities of the complex are very high in supercritical CO_2. This method is also applicable to hexavalent uranium extraction from solid UO_3 using TTA and TBP in SF-CO_2 (6).

Since UO_2 is the most common uranium oxide found in nuclear fuels, uranium extraction from UO_2 is indispensable in the nuclear fuel reprocessing and uranium waste treatment. From solid UO_2, however, uranium extraction is not possible with this method because TTA nor TBP form complexes with UO_2 directly. If we introduce acid homogeneously in SF-CO_2, we can expect UO_2 dissolution and conversion into UO_2^{2+}, which can be complexed with TBP by the charge neutralization as $TBP_2UO_2(NO_3)_2$. For example, concentrated nitric acid is widely used to dissolve UO_2 in aqueous solution forming $UO_2(NO_3)_2$.

Nitric acid is one of the three major acids of the modern chemical industry and has been known as a corrosive solvent for metals since alchemical times in the thirteenth century. The concentrated nitric acid is a mixture of HNO_3 and H_2O, and the phase diagram HNO_3-H_2O shows the presence of two hydrates, $HNO_3 \cdot H_2O$ and $HNO_3 \cdot 3H_2O$ (1).

Solubility of nitric acid in SF-CO_2 is low due to the weak solute-solvent interaction. However, if nitric acid is complexed with an organic ligand, it may become quite soluble in SF-CO_2. We have demonstrated this technique using a TBP complex of nitric acid (12).

In our experimental works, the certain volume of an anhydrous TBP (Koso Chemical, Japan) was contacted with different volumes of 60% HNO_3 (Wako Pure Chemicals, Japan) by shaking vigorously in a flask for 30 min followed by centrifugation to prepare an organic solution of TBP complex with nitric acid. Note that the prepared organic solution was a mixture of TBP, HNO_3 and H_2O and formed a single phase. The concentration of H_2O in the organic phase was measured by Karl-Fischer titration (Aquacounter AQ-7, Hiranuma, Japan). The concentration of HNO_3 in the organic phase was measured with a titrator (COM-450, Haranuma, Japan) with 0.1 molℓ^{-1} (M) NaOH solution after adding a large excess amount of deionized water.

Acidities of the TBP complex with nitric acid and equilibrated aqueous nitric acid are shown in Figure 2 for different phase ratios. In the PUREX process, acidity of 3 to 6 M is applied to dissolve spent fuels in aqueous solution. Figure 4 suggests the organic solution equilibrated with aqueous nitric acid of more than 6 M is easy to prepare, and this organic solution may dissolve UO_2.

Without SF-CO_2, we have tried to dissolve UO_2 powder with two kinds of TBP complexes with nitric acid under the normal pressure. The two composition can be described as TBP$(HNO_3)_{0.7}(H_2O)_{0.7}$ and TBP$(HNO_3)_{1.8}(H_2O)_{0.6}$. Although, with the former, UO_2 was not dissolved, with the latter well dissolved.

When we pressurized TBP$(HNO_3)_{0.7}(H_2O)_{0.7}$ with SF-CO_2, an aqueous phase appeared, and UO_2 was expected to be dissolved in the aqueous phase. The appearance of the aqueous phase can be understood as an antisolvent effect with SF-CO_2. For TBP$(HNO_3)_{1.8}(H_2O)_{0.6}$, however, no visible droplet was observed when the organic solution was mixed with SF-CO_2.

Figures 3 illustrate molecular ratios observed in the organic solution. These results suggest that there may be two different chemical structures for the organic solutions. In this chapter, we describe the TBP complex with nitric acid, TBP$(HNO_3)_x(H_2O)_y$ as TBP-HNO_3-H_2O. Similarly, TBP-HNO_3-H_2O-CO_2 denotes a high pressure mixture of TBP-HNO_3-H_2O and SF-CO_2.

Solubility of TBP Complex with Nitric Acid

We studied the solubility of TBP-HNO_3-H_2O in SF-CO_2 carefully. Figure 4 illustrates an experimental apparatus used for solubility measurements. This apparatus was prepared to operate up to 25 MPa in pressure and up to 373.15 K in temperature. Pressurized CO_2 (99.9% purity, Nippon Sanso) was introduced from a cylinder to the experimental apparatus via a syringe pump (Model 260D, ISCO) with a controller (Series D, ISCO). The temperature of the pressurized CO_2 was kept at 268.15 K through a heat exchanger placed in a cryogenic bath (CCA-1000, Eyela) before feeding to the pump. The equilibrium pressure in the cell was measured with a pressure gauge (NPG-350L, Nihonseimitsu). The internal volume of the cell is 60 cm^3. The temperature of the fluid in the cell was maintained constant with circulating water from a water bath (UA-10S, Eyela) within ±0.03 K. A magnetic stirrer (RCN-3, Eyela) was used for mixing the fluid in the cell.

14

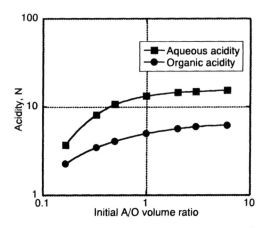

Figure 2. Acidities of TBP complex with nitric acid and equilibrated aqueous phase.

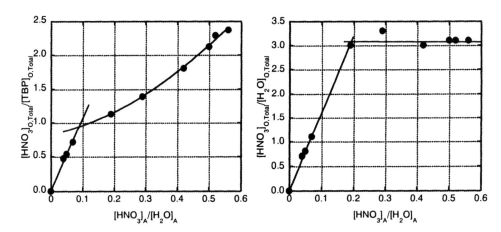

Figure 3.   Molecular ratio of (a)HNO₃ to TBP(left) and (b)HNO₃ to H₂O (right) in the TBP complex with nitric acid as a function of molecular ratio of nitric acid to water in equilibrated aqueous nitric acid.

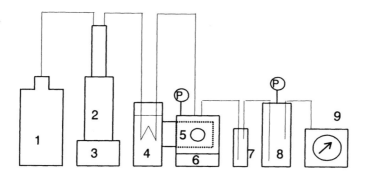

Figure 4. *A schematic diagram of the experimental apparatus for solubility measurement. 1; CO₂ cylinder, 2; syringe pump, 3; pump controller, 4; thermostated water bath, 5; view cell, 6; magnetic stirrer, 7; liquid trap, 8; gas reservoir, 9; gas volume meter.*

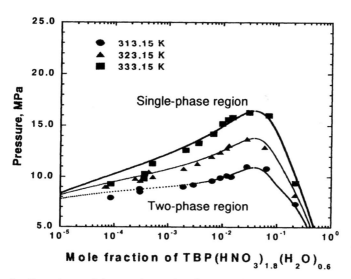

Figure 5. *Experimental data and correlated curves for transition pressures from two-phase to single-phase for a mixture of SF-CO₂ and TBP(HNO₃)₁.₈(H₂O)₀.₆ system*

For each run of the experiments, the gas in the whole line of the apparatus was replaced with CO_2 at the atmospheric pressure following a charge of a known amount of TBP-HNO_3-H_2O in the cell. The pressurized CO_2 was charged into the cell with the syringe pump and well mixed with the organic solution. In order to maintain the isothermal conditions, liquid CO_2 through the pump was heated through the heat exchanger placed in the water bath. The charging of CO_2 was continued until a single phase was formed in the view cell with pressure increase of the fluid in the cell. The solubilities of TBP-HNO_3-H_2O were determined with the bubble point method. The phase transitions resulting from pressure variations were performed and observed by direct visualization. Near the point of a single-phase formation for binary mixtures, an increment of charging pressure was set to a small value (less than 0.1 MPa), and kept the fluid in a half hour to confirm whether the fluid mixture was a single-phase equilibrium or not. A transition pressure was determined as the pressure at which the two-phase to single-phase transition took place. After it was confirmed a single-phase equilibrium, the pressure in the cell was lowered by opening the exit valve very slightly and the appearance of the two-phase mixture was checked. The total amount of the charged CO_2 gas was measured with a gas volume meter (DC-1C, Shinagawa) at a constant temperature.

We found that this organic complex is homogeneously miscible with SF-CO_2 with an arbitrary fraction when the mixture is pressurized more than 14 MPa at 323.15 K. This finding implies that we can introduce nitric acid in SF-CO_2 in order to dissolve metals including uranium and lanthanides. Figure 5 illustrates experimental data obtained for the transition pressure from two phases to a single phase at different temperatures. In this figure, the upper part is the single-phase region and the lower part is the two-phase region. There exists the maximum point of the locus of the transition pressure, with increasing the concentration of TBP complex near 0.03 mole fraction of TBP complex under isothermal condition. The similar phase behavior has been reported for TBP and CO_2 system (8). However, the maximum transition pressure reported for the mixture of TBP and CO_2 was 11 MPa at 323.15 K, which was lower than the value we obtained for the mixture of SF-CO_2 and TBP$(HNO_3)_{1.8}(H_2O)_{0.6}$. The critical opalescence was observed for this system; this means that there exists a pseudocritical point of this mixture where the system goes through a smooth change from two phase coexistence to a one phase state, without a true phase transition. The correlated curves plotted in Figure 5 were obtained by using the same approach described in the literature (14) and the detail is given in Ref. (16)

Uranium Extraction from Uranium Oxide with a High Pressure Mixture of TBP-HNO_3-H_2O-CO_2

Uranium extraction from UO_2 has been demonstrated using a high pressure mixture of TBP-HNO_3-H_2O-CO_2 (10). The UO_2 powder was prepared by grinding mechanically the UO_2 nuclear fuel pellet using a vibrating sample mill (TI-100,

Heiko Co., Japan). The UO_2 powder was fractionated using a sieve of 500 mesh and the grain size of the UO_2 powder was found to be 1–5 μm by microscopic observation. The O/U ratio of the UO_2 powder was determined to be 2.005±0.003 by spectrophotometry after the dissolution of the sample with strong phosphoric acid (13). The oxide powder of UO_2 (520 mg) was taken in the reaction vessel, and uranium was dissolved and extracted with TBP-HNO_3-H_2O-CO_2 dynamically by flowing SF-CO_2. Several runs of the dynamic extraction were repeated by changing pressure, temperature and composition of TBP-HNO_3-H_2O-CO_2. Samples were collected in a series of bottles during a dynamic extraction of 1 to 2 hours. Back extraction of uranium in the collected sample was performed by shaking the collected sample with deionized water. The uranium content was analyzed by Inductively Coupled Plasma Atomic Emission Spectroscopy (ICP-8000E, Shimazdu, Japan).

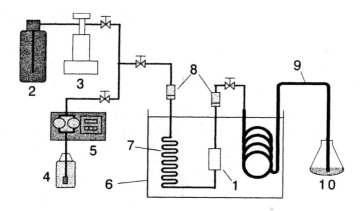

Figure 6. A schematic diagram of the experimental apparatus for dissolution and extraction experiment. 1; reaction vessel, 2; CO$_2$ cylinder, 3; syringe pump, 4; ligand container, 5; plunger-type pump, 6; thermostated water bath, 7; pre-heating coil, 8; filter, 9; restrictor, 10; collection vessel.

Figure 7 illustrates an example of dissolution curve for the UO_2 powder. The rate of the dissolution of UO_2 increased when the content of HNO_3 in the HNO_3-TBP complex was higher. A half life, t_h, defined as an extraction time where 50% of uranium is extracted was evaluated for each experiment, and a reciprocal number of t_h, which is proportional to the dissolution rate, was plotted in Figure 8. The correlated curve shows the dissolution rate is proportional to $([HNO_3]/[H_2O])^{1.3}$. This dependency is smaller than the value (2.3) reported for UO_2 dissolution in aqueous nitric acid (15), and it suggests the chemical mechanism of UO_2 dissolution in TBP-HNO_3-H_2O-CO_2 system is somewhat different from that in aqueous nitric acid.

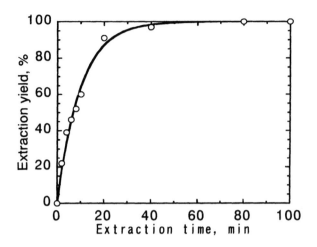

Figure 7 Dissolution curve for UO₂ with TBP(HNO₃)₁.₈(H₂O)₀.₆ at 13 MPa

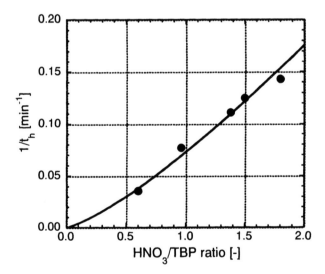

Figure 8. Correlation between the dissolution rate and the HNO₃/TBP ratio of the TBP complex with nitric acid

Solubility of TBP Complex with Uranyl Nitrate in SF-CO$_2$

Figure 9 illustrates experimental data obtained for the transition pressure from two phases to a single phase for SF-CO$_2$ and UO$_2$(NO$_3$)$_2$(TBP)$_2$ at 323.15 K In this figure, mole fraction was calculated based on TBP concentration and the upper part is the single-phase region and the lower part is the two-phase region. There exists the maximum point of the locus of the transition pressure, with increasing the concentration of TBP complex at near 0.03 mole fraction of TBP complex under isothermal conditions. The phase behavior is very similar to that observed for SF-CO$_2$ and TBP complex with nitric acid, and the maximum pressure for the transition was also 13 MPa, which is the same value observed in the mixture of SF-CO$_2$ and TBP(HNO$_3$)$_{1.8}$(H$_2$O)$_{0.6}$. This fact implies that UO$_2$(NO$_3$)$_2$(TBP)$_2$, which is a product of UO$_2$ or U$_3$O$_8$ dissolution with TBP(HNO$_3$)$_{1.8}$(H$_2$O)$_{0.6}$ in CO$_2$, does not precipitate after the dissolution if the system pressure is maintained at more than 13 MPa.

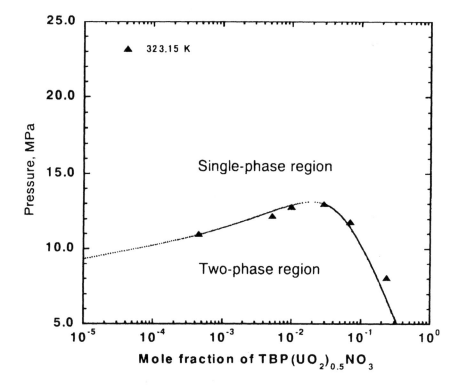

Figure 9. Experimental data and correlated curves for transition pressures from two-phase to single-phase for a mixture of SF-CO$_2$ and UO$_2$(NO$_3$)$_2$•2TBP system

Application to Nuclear Fuel Reprocessing and Uranium Waste Treatment

Applicability of SFE to nuclear fuel reprocessing has been proposed by Smart and Wai et al. (17, 18). We have developed a new process which employs a high pressure mixture of TBP-HNO$_3$-H$_2$O-CO$_2$ as is described in this chapter and this approach has indicated a very efficient extraction of uranium from UO$_2$. Now, the nuclear industries have paid attention to the applications of SFE to future processes. In Japan, we have started a four-year project with nuclear plant construction companies to demonstrate uranium and plutonium extraction from a mixed oxide fuel using the high pressure mixture. On the other hand, uranium and plutonium will be extracted from the irradiated nuclear fuel with TBP(HNO$_3$)$_{1.8}$(H$_2$O)$_{0.6}$ in the same project.

Another successful application under investigation is decontamination of the uranium waste, which is accumulating in nuclear fuel fabrication plants. Tomioka et al. demonstrated potential uranium decontamination for a simulated waste (8). In their experiments, the synthetic solid waste sample was prepared as a mixture of ca. 20 to 200 mg of UO$_2$ or U$_3$O$_8$ powders and 50 g of standard sea sand (20-30 mesh, Wako, Japan). The sample was placed in the reaction vessel of 50 cm^3. After a clean-up procedure with a high pressure mixture of TBP-HNO$_3$-H$_2$O-CO$_2$, UO$_2$ and U$_3$O$_8$ remaining on the sand sample were 0.3 mg (decontamination factor, DF=350) or 0.01 mg (DF=10000), respectively (8). This process is named as Supercritical Fluid Leaching (SFL). The solid waste after the SFL treatment was dry and contained low concentrations of TBP and HNO$_3$. One technical problem we have encountered so far is slow mass transfer of the TBP complex with UO$_2$(NO$_3$)$_2$ from the material surface into SF-CO$_2$. It was already demonstrated a chemical reaction from UO$_2$ to UO$_2$(NO$_3$)$_2$ with TBP-HNO$_3$-H$_2$O-CO$_2$ is very efficient, but (TBP)$_2$UO$_2$(NO$_3$)$_2$ seems very sticky on the surface of the sand etc. This may be solved by application of ultrasound. The authors verified the effectiveness of ultrasound application for uranium decontamination using TBP-HNO$_3$-H$_2$O-CO$_2$. Rate constants for uranium decontamination were more than ten times enlarged by applying ultrasound (11). This kind of assisting technology seems useful for in the future industrial processes.

Lanthanide Extraction with a High Pressure Mixture of TBP-HNO$_3$-H$_2$O-CO$_2$

Lanthanide metals were also successfully extracted from their oxides with a high pressure mixture of TBP-HNO$_3$-H$_2$O-CO$_2$ (12). We applied this technique to extract europium from the additives contained in fluorescence light tubes. The additives were recovered as white powder by crashing the glass tubes. An experimental study was performed with a similar apparatus as shown in Figure 6. The results of extraction of europium are given in Table I as well as the experimental conditions. In Ref. (12), the recovery yield approached 100% easily, but relatively low recoveries were obtained for the sample from the fluorescence light tubes. The major reason of the low recovery yields is due to insolubility of the solid matrix with nitric acid. A separate experiment with aqueous concentrated nitric acid (13.6 M) resulted in only 50% europium recovery from the powder. When we applied sulfuric acid to leach the powder, all powder dissolved in the aqueous solution and 100% of europium was recovered. Further study is required to recover lanthanides from the used fluorescence light tubes.

Table I. Extraction of Europium from the Crushed Powder of Fluorescence Light Tubes

Composition of TBP-HNO$_3$-H$_2$O	Recovery yield / %
TBP(HNO$_3$)$_{0.7}$(H$_2$O)$_{0.7}$	20.9 ± 0.1
TBP(HNO$_3$)$_{1.0}$(H$_2$O)$_{0.4}$	46.2 ± 0.1
TBP(HNO$_3$)$_{1.8}$(H$_2$O)$_{0.6}$	31.4 ± 0.1

Note: Extraction pressure: 15 MPa, temperature: 323 K, sample weight: 0.5 g, employed volume of TBP-HNO$_3$-H$_2$O: 20 cm^3, employed volume of liquid CO$_2$ (at 268 K): 100 cm^3, extraction time: 1 hour, extraction condition: dynamic extraction.

Acknowledgment

This work was partially granted by the Institute of Applied Energy, Japan. We acknowledge the assistance of Masataka Suzuki, Osamu Tomioka and Samir A. El-Fatah who performed SFE works.

References

1. Greenwood, N. N.; Earnshaw, A.; *Chemistry of the Elements;* Butterworth-Heinemann: Woburn, MA, 1997; p 1341.
2. Benedict, M.; Pigford T.; Levi, H.; *Nuclear Chemical Engineering;* McGraw-Hill Publishing: Columbus, OH, 1982; p 1008.
3. *The Nuclear Fuel Cycle;* Wilson, P. D., Ed.; Oxford Science Publications: Oxford, UK, 1996; p 323.
4. Cochran, R. G.; Tsoulfanidis, N.; Miller, W. F.; *Nuclear Fuel Cycle: Analysis and Management;* American Nuclear Society: La Grange Park, IL, 2000; p 382.
5. Lin, Y.; Brauer, R. D.; Laintz K. E.; Wai, C. M. *Anal. Chem.* **1993**, 65, 2549.
6. Lin, Y.; Wai, C. M.; Jean F. M.; Brauer, R. D. *Environ. Sci. Technol.* **1994**, 28, 1190-1193.
7. Wai C. M.; Wang, S. *J. Chromatography A* **1997**, 785, 369-383.
8. Tomioka, O; Meguro, Y.; Iso, S.; Yoshida, Z.; Enokida Y.; Yamamoto, I. *J. Nucl. Sci. Technol.* **2001**, 38, 461-462.

9. Samsonov, M. D.; Wai, C. M.; Lee, S. C.; Kulyako, Y.; Smart, N. G. *Chem. Commun.* **2001**, 1868-1869.
10. Tomioka, O.; Meguro, Y.; Yoshida, Z.; Enokida Y.; Yamamoto, I. *J. Nucl. Sci. Technol.* **2001**, 38, 1097-1102.
11. Enokida, Y.; El-Fatah, S. A.; Wai, C. M. *Ind. Eng. Chem. Res.* **2002**, 41, 2282-2286.
12. Tomioka, O.; Enokida, Y.; Yamamoto, I. *J. Nucl. Sci. Technol.* **1998**, 35, 513-514.
13. Kihara, S.; Adachi, T.; Hashitani, H. *Fresenius Z. Anal. Chem.***1980**, 303, 28-30.
14. Joung S. N. et al. *J. Chem. Eng. Data* **1999**, 44, 1034-1040.
15. Asano, Y.; Kataoka, M.; Tomiyasu, H.; Ikeda, Y. *J. Nucl. Sci. Technol.* **1996**, 33, 152-162.
16. Enokida, Y; Yamamoto, I. *J. Nucl. Sci. Technol.* **2002**, in press.
17. Smart, N. G.; Wai, C. M.; Phelps C. *Chem. Britain.* **1998**, 34, 34-36.
18. Wai, C. M. in *Hazardous and Radioactive Waste Treatment Technologies Handbook;* Oh, C. H., Ed.; 5.1; CRC Press: Boca Raton, FL, 2001, pp. 5.1-3-23.

Chapter 3

Supercritical Fluid Extraction of Actinides and Heavy Metals for Environmental Cleanup: A Process Development Perspective

Yuehe Lin[1] and Neil G. Smart[2]

[1]Pacific Northwest National Laboratory, Richland, WA 99352
[2]BNFL, Sellafield, Cumbria CA14 1XQ, United Kingdom

The extraction of heavy metal ions and actinide ions is demonstrated using supercritical carbon dioxide (CO_2) containing dissolved protonated ligands, such as β-diketones and organophosphinic acids. High efficiency extraction is observed. The mechanism of the extraction reaction is discussed and, in particular, the effect of addition of water to the sample matrix is highlighted. In-process dissociation of metal-ligand complexes for ligand regeneration and recycle is also discussed. A general concept for a process using this technology is outlined.

Introduction

Supercritical fluids (SFs) are being used increasingly as extraction solvents due to increased restrictions on the use of traditional solvents, particularly chlorinated solvents, that are a result of environmental legislation. Carbon dioxide (CO_2) has been the solvent of choice due to its low toxicity, relatively low cost, convenient critical properties, and ease of recycling. Supercritical (SC) CO_2 has been found to be a particularly useful solvent for extraction of organic compounds at both analytical and process scales (1). Direct extraction of metal ions by supercritical CO_2 is highly inefficient because of the charge neutralization requirement and the weak solute-solvent interaction. However, when metal ions are chelated with organic ligands, they become quite soluble in SC CO_2 *(2-6)*. Conversion of metal ions into metal chelates can be performed by two methods. One method is on-line chelation, which is performed by first dissolving ligands into SC CO_2 which is then directed through the sample matrix. Another method is *in situ* chelation, which involves adding the ligands directly to the sample matrix prior to the supercritical fluid extraction (SFE). Both methods have been found to be successful for metal ion extraction using SC CO_2 *(2-24)*.

According to the literature *(2-15)*, the following factors are important in determining the effectiveness of metal extraction into supercritical CO_2: (1) solubility of the chelating agent in SC CO_2, (2) solubility and chemical stability of the metal chelate in SC CO_2, (3) density of the SC CO_2, (4) chemical form of the metal species, (5) water content and pH of samples, and (6) sample matrix. SFE studies using a variety of organic chelating agents for extraction of metal ions have achieved efficient extraction. It has been observed that the introduction of fluorine-substituted groups to the ligands *(2-8,10-13)* or using synergistic two-ligand systems *(6-8,10-13)* can enhance the efficiencies of extraction into supercritical fluids. The method of this enhancement effect is thought to be three fold: (1) increased solubility of the fluorine-substituted ligands and metal complexes in SC CO_2 over non-fluorine-substituted analogues, (2) increased solubility and chemical stablity of metal complexes due to adduct formation, and (3) the ability of the fluorine substituted ligands and synergistic two-ligand systems to extract metal ions from matrices with relatively acidic conditions.

In this chapter, we outline the current understanding in the field of supercritical fluid extraction of metal ions. Of particular interest is the extraction of heavy metals and actinides from environmental matrices, which may impact human health and the environment. This discussion will include an overview of the fundamental chemistry, examples of the extraction of actinide and heavy metals, and an outline of the proposed industrial process.

The Molecular Concept of Supercritical Fluid Extraction of Metals

Consideration of SFE processes must start at the molecular level, to understand the parameters that are most important to the extraction process. As a simple case, consider the extraction of metal ions via complexation with a protonated ligand into a SF. The following reaction series is illustrative of such a process at the molecular level:

$$HL_{(S)} + SF \quad \leftrightarrows \quad HL_{(SF)} \qquad (1)$$
Ligand (HL) Dissolution

$$xHL_{(SF)} + M^{x+} \quad \leftrightarrows \quad ML_{(x)(S)} + H^+ \qquad (2)$$
Metal ion complexation

$$ML_{(x)(S)} + SF \quad \leftrightarrows \quad ML_{(x)(SF)} \qquad (3)$$
Metal complex dissolution in SF

Experimental results, with both actinides and heavy metals, have demonstrated that the presence of water aids the above reaction scheme *(6, 15)*. It is believed that the water is required to hydrate the metal ion prior to complexation and also aids in deprotonation of the acidic ligand. In SFE from environmental matrices, the presence of water may also aid the breaking of bonds to the binding sites on matrices such as soil, clay and other minerals present in the matrix. This overall reaction scheme is illustrated conceptually in Figure 1.

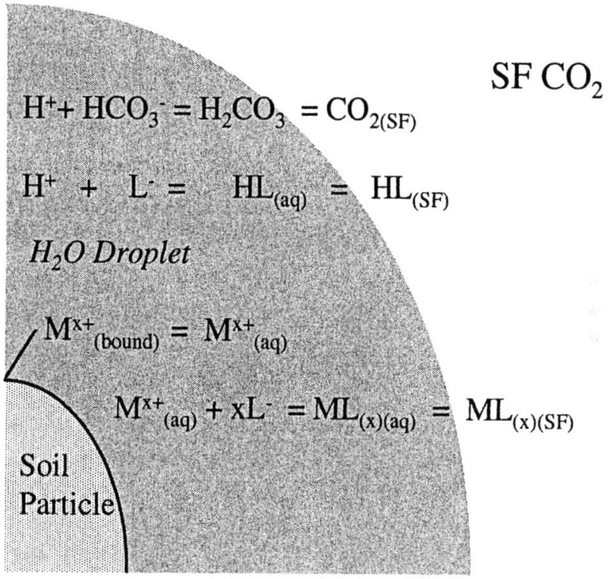

Figure 1. Molecular reaction scheme for metal extraction into SF using protonated ligands.

Examples of this reaction scheme have been demonstrated for extraction of actinides using the β-diketone ligand family and for heavy metals using organophosphinic acid ligand family (15). Extraction of heavy metals is illustrated for the range of organophosphinic acid ligands listed in Table 1. Table 1 shows that SFE of metals using the Cyanex 301, Cyanex 302 and D2EHTPA ligands is highly effective.

Table 1. Percent Extracted (% E) and Collected (% Coll.) for Cu^{2+}, Pb^{2+}, Zn^{2+} and Cd^{2+} Ions from a Cellulose Support using SC CO_2

Ligand	Cu^{2+} % E	Cu^{2+} % Coll.	Pb^{2+} % E	Pb^{2+} % Coll.	Zn^{2+} % E	Zn^{2+} % Coll.	Cd^{2+} % E	Cd^{2+} % Coll
None	2	2	2	0	3	2	1	0
Cyanex 301	82	81	100	85	100	89	95	79
Cyanex 302	98	93	60	56	32	50	85	77
Cyanex 272	50	41	52	11	81	77	30	22
D2EHTPA	98	100	100	89	97	76	99	90

Conditions: 200 atm; 60 °C A. 350 mg portion of ligand was used in each case. The results represent the mean value obtained from duplicate runs under 10 min

To demonstrate the chelation/SFE technology for environmental remediation applications, extraction of uranium-contaminated soil samples collected from a spent uranium mine in the northwest United States was investigated using the synergistic ligand system of hexafluoroacetylacetone (HFA) plus trioctylphosphine oxide (TOPO). The experimental procedures used for uranium extraction from soil and Neutron Activation Analysis (NAA) of uranium in soil were the same as reported previously (4,7). Results obtained from this study, summarized in Table 2, were promising. With 10 minutes of static extraction followed by 20 minutes of dynamic extraction, 80-88% uranium was extracted from the wet soil samples. A significant increase in extraction effectiveness was observed for the mine soil samples when wet as compared to dry samples, supporting the above reaction mechanism that uses water to enhance ligand deprotonation and release of metal ions from mineral binding sites. A small amount of water might also dissolve in SC CO_2 and served as a solvent modifier to increase the solubility of metal chelates.

Table 2. Extraction of Uranium from Contaminated Soils with Supercritical CO_2 (5% methanol) Containing HFA and TOPO at 60°C and 200 atm

Sample*	Initial U Conc. (mg/g)	Extracted **(%)	
		Dry	Wet
#1	0.14	61	88
#2	0.48	57	80
#3	0.66	54	81

* Soil samples collected from different locations in a spent U mine.

** Average data of three runs

Conditions: 100 mg soil was used for extraction.; 100 μL HFA + 50 mg TOPO; 10 min static extraction followed by 20 min dynamic extraction for each sample. In "dry" condition, no water was added. In "wet" condition, 10 μL water was added to the soil sample prior to experiment.

The Chemical Concept of Supercritical Fluid Extraction of Metals

The extraction of metals using SFE presents a number of difficulties not found when extracting organic compounds. The SF must be able to dissolve both ligand and ligand-metal complexes in quantities sufficient to achieve extraction in a reasonable time scale. The ligand must be able to access and react with the metal within the sample matrix. Finally, once dissolved in the SF, the metal complex must not decompose significantly or metal will be lost from the SF. These processes are illustrated in Figure 2.

SFE processes typically involve mass transfer between a supercritical fluid, such as CO_2, and a solid or liquid phase matrix under conditions of high pressure and temperature. Slight changes in temperature or pressure of the system can cause large changes in the density of the solvent and consequently the solvent's ability to dissolve heavy, nonvolatile compounds from the sample matrix. Nonvolatile compounds can be extracted by proper manipulation of the system pressure. A reduction in pressure, generally to a pressure below the solvent critical pressure, results in the complete precipitation of the solute.

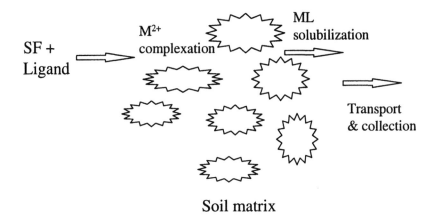

Figure 2. Schematic of processes involved in the SFE of metals from solid environmental matrices

Example results for the extraction of heavy metals from environmental matrices into supercritical CO_2 are shown in Figures 3a and 3b. A heavy metal-contaminated soil (primarily lead and zinc contamination) obtained from mine waste produced in the Cour D'Alene, Idaho mining district was studied. The samples were subjected to repeated extraction cycles using conditions of 200 atm, 300 atm, or 400 atm, all at 60°C. Approximately 0.5 g of sample was extracted in each case and each extraction was allowed to proceed for 20 minutes static and 20 minutes dynamic. Extraction at higher pressure significantly increases the amount of both lead and zinc that is recovered. Initial extraction cycles remove relatively large amounts of metal, while later cycles extract much less. This pattern indicates that the extraction is limited by the rate of reaction of the ligand with the metal rather than the solubility of the complexed metal in the SC CO_2. Extraction profiles of this type have been observed for the removal of organic compounds from environmental matrices, such as soils. After the final SFE cycle, the samples were subjected to an equivalent of the US EPA TCLP procedure *(25)*. The amount of leachable metal is greatly reduced as compared to untreated samples, indicating that the metals removed using the SFE process are those with the greatest potential for leaching into aqueous streams. At conditions of 400 atm and 60°C, the concentration of lead in a TCLP leachate from SFE-treated samples was 12.4 ppm, which is close to the regulatory standard of 10 ppm.

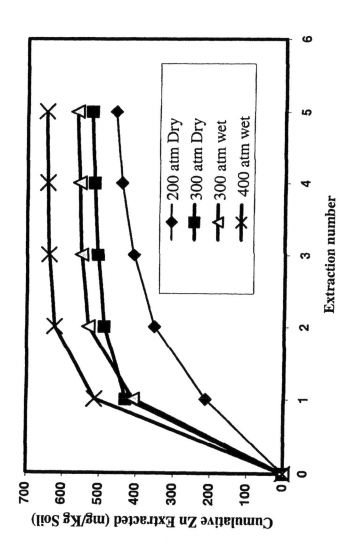

Figure 3a. Cumulative mass of Zn extracted from contaminated soils using SC CO$_2$ with Cyanex 302 ligand at 60°C and varying pressures and moisture additions. A quantity of 0.5 g of Cyanex was added per extraction cycle. 'Wet' samples were obtained by addition of 10 µL of water to the soil matrix prior to first extraction.

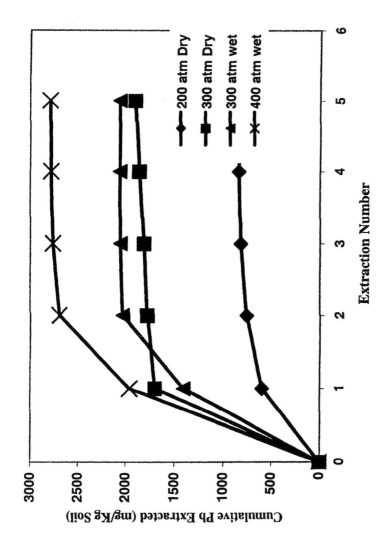

Figure 3b. Cumulative Pb extracted from contaminated soils using SC CO$_2$ with Cyanex 302 ligand at 60°C and varying pressures and moisture additions. A quantity of 0.5 g of Cyanex was added per extraction cycle. 'Wet' samples were obtained by addition of 10 μL of water to the soil matrix prior to first extraction.

In-Process Dissociation of Metal-Ligand Complexes and Regeneration/ Recycle of Ligand

For industrial-scale application of chelation/SFE technology, the ability to recycle the solvent (e.g., CO_2) and the ligand is important for reducing operational costs. In-process dissociation of extracted metal-ligand complexes can regenerate the ligands. Subsequent recycle of the CO_2 solvent and dissolved ligands can be performed without depressurizing the system. In-process regeneration and recycle can make the extraction process more convenient and economic. Recently, we developed two technologies for dissociation of metal-ligand complexes and regeneration of ligand (26,27), both of which have been demonstrated to be highly effective for on-line dissociation of metal-ligand complexes.

The first method causes dissociation of extracted metal-ligand complexes in the SF by reduction of the metal-ligand complex using hydrogen in a specially-designed reactor (26). The corresponding chemical equation is:

$$ML_{x\,(SF)} + H_2 \xrightarrow{\text{heat}} M\,(s) + x\,HL_{(SF)}$$

During the dissociation of the metal-ligand complexes, metal was deposited on a heated plate and the ligand remained in the SF for subsequent recycle.

In the second method, extracted metal-ligand complexes in SF were dissociated in a back-extraction device containing nitric acid solution by a reverse reaction of chelation/extraction (13,27), which is generically shown as follows:

$$ML_{x\,(SF)} + x\,H^+ \rightarrow M^{x+}_{(aq)} + x\,HL_{(SF)}$$

A process flow diagram of the SFE system with an on-line back-extraction device for ligand regeneration is shown in Figure 4. The liquid back-extraction vessel was modified from a commercial SFE cell (Dionex, 1.0 cm i.d. and 13 cm length) that has a volume of 10 mL. The 1/16-inch stainless steel inlet tubing on the cell was extended to the bottom of the vessel cavity, thereby forcing the supercritical CO_2 to flow through the nitric acid solution in the vessel before exiting through the outlet tubing at the top. Metal complexes in the SF were decomposed upon contact with the nitric acid solution. Metal ions were trapped in the nitric acid solution and ligands remained dissolved in the SF phase, which can subsequently be recycled.

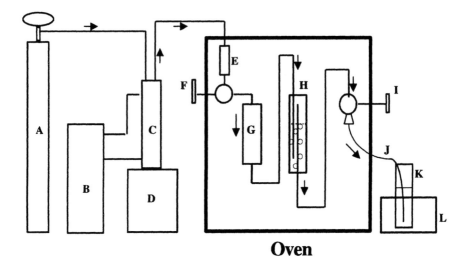

Oven

*Figure 4. SFE system with an on-line back-extraction device.
(A) CO₂ cylinder; (B) circulator coolant; (C) liquid CO₂ reservoir; (D) Haskel
pump; (E) temperature equilibration device; (F) inlet valve; (G) extraction
vessel; (H) liquid vessel for on-line back-extraction; (I) outlet valve; (J)
restrictor; (K) collection vessel; (L) water bath*

The Full-Scale Process Concept

Although the SFE/ligand technology has only been tested at small scale for the extraction of metal ions, it is useful to consider how the full-scale process would be configured. This provides an understanding of the information required to advance this new technology area. One possible configuration for a process utilizing supercritical fluid technology to extract metals from environmental matrices is illustrated in Figure 5.

In this conceptual process, the main process vessel is packed with contaminated material (i.e., soil). The supercritical fluid, containing dissolved ligands with affinity for the metals of interest, would pass through the soil. Metal ions would be extracted into the SF as a metal-ligand complex and subsequently transported into a metal-ligand separation vessel. In the separation vessel, the metal ion would be removed from the system by decomplexation, thereby regenerating the used ligand.

The fundamental issue is that the following change is achieved:

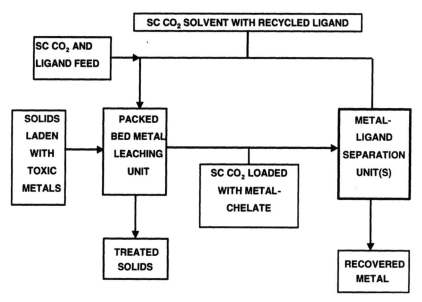

Figure 5. Conceptual process diagram for a SFE system to remove metal ions from environmental matrices.

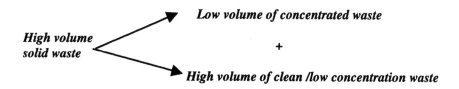

Figure 6. Overall treatment concept for SFE of metal-contaminated soil.

The next step towards successful development of the SF extraction process is to move from small-scale investigations to bench scale feasibility studies. Bench-scale data will provide information to evaluate process design issues and to establish the utility of this technology. Key components of this technology that must be evaluated are: (a) ligand solubility in SC CO_2, (b) metal ion extraction into SC CO_2, (c) metal chelate solubility in SC CO_2, and (d) metal ion recovery from a metal-laden SC CO_2 phase. The overall goal of the SFE process is the concentration of the contaminants, as depicted in Figure 6.

Conclusions

The mechanism of extraction of metal ions such as actinides and heavy metals is well understood at the molecular level. For extraction of metals from environmental systems, water content is found to be an important parameter in the reaction scheme. Extraction from soils and other solid matrices was demonstrated at a small scale with similar efficiency to that of the US EPA TCLP test. The main areas for future development in this technology are the issues surrounding process scale up.

Acknowledgements

The Authors would like to thank Professor C.M. Wai at University of Idaho for helpful discussion and Mr. Christian Johnson at PNNL for reviewing the manuscript. Pacific Northwest National Laboratory (PNNL) is operated by Battelle Memorial Institute for the US DOE under Contract DE-AC06-76RLO 1830.

Refecences

1. Hawthorne, S.B. *Anal. Chem.*, **1990**, *62*, 633A
2. Laintz, K.E.; Wai, C.M.; Yonker, C.R.; and Smith, R.D. *J. of Supercritical Fluids*, **1991**, *4*,194-198
3. Laintz, K.E.; Wai, C.M.; Yonker, C.R.; Smith, R.D. *Anal Chem.* **1992**, *22*, 2875
4. Lin, Y.; Brauer, R.D.; Laintz, K.E.; Wai, C.M. *Anal. Chem.*, **1993**, *65*, 2549-2551
5. Wai, C.M.; Lin, Y.; Brauer, R.D.; Wang, S.; Beckert, W.F. *Talanta*, **1993**, *40*,1325-30.
6. Lin, Y.; Smart, N.G.; Wai, C.M. *Trends Anal. Chem.* **1995**; *14(3)*:123-33

7. Lin, Y.; Wai, C.M.; Jean, F.M.; Brauer, R.D. *Environ. Sci. Technol.* **1994**, *28*, 1190-1193.

8. Lin, Y.; C.M. Wai, *Anal. Chem.* **1994**,*66*, 1971-1975

9. Lin, Y.; Smart, N.G.; Wai, C.M. *Environ. Sci. Technol.*, **1995**, *29*, 2706-2711.

10. Wai, C.M.; Lin, Y.; Ji, M.; Toews, K.L.; Smart, N.G. "Extraction and Separation of Uranium and Lanthanides with Supercritical Fluids," Chapter 23, pp 390-400 in "*Metal Ion Separation and Preconcentration: Progress and Opportunities*"; A. H. Bond and M.L. Dietz; Eds; ACS Symposium Series 716; American Chemical Society: Washington, DC, 1998; Oxford University Press; 1999.

11. Lin, Y.; Wu, H.; Smart, N.G.; Wai, C.M. *J. Chromatogr. A*, **1998**, *793*, 107-113.

12. Lin, Y.; Wu, H.; Smart, N.G.; Wai, C.M. *Talanta*, **2000**, *52*, 695-701

13. Lin, Y.; Wu, H.; Smart, N.G.; Wai, C.M. *Sep. Sci. Tech.* **2001**, *36*, 1149-1162

14. Elshani, S.; Smart, N.G.; Lin, Y.; Wai, C.M. *Sep. Sci. Tech.* **2001**,*36*,1197-1210

15. Smart, N.G.; Carleson, T.E.; Elshani, S.; Wang, S.; and Wai, C.M. *Ind. Eng. Chem. Res.* **1997**, *36*, 1819-1826.

16. Clifford, A.A.; Zhu, S.; Smart, N.G.; Lin, Y.; Wai, C.M.; Yoshida, Z.; Meguro, Y.; Iso, S. *J. Nucl. Sci. Technol.* **2001**, *38*, 433-438

17. Addleman, R.S.; Carrott, M.J.; Wai, C.M. *Anal. Chem.* **2000**, *72*, 4015-4021

18. Wai, C.M.; Waller,B. *Ind. Eng. Chem. Res.* **2000**, *39*, 4837-4841

19. Erkey, C. *J. Supercritical fluids*, **2000**, *17*, 259-287

20. Li, J.; Beckman, E.J.; *Ind. Eng. Chem. Res.* **1998**, *37*, 4768

21. Mincher, B.J.; Fox, R.V.; Holmes, R.G.G; Robbins, R.A.; Boardman, C. *Radiochim. Acta*, **2001**, *89*, 613-617

22. Kersch, C.; van Roosmalen, M.J.E.; Woerlee, G.F.; Witkamp, G.J. *Ind. Eng. Chem. Res.* **2000**, *39*, 4670-4672

23. Roggeman, E.J.; Scurto, A.M.; Brennecke, J.F. Ind. Eng. Chem. Res. **2001**, 40, 980-989

24. Tomioka, O.; Enokida, Y.; Yamamoto. I. *Sep. Sci. Tech.* **2002**, *37*, 1153-1162

25. U.S. Environmental Protection Agency, Toxicity Characteristics Leaching Procedure (TCLP), Vol. 1C, Ch.8, Sec. 8.4, 7/92, Method 1311

26. Wai, C.M.; Hunt, F.H.; Smart, N.G.; Lin, Y. *U.S. Patent* No 6,132,491, 2000

27. Wai, C.M.; Smart, N.G.; Lin, Y. *U.S. Patent* No 5,792,357, 1998

Chapter 4

Supercritical Fluid Extraction of Plutonium and Americium from Soil Using β-Diketone and Tributyl Phosphate Complexants

Robert V. Fox and Bruce J. Mincher

Idaho National Engineering and Environmental Laboratory,
Idaho Falls, ID 83415

Supercritical fluid extraction (SFE) of plutonium and americium from soil was demonstrated using supercritical fluid carbon dioxide solvent augmented with organophosphorus and beta-diketone complexants. Soil from a radioactive waste management site in Idaho was spiked with plutonium and americium, chemically and radiologically characterized, then extracted with supercritical fluid carbon dioxide at 2,900 psi and 65 °C. The organophosphorus reagent tributyl phosphate (TBP) and the beta-diketone thenoyltrifluoroacetone (TTA) were added to the supercritical fluid as complexing agents. A single 45 minute SFE with 2.7 mol% TBP and 3.2 mol% TTA provided as much as 88% ± 6 extraction of americium and 69% ± 5 extraction of plutonium. Use of 5.3 mol% TBP with 6.8 mol% of the more acidic beta-diketone hexafluoroacetylacetone (HFA) provided 95% ± 3 extraction of americium and 83% ± 5 extraction of plutonium in a single 45 minute SFE at 3,750 psi and 95 °C. Sequential chemical extraction techniques were used to chemically characterize soil partitioning of plutonium and americium in pre-SFE soil samples. Sequential chemical extraction techniques demonstrated that spiked plutonium resides primarily (69% ± 2) in the sesquioxide fraction with minor amounts being absorbed by the oxidizable fraction (18% ± 1) and residual fractions (8% ± 1). Post-SFE soils subjected to sequential chemical extraction characterization demonstrated that 97% ± 3 of the oxidizable, 79% ± 2 of the sesquioxide and 80% ± 2 of the residual plutonium could be removed using SFE. These preliminary results show that SFE may be an effective solvent extraction technique for removal of actinide contaminants from soil.

Introduction

The United States Department of Energy (DOE) manages approximately 1.9 billion cubic meters of radionuclide contaminated environmental media and 4.1 million cubic meters of stored, contaminated waste at 150 different sites located in 30 different states (*1, 2*). This environmental legacy is a result of the massive industrial complex responsible for defense related and non-defense related research, development and testing of nuclear weapons, nuclear propulsion systems and commercial nuclear power systems. Cleaning up the environmental legacy is expected to cost several hundred billion dollars over the next 5 to 7 decades. To reduce costs and speed remediation efforts the DOE has invested in waste treatment and environmental remediation research.

Remediation of solid and liquid environmental media contaminated with actinides and fission products is a challenging task. The challenge calls for ultimately rendering the contaminated sites safe. One strategy is removal of contaminants from solid and liquid media to concentrations which are below release criteria; preferably without generation of considerable secondary waste streams and without denaturing the media such that it can be returned to the environment. Other factors which must be considered include cost, safety, long-term impact and public acceptance. Physical and chemical separation techniques for removal of actinides and fission products from contaminated soil and water have met with varying success (*3–6*). The most promising *in-situ* soil treatment techniques focus on sequestration of radionuclides and methods for keeping surface water and groundwater away from the source term (e.g., barriers, grouting, etc.). *In-situ* treatment of radionuclide contaminated groundwater and aquifer zones has been approached through use of selective barriers or selective adsorbents which pass water but sequester metals. *Ex-situ* treatment of radionuclide contaminated water primarily involves pump-and-treat, where treatment can be any number of chemical or physical methods which remove radionuclides from water (e.g., chemical precipitation, ion exchange, adsorption, filtration, etc.). *Ex-situ* soil treatment techniques have been investigated but there are no effective methods developed which can remove recalcitrant and strongly adsorbed radionuclides from the soil without significant loss of soil mass or denaturation of the soil. The best technique demonstrated to date has been soil collection followed by above ground isolation and storage. Physical separation technologies, such as particle size segregation or high gradient magnetic separation techniques, have been demonstrated to be effective only in certain cases. Whereas those techniques provide some volume minimization relief they fail at separating strongly sorbed metal species from the soil host, or rely upon the assumption that the bulk of the contaminant resides with a particular size fraction. Chemical treatment and soil washing techniques have been investigated and continue to undergo further research and development targeted at enhancing extraction efficiency and reducing secondary waste streams.

Metal complexation followed by supercritical fluid extraction (SFE) is a relatively new solvent extraction technique which couples the energy efficiency of a supercritical fluid process with the extractive power of more than 5 decades of research in extractive radiochemistry (7). The technique is performed by dissolving a metal complexing agent in supercritical fluid carbon dioxide. The augmented solvent is then passed through the solid or liquid matrix containing the radionuclide. A chemical reaction occurs between the radionuclide and the metal complexing agent. An organometallic complex is formed which itself is soluble in the supercritical fluid. As the solvent flows through the matrix the organometallic complex is swept out. A downstream reduction in pressure effectively precipitates and isolates the metal complex. The solvent is recyclable and the complexing agent can also be regenerated if desired.

Metal-complexation/SFE using carbon dioxide has been successfully demonstrated for removal of lanthanides, actinides and various other fission products from solids and liquids (8-18). Direct dissolution of recalcitrant uranium oxides using nitric acid and metal-complexing agents in supercritical fluid carbon dioxide has also been reported (19-25). In this paper we explored supercritical fluid extraction of sorbed plutonium and americium from soil using common organophosphorus and beta-diketone complexants. We also qualitatively characterize actinide sorption to various soil fractions via use of sequential chemical extraction techniques.

Experimental

Clean soils were chosen from geographical locations at and near the Radioactive Waste Management Complex (RWMC) at the Idaho National Engineering and Environmental Laboratory (INEEL). The RWMC at the INEEL served as a DOE burial ground from the early 1950s through to the 1970s and contains numerous pits, trenches and soil vaults wherein soil contaminated with actinides and fission products is known to exist. Soil samples were air dried, sieved to 50 mesh, then partitioned into 100 g batches. A slurry was made by adding demineralized water to the 100 g soil batches and the slurry was then spiked with varying amounts of an aged stock solution which contained approximately 26 μCi/mL ^{239}Pu and approximately 0.2 μCi/mL ^{241}Am in 8 M nitric acid. Spikes were added in small (<0.1 ml) aliquots with 1 minute stirring intervals between additions. Plutonium and americium nitrates were spiked onto the clean RWMC soils at activities ranging from 100 nCi/g (3.7 X 10^3 Bq/g) to as high as 1000 nCi/g (3.7 X 10^4 Bq/g). The samples were then air dried and radiologically characterized using both gamma-ray spectroscopy and alpha spectrometry techniques.

Plutonium and americium were extracted from spiked INEEL soil samples using supercritical fluid carbon dioxide augmented with commonly known beta-diketones and neutral oxygen donor ligands. Ten gram batches of the spiked soils were loaded into a supercritical fluid extraction vessel made from stainless steel high pressure tubing. The extraction vessel was approximately 17 inches long with a ½ inch o.d. Supercritical carbon dioxide was passed over the soil samples at a pressure of 2,900 psi, a temperature of 65 °C, and a flow rate of 3.5 milliliters per minute for approximately 45 minutes while temperature and pressure equilibrated. Once the system had equilibrated tributyl phosphate (TBP) and thenoyltrifluoroacetone (TTA) were added to the supercritical fluid solvent stream at a constant flow rate and the augmented solvent was then passed over the soil sample for another 45 minutes. Numerous extractions were performed on different soil samples where the TBP concentration was ranged from 0.27 to 2.7 mol% and the TTA concentration was ranged from 0.32 to 3.2 mol%. Different extraction parameters were changed (temperature, pressure, addition of solvent modifiers, extraction time) to determine their effects on extraction efficiency. SFE was also performed on spiked soil samples using TBP and the more acidic beta-diketone hexafluoroacetylacetone (HFA). For those extractions the TBP concentration was ranged from 2.7 to 5.3 mol% and the HFA concentration was ranged from 3.5 to 6.8 mol%. Various extraction parameters (temperature, pressure, addition of solvent modifiers) were also changed to determine their effects on extraction efficiency.

A flow schematic of the SFE system is shown in Figure 1. A photograph of the extraction system in a radiological control hood is shown in Figure 2, and a photograph of the actinide-ligand extract is shown in Figure 3.

Post-SFE soils and the liquid extract were radiologically characterized using both gamma-ray spectroscopy and alpha spectrometry techniques. Sequential chemical extractions were performed on both pre-SFE and post-SFE soils to obtain information related to the percent plutonium and americium removed from the various soil fractions. Sequential chemical extraction analysis was performed on 1 g batches of soils. Sequential chemical extraction techniques were the same as those employed by Litaor and Ibrahim (26) and Tessier et al (27). Both techniques were employed for comparative purposes on pre-SFE samples, but only the technique of Litaor et al was used to characterize plutonium and americium in the post-SFE samples.

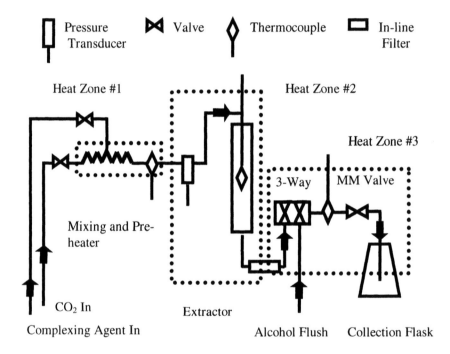

Figure 1. Flow schematic of the SFE system.

Figure 2. Photograph of the SFE System in a radiological control hood.

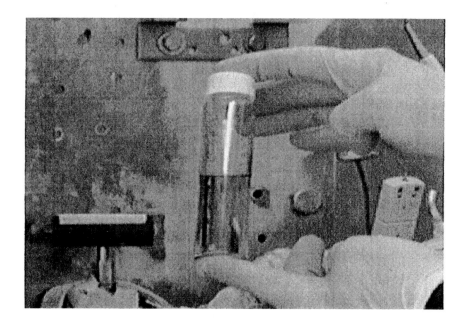

Figure 3. Photograph of the actinide-ligand extract.

That technique was viewed as the most conservative. Sequential extraction techniques developed and reported by Schultz *et al* (*28, 29*) were also tried. The technique reported by Schultz provided unsatisfactory results for these samples because the bulk of the plutonium was not removed by the reagents and thus became incorrectly assigned to the residual fraction.

Results

Sequential chemical extraction techniques are widely published in the literature and useful for determining the geochemical fractionation of metals in soils and sediments (*27–29*). Even though some methods may suffer from limitations (e.g., re-adsorption) the data are useful for assessing the conditions under which bound metal can be released from the soil.

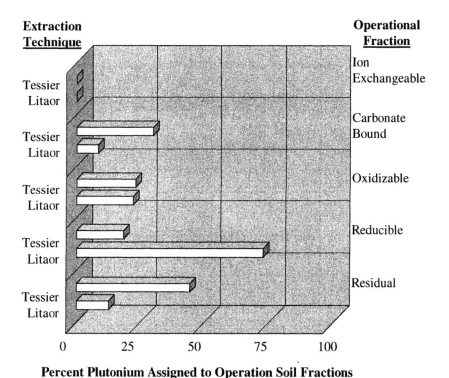

Percent Plutonium Assigned to Operation Soil Fractions

Figure 4.   Sequential chemical extraction results for plutonium on pre-SFE soil.

Sequential chemical extractions on pre-SFE soils, using two different procedures for comparison, were conducted. The results for plutonium are found in Figure 4. The data indicate there is variability in the results between different procedures. The procedure of Litaor and Ibrahim was viewed as the most conservative because it tended to extract plutonium from all phases leaving approximately 12.8% ± 1 assigned to the residual. The sequential chemical extraction results for americium on pre-SFE soil are found in Figure 5. Results indicate americium is primarily associated with the reducible soil fraction. Comparison of the data from the two different sequential chemical extraction techniques shows similar results.

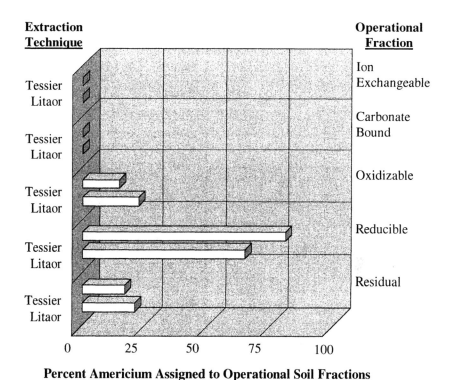

Percent Americium Assigned to Operational Soil Fractions

Figure 5. Sequential chemical extraction results for americium on pre-SFE soil.

From experimentation with supercritical fluid extraction it is known that neither plutonium nitrate or oxides nor americium nitrate or oxides are soluble in carbon dioxide without the aid of soluble complexing agents. Even with the aid of selected complexing agents plutonium extraction efficiency from spiked

INEEL soils was less than 20% when 2.7 mol% tributyl phosphate was used as the sole complexing agent in a supercritical fluid extraction. Neither plutonium nor americium were extracted when various beta-diketones were used as sole complexing agents. However, when a dual ligand system comprised of 2.7 mol% tributyl phosphate and 3.2 mol% thenoyltrifluoroacetone was used in a synergistic manner, as much as 69% ± 5 of the plutonium and 88% ± 6 of the americium could be extracted from spiked INEEL soils. The addition of water as a co-solvent modifier was found to have minor beneficial effect on extraction efficiency and instead proved to create column plugging problems. Ethanol as a co-solvent modifier served to nearly double the extraction efficiency over the initial conditions; thus, providing a beneficial effect. However, addition of co-solvents did not provide nearly as great of benefit as did saturating the critical fluid phase with the extracting agents. Increasing the extraction run time, increasing pressure and performing multiple extractions on the same sample had minor beneficial effects and were thus deemed unworthy of further pursuit. The SFE results for the TBP - TTA system are given in Table I.

Table I. SFE Results Using the TBP - TTA System in $scCO_2$

	% Extracted	
Conditions	Pu	Am
Initial Extraction Conditions: 2,900 psi, 65 °C 45 min., 3.5 ml/min, [TBP] = 0.27 mol%, [TTA] = 0.33 mol%	17 ± 4	32 ± 5
Initial Conditions Plus Modifiers 2 mol% water	17 ± 1	21 ± 1
5 mol% ethanol	45 ± 8	59 ± 4
Initial Conditions Plus Increase Extraction Time From 45 minutes to 4 hours (240 min.)	17 ± 1	26 ± 1
Initial Conditions Plus Increase Extraction Pressure From 2,900 psi to 3,750 psi	29 ± 2	51 ± 1
Initial Conditions Plus Increase TBP – TTA Concentration [TBP] = 2.7 mol%, [TTA] = 3.2 mol%	69 ± 5	88 ± 6

When 3.5 mol% of the acidic beta-diketone hexafluoroacetylacetone was used in combination with 2.7 mol% tributyl phosphate, 76% ± 2 of the plutonium and 93% ± 1 of the americium could be extracted in a single 45 minute extraction at 65 °C. When the ligand concentrations were increased to 6.8 mol% hexafluoroacetylacetone and 5.3 mol% tributyl phosphate, and the extraction temperature was increased to 95 °C, a maximum of 83% ± 5 plutonium and 95% ± 3 americium were extracted. The SFE results for the TBP - HFA system are given in Table II.

Table II. SFE Results Using the TBP - HFA System in $scCO_2$

Conditions	% Extracted Pu	Am
Initial Extraction Conditions: 2,900 psi, 65 °C, 45 min., 3.5 ml/min, [TBP] = 2.7 mol%, [HFA] = 3.5 mol%	76 ± 2	93 ± 1
Initial Conditions Plus Add Modifier 5 mol% ethanol	74 ± 10	90 ± 3
Initial Conditions Plus Increase Temperature, Pressure and Complexing Agent Concentration 3,750 psi, 95 °C, [TBP] = 5.3 mol%, [HFA] = 6.8 mol%	83 ± 5	95 ± 3

Experimental results from sequential chemical extractions performed on post-SFE soils (Figure 6) using the TBP - HFA system show that most (86% ± 2) of the remaining plutonium not removed by SFE is associated with the reducible (sesquioxide) mineral fraction of the soil. Approximately 10% of the remaining plutonium is partitioned into the residual fraction and the balance (~4%) is found in the oxidizable and carbonate bound fractions.

Unlike the results found by Loyland et al (30) for uranium, we observed no repartitioning of either plutonium or americium amongst the various operational soil fractions after SFE. In all SFE experiments conducted in this work a net reduction of both plutonium and americium activity was observed across all soil fractions. Additionally, as the ligand system was changed from the less acidic beta-diketone TTA to the more acidic beta-diketone HFA the extraction efficiency of plutonium from the sesquioxide and residual fractions increased dramatically. Thus, it is likely that a more effective, and possibly complete, extraction could be performed if a reagent were added in small quantities to the extraction mixture to specifically attack the sesquioxide fraction.

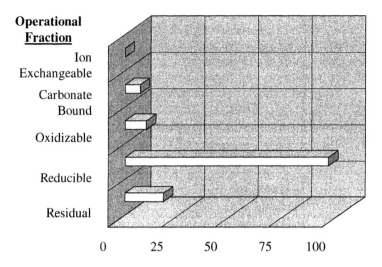

Percent Plutonium Assigned to Operational Soil Fractions

Figure 6. Sequential chemical extraction results for plutonium on post-SFE soil.

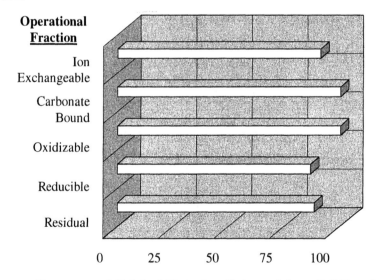

Percentage of Original Plutonium Extracted from Soil Fraction

Figure 7. Sequential chemical extraction results for plutonium on post-SFE soils.

In Figure 7 the sequential chemical extraction results on post-SFE soils are plotted as a percentage of plutonium extracted from each phase based on the original amount of plutonium assigned to that operational soil fraction in the pre-SFE samples. When the data are plotted in this manner it is easier to see that plutonium was removed from all soil fractions including the more recalcitrant/strongly sorbed residual fraction. Even though the data for the ion exchangeable, carbonate bound and oxidizable fractions points towards high extraction efficiency it should be noted that the original amount of plutonium partitioned into those phases accounted for only approximately 22% of the initial spike.

Conclusion

SFE results and sequential chemical extraction characterization of pre- and post-SFE soils suggest that the best gains in supercritical fluid extraction efficiency can be made by designing a supercritical fluid extraction process which targets the reducible mineral fraction of the soil. These preliminary results also indicate that metal-complexation followed by supercritical fluid extraction holds potential as being a highly effective, safe and energy efficient process for removal of strongly adsorbed and recalcitrant radionuclide species from solid environmental media.

Acknowledgment

The authors wish to thank the staff of the Radiation Measurements Laboratory at the INEEL. This work was funded in part by BNFL Group. Work was performed under contract number DE-AC07-99ID13727.

References

1. *The 1996 Baseline Environmental Management Report*; U.S. Department of Energy, Office of Environmental Management, U.S. Government Printing Office: Washington, DC, 1996; DOE/EM-0290.
2. *Linking Legacies: Connecting the Cold War Nuclear Weapons Processes to their Environmental Consequences*; U.S. Department of Energy, Office of Environmental Management, U. S. Government Printing Office: Washington, DC, 1997, DOE/EM-0319.
3. Hamby, D. M. *Sci. Tot. Env.* **1996**, *191*, 203-224.

48

4. Ijaz, T.; Rutz, E. E.; Eckart, R.; Church, B. W. *Rad. Waste Management and Env. Rest.* **1999**, *22*, 49-77.
5. Mulligan, C. N.; Yong, R. N.; Gibbs, B. F. *Engineering Geology* **2001**, *60*, 193-207.
6. Peters, R. W. *J. Haz. Mat.* **1999**, *66*, 151-210.
7. Darr, J. A.; Poliakoff, M. *Chem. Rev.* **1999**, *99*, 495-541.
8. Lin, Y.; Wai, C. M.; Jean, F. M.; Brauer, R. D. *Environ. Sci. Technol.* **1994**, *28*, 1190-1193.
9. Laintz, K. E.; Tachikawa, E. *Anal. Chem.* **1994**, *66*, 2190-2193.
10. Shamsipur, M.; Ghiasvand, A. R.; Yamini, Y. *J. Supercritical Fluids* **2001**, *20*, 163-169.
11. Wai, C. M. *Anal. Sci.,* **1995**, *11*, 165-167.
12. Lin, Y.; Brauer, R. D.; Laintz, K. E.; Wai, C. M. *Anal. Chem.* **1993**, *65*, 2549-2551.
13. Saito, N.; Ikushima, Y.; Goto, T. *Bull. Chem, Soc. of Japan* **1990**, *63*, 1532-1534.
14. Laintz, K. E.; Wai, C. M.; Yonker, C. R.; Smith, R. D. *Anal. Chem.* **1992**, *64*, 2875-2878.
15. Wai, C. M.; Lin, Y.; Brauer, R. D.; Wang, S. *Talanta,* **1993**, *40*, 1325-1330.
16. Wang, J.; Marshall, W. D. *Anal. Chem.* **1994**, *66*, 1658-1663.
17. Erkey, C. *J. Supercritical Fluids* **2000**, *17*, 259-287.
18. Mincher, B. J.; Fox, R. V.; Holmes, R. G. G.; Robbins, R. A.; Boardman, C. *Radiochim. Acta* **2001**, *89*, 613-617.
19. Samsonov, M. D.; Trofimov, T. I.; Vinokurov, S. E. In Proceedings International Solvent Extraction Conference, ISEC 2002, Cape Town, South Africa, March 17-21, 2002, 1187.
20. Shadrin, A.; Murzin, A.; Starchenko, V. In Proceedings Spectrum'98, Denver, Colorado, USA, September 13-18, 1998, 94.
21. Shadrin, A.; Murzin, A.; Smart, N. G. In Proceedings 5th Meeting on Supercritical Fluids, Nice, France, March 23-25, 1998, 155.
22. Shadrin, A.; Murzin, A.; Babain, V. In Proceedings 5th Meeting on Supercritical Fluids, Nice, France, March 23-25, 1998, 791.
23. Tomioka, O.; Enokida, Y.; Yamamoto, I. *Progress in Nuclear Energy* **2000**, *37*, 417-422.
24. Murzin, A.; Babain, V. A.; Shadrin, A. Y.; Smirnov, I. V.; Lumpov, A. A.; Gorshkov, N. I.; Miroslavov, A. E.; Muradymov, M. Z. *Radiochemistry* **2001**, *43*, 177-182.
25. Trofimov, T. I.; Samsonov, M. D.; Lee, S. C.; Smart, N. G.; Wai, C. M. *J. Chem. Tech. Biotech.* **2001**, *76*, 1223-1226.
26. Litaor, M. I.; Ibrahim, S. A. *J. Environ. Qual.* **1996**, *25*, 1144-1152.

27. Tessier, A.; Campbell, P. G. C.; Bisson, M. *Anal. Chem.* **1979**, *51*, 844-851.
28. Schultz, M. K.; Burnett, W. C.; Inn, K. G. W.; Smith, G. *J. Radioanalytical and Nuclear Chemistry* **1998**, *234*, 251-256.
29. Schultz, M. K.; Burnett, W. C.; Inn, K. G. W. *J. Environ. Radioactivity* **1998**, *40*, 155-174.
30. Loyland, S. M.; Yeh, M.; Phelps, C.; Clark, S. B. *J. Radioanal. Nucl. Chem.* **2001**, *248*, 493-499.

Chapter 5

Solubility and Modifications of Metal Chelates in Supercritical Carbon Dioxide

Bernd W. Wenclawiak, H. Beer, A. Ammann, and A. Wolf

Analytical Chemistry, Department of Chemistry, University of Siegen,
Adolf Reichwein Strasse 9, D–57078, Siegen, Germany

The solubilities of chelates with palladium, rhodium, lead and copper have been measured with static and dynamic spectroscopic methods at different temperatures and different pressures. The influence of different ligands, ligand modifications and of the metal ions coordination sphere on the solubility were studied and compared to the influence of pressure and temperature of the supercritical carbon dioxide (scCO$_2$). In a series of C$_2$, C$_4$, C$_6$, C$_8$ copper dithiocarbamates a maximum solubility was measured with the butyl substituents.

In recent years the use of supercritical fluids (ScF) for analytical and process scale extraction has increased dramatically in effort to reduce the amount of organic solvents used. For a wide variety of low-polarity solutes pure scCO$_2$ can quantitatively extract organics from a wide variety of matrices. (1-3) Because of practical considerations of low toxicity, high purity, and good ability to solvate a range of organics, scCO$_2$ has received the most use for analytical scale extractions. Many methods to dissolve and extract metals as chelates into scCO$_2$ have been described and different ways to determine chelate solubility data in scCO$_2$ have been reported (4-27).

Modern catalytic exhaust converters used in cars contain the precious metals palladium and rhodium in the gram level. (28) Common techniques of recycling heavy metals use the transformation of the metals into chloro complexes which are extracted by means of organic solvents. In order to evaluate the replacement of organic solvents by $scCO_2$ by a different extraction process the solubility of different chelates of palladium(II), rhodium(III), lead(II) and copper(II) was investigated.

There are principally two ways of modifying the solubility of metal chelates such as dithiocarbamates or beta-diketonates:

1. Change of residual groups - e.g. replacement of hydrogen by fluorine or increase the carbon chain length. (6,29)
2. Adding of a synergism or additional modifier to the chelates – e.g. TBP or other. (5)

We have systematically studied this behaviour for a group of palladium, rhodium, lead and copper chelates. Linear and branched dialkyldithiocarbamates (dtc) as S,S-type, 2,2,6,6-tetramethyl-3,5-heptanedionate (thd) as O,O-type and methylglycolates (mtg) as S,O-type were selected. The diisopropyldithiocarbamate (DPDTC) ligand was used as one dtc representative to investigate the influence of fluid density on the solubility of the dtc chelates. The influence of alkylchain length ($R=C_2-C_8$) of dialkyldithiocarbamates on the solubility of palladium(II), rhodium(III) and copper(II) chelates was measured at constant conditions with two different methods. We observed that branching and length increase of dtc-ligand alkylchain increases the solubility of the respective palladium, rhodium and copper dtc-chelate but only to a certain content. After a solubility maximum the solubility decreases again. The optimum solubility showed dtc with linear or branched C_4 or C_5 alkyl chains. The ligand thd, belonging to the beta-diketon group, was used to measure the influence of the central ion coordination number and therefore the chelates spherical structure on the solubility of the palladium(II)- and rhodium(III) chelates in $scCO_2$. As another type of ligand we tested methylglycolate and measured the solubility of palladium(II), rhodium(III) and lead(II) methylglycolate at different conditions.

We can compare independent results measured with different apparatuses: Most working groups use a static-spectroscopic or dynamic-spectroscopic method with offline quantification to determine chelate solubilities. We present a new dynamic-spectroscopic method with online quantification compared to a static-spectroscopic method. Both methods depend on photometrical absorption measurements of chelates dissolved in $scCO_2$.

To increase the understanding of factors determining the solubility of metal chelates in ScF we will report our concept and recent results. We discuss the influence of ligand structure modifications and the influence of the $scCO_2$ density on the solubility of the chosen chelates.

Methods

Apparatus for Dynamic Spectroscopic Online Measurements

The center of the dynamic-spectroscopic apparatus consists of a modified Suprex MPS 225 (Suprex Corp. Pittsburgh, PA.) system as shown in Figure 1.

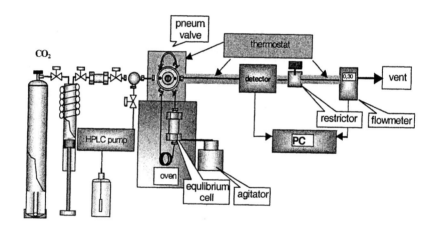

Figure 1. Apparatus for online dynamic-spectroscopic measurement of chelate solubility

A 250 mL syringe pump is used to pressurize CO_2. All heated components, except the pneumatic driven valve and the capillary connected to the detector are placed in an oven. The pressure of the $scCO_2$ and the temperature of the oven, detection cell and of the restriction system can be programmed via a control panel. The CO_2 we used was heated in a 1.0 m preheating coil of 1.6 mm x 762 μm i.d. (1/16 in. x 0,03 in.) stainless steel tubing and then passes through the saturation cell (V=1,027 mL, 14,5 mm x 9,5 mm i.d. SFE cell, Keystone

Scientific, Bellfonte, PA) containing approx. 100 mg of the chelate. The saturation cell with the chelate was equilibrated to the desired temperature before CO_2 was filled in. The preheating coil was necessary to ensure that the CO_2 is at operating temperature prior to entering the saturation cell. After pressurizing the cell the pneumatic valve is switched to the "load" position where the ScF moves directly from the pump to the detector. In this position, the flow rate is adjusted to a value of 0.3 mL/min while the chelate dissolves in the fluid (see Figure 2).

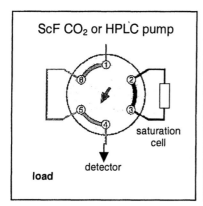 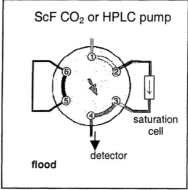

Figure 2: Positions of the pneumatic valve.

The dissolution of the chelates is accelerated by shaking the cell with a mixer motor (Cenco Intrumenten, Netherlands). After equilibration the valve is switched for 500 ms to a position where the saturation cell is flooded and an aliquot of the saturated solution is transferred to a "Spectra Focus" (Spectra Physics) forward scanning absorbance detector where the absorbance of the chelate aliquot is measured time-resolved (Figure 3). The flow rate, measured by an Aalborg mass flow meter is recorded in parallel with the signal of the UV detector. The spectra of five extractions were recorded and afterwards the chelate is eluted from the steel capillary and restrictor by pumping ethanol through it. The ethanol of the extract is removed and the chelate is treated with hydrochloric acid (1:5 v:v). This solution is analyzed with an ICP-OES (Leeman Labs, Inc.). To calibrate the system a correlation between known amounts of dissolved chelates and the area of the time resolved chelate absorption was derived. With the analyzed quantity of five extractions, the known switching time of valve and the measured flow rate, the concentration of the extracted solution can be calculated. This calculated concentration is set in relationship to the area under the spectra measured. Therefore a correlation between the area of the on line absorption area of the chelate aliquots and the solubility was found.

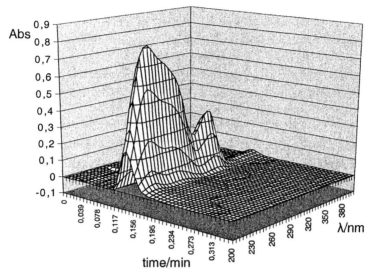

Figure 3. Representative time-resolved absorption measurement of a Cu(thd)₂ aliquot passing the scanning Spectrafocus UV detector.

Apparatus for Static Solubility measurements

The second apparatus to measure chelate solubilities in scCO₂ used here already is described in the work of Carrott and Wai. Details to this method can be found there. (*16*) With this static method the UV absorption of a saturated chelate solution is measured directly in a capillary tubing connected to the fiber optic (Polymicro Technologies, Phoenix, Arizona) of a Cary 1E UV-VIS-Spektrometer (Varian Instruments, Sugarland, Texas).

Classification and Characterization of Chelates

Preparation of metal complexes

All metal beta-diketonates and methylglycolates were prepared according to methods described in the literature. (30-31) A detailed way to get long alkyl chained dtc of palladium and rhodium is described in the thesis of Ammann. (*32*) In Figure 4 the structure of the ligands is shown. In Table I the systematic names of dtc ligands we have chosen is given.

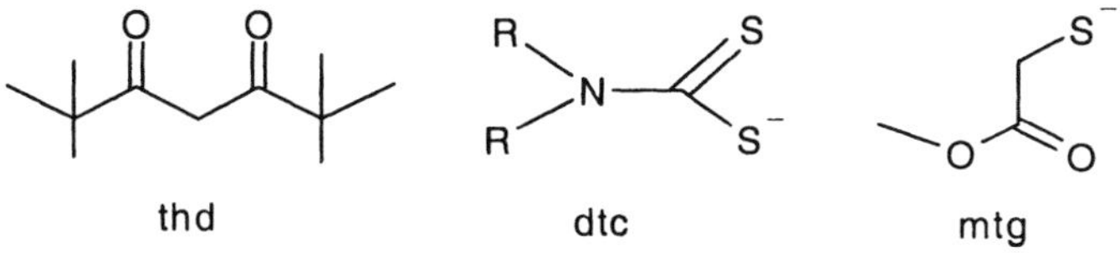

thd   dtc   mtg

Figure 4. Structure of 2,2,6,6-tetramethyl-3,5-heptanedion (thd), dialkyldithiocarbamate (dtc) and methylglycolate (mtg) ligands with R representing different alkyl chains

Table I. Dtc ligands used in this study

formula	chemical name
$[S_2CN(C_2H_5)_2]^-$	Diethyldithiocarbamate (DEDTC)
$[S_2CN(C_3H_7)_2]^-$	Diisopropyldithiocarbamate (DPDTC)
$[S_2CN(C_4H_9)_2]^-$	Dibutyldithiocarbamate (DBDTC)
$[S_2CN(C_6H_{13})_2]^-$	Dihexyldithiocarbamate (DHDTC)
$[S_2CN(C_8H_{17})_2]^-$	Dioctyldithiocarbamate (DODTC)

All chelates have been characterized by means of elementary analysis and melting points (see Table II). Dtc chelates also have been characterized by FAB-mass spectrometry and ^{1}H-NMR. (*32*) The structures of the Pd(DPDTC)$_2$ and of Pd(thd)$_2$ were solved by x-ray diffraction measurements. Further information to crystal structures and spectroscopic data can be achieved from the authors. (*31-34*).

Results of Dynamic Solubility Measurements

Solubility of Pd(thd)$_2$ and Rh(thd)$_3$

The solubilities of chelates presented in this chapter have been measured with the dynamic spectroscopic method. In Figure 5 the solubility data of Pd(thd)$_2$ and Rh(thd)$_3$ measured with this system are presented. In this diagram the molar solubilities are plotted against the reduced density of the scCO$_2$ at a temperature of 70 °C. The solubility of Pd(thd)$_2$ is approximatelya factor of 50 higher than the solubility of Rh(thd)$_3$. This is quite surprising because in general the solubility of chleates with coordination number Z=3 (for example Cr(thd)$_3$)

Table II. Melting points of dithiocarbamates with different metals
*not determined

chelate	Melting point °C	chelate	Melting point °C
Pd(DEDTC)$_2$	225-226	Pd(DHDTC)$_2$	65-66
Rh(DEDTC)$_3$	>235	Rh(DHDTC)$_3$	176-177
Cu(DETDC)$_2$	189-190	Cu(DHTDC)$_2$	49-56
Pd(DPDTC)$_2$	>300	Pd(DODTC)$_2$	<30
Rh(DPDTC)$_3$	>235	Rh(DODTC)$_3$	165-169
Cu(DPTDC)$_2$	not determined	Cu(DOTDC)$_2$	<30
Pd(DBDTC)$_2$	105-106		
Rh(DBDTC)$_3$	not determined		
Cu(DBTDC)$_2$	60-61		

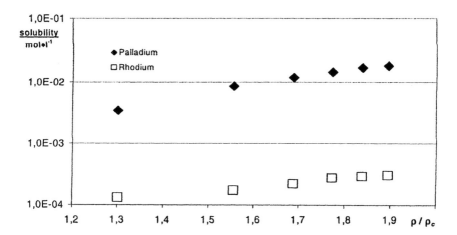

Figure 5. Solubility of Pd(thd)$_2$ and Rh(thd)$_3$ at 70 °C plotted versus reduced density (ρ/ρ_c).

show higher solubilities in scCO$_2$ than chelates with coordination number Z=2 (for example Cu(thd)$_2$). This shows that the planar quadratic structure of palladium(II) chelates causes an extraordinary solute behavior.

In Figure 6 the solubility of Pd(thd)$_2$ at different temperatures and different pressures is shown. At 150 bar we can observe that an increase of temperature leads to a reduced solubility at constant pressure. This behavior changes above 200 bar. The solubility increases with increasing temperature. A similar behavior at low temperatures was observed for Rh(thd)$_3$.

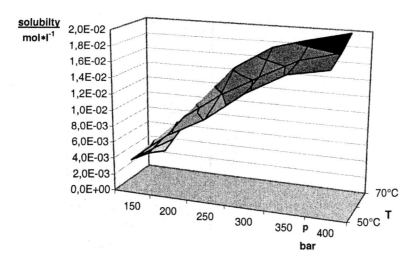

Figure 6. Solubility of Pd(thd)$_2$ at different pressures and temperatures.

Solubility of Pd(mtg)$_2$ and Rh(mtg)$_3$

In Figure 7,8 the solubilities of Pd(mtg)$_2$ and Rh(mtg)$_3$ at different densities of scCO$_2$ are shown. Here the solubility of Rh(mtg)$_3$ is slightly higher compared to the solubility of Pd(mtg)$_2$. The solubilities of both chelates is around 10^{-4} mol/L. This is about two orders of magnitude lower than that of Pd(thd)$_2$. We attribute this to the minor size of the mtg ligand with only one methoxy-group determining the polarity of the whole chelate. The influence of the spherical geometry of the chelate becomes less important on polarity and therefore the difference of the solubility between rhodium and palladium is only small. Also the shielding of the central ion is probably significantly smaller as compared for example to thd-chelates. In Figure 8 is a plot of Rh(mtg)$_3$. As can be seen at 150 bar solubility decreases with increasing temperatures. Fig. 6, 8, 9 depict one phenomenon: at our lowest studied pressure there is a different temperature effect than at higher pressures. This in fact is not new, but not yet sufficiently studied with chelates. However a better apparatus, e.g. with better temperature and pressure control at lower levels, is necessary. Our apparatus was not suited for this.

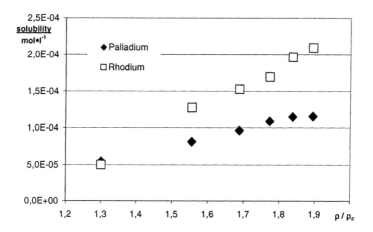

Figure 7. Solubility of Pd(mtg)$_2$ and Rh(mtg)$_3$ at 70 °C plotted versus reduced density (ρ/ρ_c).

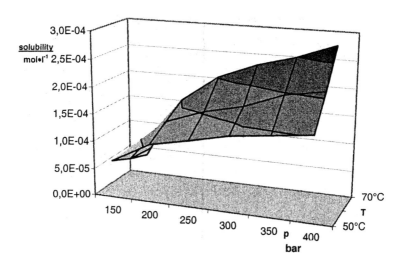

Figure 8. Solubility of Rh(mtg)$_3$ at different pressures and temperatures.

Solubility of Pd(DPDTC)₂, Rh(DPDTC)₃ and Pb(DPDTC)₂

In Figure 9 the measured solubilities of the diisopropyldithiocarbamates (DPDTC) at 70 °C are presented. The solubility of the rhodium chelate is approximately a factor of 10 higher than the palladium chelate. The highest solubility shows Pb(DPDTC)₂ . Consequently, the higher solubility of the Pb(DPDTC)₂ can only be attributed to structural differences in the metals coordination sphere.

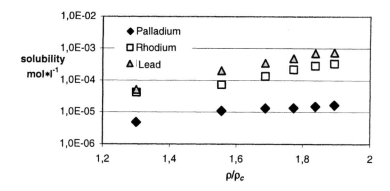

Figure 9. Solubility of Pd(DPDTC)₂ , Rh(DPDTC)₃ and Pb(DPDTC)₂ at 70 °C plotted versus reduced density (ρ/ρ_c).

To confirm the observed solubility differences between palladium(II)- and lead(II) DPDTC-chelates further theoretical considerations to describe the interaction between $scCO_2$ and the chelates have to be done. Furthermore, the influence of geometry, size and molecular mass of the chelates on the solubility have to be studied in detail.

We suppose that the planar quadratic coordination of the palladium(II) chelates allow CO_2 molecules or other small molecules like traces of the synthesis solvent to be coordinated to the Pd(II) centre and therefore change the polarity /solubility of the chelate. Voluminous and flexibel residual ligand substitutes protect the central palladium ion against interaction of small molecules and we can observe a drastic increase of solubility. Otherwise, if the coordination sphere of the chelate centre ion is octahedral or tetrahedral, other molecules (e.g. the solvent) cannot interact with the metal centre easily.

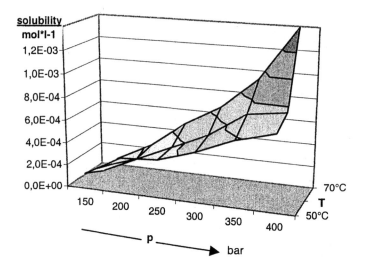

Figure 10. Solubility of Pb(DPDTC)₂ at different pressures and temperatures.

Results of Static Solubility Measurements

In this chapter we present one example for the influence of the dtc alkyl chain length on chelate solubility in $scCO_2$. As one representative compound for all dialkyldithiocarbamates we demonstrate the results for $Cu(DEDTC)_2$, $Cu(DBDTC)_2$, $Cu(DHDTC)_2$ and $Cu(DODTC)_2$.

The p and T conditions chosen for solubility measurements with the dynamic method were limited by the temperature stability of the fiber optic system. The maximum temperature was 70 °C at a maximum pressure of 250 atm. In Figure 11 eight p/T- conditions for solubility determinations are presented.

To describe the influence of temperature on the solubility at constant density of the ScF CO_2 we have chosen four different conditions. Under these conditions the polarity of the solvent is constant, e.g. only the solvent temperature has an influence on the chelate solubility. The quantification of dissolved copper chelate was performed by measuring the absorption spectrum directly in the saturated $scCO_2$ solution. Extinction coefficients to quantify the dissolved chelate amounts have been determined previously in n-hexan solution. In Figure 12 the absorption spectrum of $Cu(DHDTC)_2$ in $scCO_2$ is shown. The absorption maxima of the copper chelates in $scCO_2$ were hypsochrom shifted by 5 nm to 428 nm compared to the maxima in n-hexane solution.

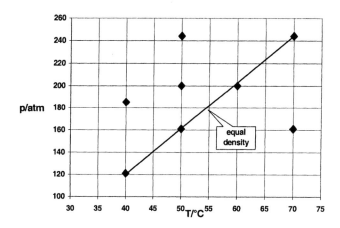

Figure 11. Conditions for solubility measurements with the static method.

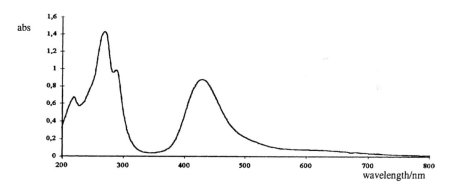

wavelength/nm

Figure 12. Typical UV-VIS absorption spectrum of Cu(DHDTC)$_2$ in scCO$_2$.

In Figure 13 average solubility values for copper(II)-dialkydithiocarbamates of at least three single measurements at different conditions are presented. The selected conditions are according to Figure 11.

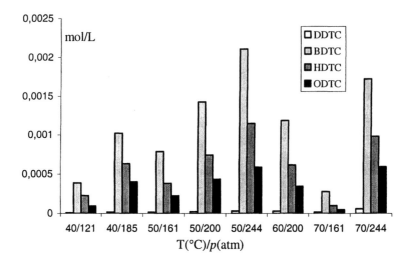

Figure 13. Solubilities of Cu(II)-dialkyldithiocabamates with different residual substituents measured with the static method at given conditions

These results show that the solubility of dialkyldithiocarbamates primarily depends on the length (and shape) of the ligands' residual alkyl chain. The influence of pressure and temperature is only small as compared to structural factors. In general high pressures and high temperatures show higher solubilities of the dialkyldithiocarbamates. The solubility of $Cu(DEDTC)_2$ never exceeds 3.3 x 10^{-5} molL^{-1} according to conditions given in Figure 11 and is never lower than 1.10 x 10^{-5} molL^{-1}. The solubility of $Cu(DBDTC)_2$ in contrast is at 2.10 x 10^{-3} molL^{-1} about two orders of magnitude higher. However the differences in solubility for the conditions selected are much more dependent for butyl DTC (one order of magnitude) than for ethyl DTC (only 3 fold). Further increase of the residual alkyl chain decreases the solubility steadily. The existence of a solubility maximum at a certain length or shape of the alkyl chains so far was not considered. We assume that the special structure of dialkyldithiocarbamate alkyl chains with R=C_4H_9 or R=C_5H_{11} function as a spherical shield and therefore prevent CO_2 molecules from interacing with the metal ion. The influence of residual alkyl chain length on compounds in scCO$_2$ has already been proven by some other authors. Schneider et al. showed that alkyl substituted anthraquinone dyes have highest solubilities in scCO$_2$ when they are substituted by the pentyl group. The octyl- and ethylhomologues showed significant lower solubilities. (*35-36*) Glennon et al. as well as Eastoe et al. reported similar results for their investigations. (37-38) Similar results for homologue arsenic(III), palladium(II), nickel(II) and rhodium(III) dialkyldithiocarbamates recently have been proven by experiments with different methods in our working group. (*33-34*)

The contribution of volatility on solubility at constant density and different temperatures for Cu(BDTC)$_2$ is shown in Figure 14. It is obvious that the solubility increases although the solvent properties are the same (same density at different temperatures) . ·

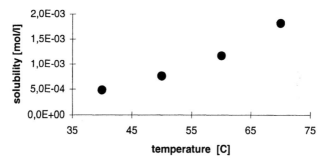

Figure 14. Solubility of Cu(BDTC)$_2$ at different temperatures and constant density of ScF CO$_2$.

Conclusions

We demonstrated that systematic modification of the ligand structure is a powerful way to optimize solubilities of chelates in ScF. The use of fluorinated, expensive, or environmental harmful ligands should be reconsidered in light of economic recovery rates of metals via supercritical fluid extraction. Dialkyldithiocarbamates are one group of ligands for appropriate design of suitable ligands to extract efeectively different metals with supercritical fluids. We have shown that dialkyldithiocarbamates with R=C$_4$-C$_5$ possess the potential for different applications e.g. to extraction and recovery of toxic heavy metals or platinum group elements from different matrices. A further advantage of the dialkyldithiocarbamates is the low price, easy handling and almost unlimited availability. Nevertheless, further experiments have to be done to describe the influence of the ligand structure on the solubility of metal chelates. Dialkyldithiocarbamate or ß-diketonate ligands containing branched or nonsymmetrical alkyl chains so far are described only in a few cases. Finally, our results demonstrate that it is not trivial to foresee the solubility behavior of metal chelates in all cases. The metals' coordination chemistry, molecular mass and the ligands' structure under supercritical conditions as well as thermodynamic considerations have to be discussed in detail and therefore further metal chelates solubility data are necessary.

For the chelates studied under dynamic extraction conditions we conclude that palladium(II) chelates show high solubilites in $scCO_2$ only if the ligand contains bulky alkyl chains or additional coordinated solvent molecules which can fill the empty space above and under the coordination plain of the chelate and therefore cover as "secondary ligands" the metal centre and thus influence the interaction with other molecules like CO_2. Ligands with short alkyl chains like mtg or DPDTC ligand in contrast show relatively low solubilities. At pressures above 150 bar solubilities can be increased more effectively by increasing the temperature than increasing the pressure.

Acknowledgments

Acknowledgment is made to the Donors of the Petroleum Research Fund, administered by the American Chemical Society, for partial travel support to B.Wenclawiak.

We wish to thank the German Science foundation (DFG) for a grant: We 1073/13/1-3.

References

1. Chester, T.L.; Pinkston, J.D.; Raynie, D.E. *Anal. Chem.* **1998**, *70*, 301R-319R.
2. Taylor, L.T.: *"Supercritical Fluid Extraction"*, John Wiley & Sons Inc., New York, **1996.**
3. Hawthorne, S.B. *Anal. Chem.* **1990**, 62, *11.*
4. Wai, C.M.;Waller, B. *Ind. Eng. Chem. Res.* **2000**, *39*, 4837-4841.
5. Wai, C.M.;Wang, S. *Anal. Chem.* **1996**, *68*, 3516-3519.
6. Smart, N.G.;Carleson, T.; Kast, T.;Clifford, A.A.;Burford, M.D.;Wai, C.M. *Talanta* **1995**, 44, 137-135.
7. Lin, Y.; Smart, N.G.; Wai, C.M. *Environ. Sci. Technol.*, *29*, 2706.
8. Toews, K.L.; Shroll, R.M.;Wai, C.M.;Smart, N.G. *Anal.Chem.*, **1995**, *67*, 4040.
9. Wai, C.M.; Wang, S.;Liu, Y.; Lopez-Avila, V.; Beckert, W.F *Talanta,* **1996**, *43*, 2083-2091.

10. Lin Y.; Wai, C.M *Anal. Chem.* **1994**, *66*, 1971-1975.
11. Laintz, K.E; Wai, C.M.; Yonker, C.R.; Smith, R.D. *Anal. Chem.* **1992**, 64, 311-315.
12. Smart, N.G.; Carleson, T.; Kast, T., Clifford, A.A.; Burford, M.D.; Wai, C.M *Talanta* **1997**,*44*, 137.
13. Wang, J.J. Yu, C.M. Wai, *Anal. Chem* **1996.**, **68**, .3516.
14. Smart, N.G.; Carleson, T.E.; Elshani, S.; Wang S.; Wai, C.M. *Ind. Eng. Chem. Res.* **1997**, *36*, 1819-1826.
15. Carrott, M.J.; Waller, B.E.; Smart, N.G.; Wai, C.M *Chem. Commun.* **1998**, 373-374.
16. Carrot, M.J.; Wai, C.M *Anal. Chem.*, **1998**, *70*, 2421-2425.
17. Ashraf-Khorassani, M.; Combs, M. T.; Taylor, L.T *Talanta* **1997**, *44*, 755-763.
18. Cross, W., Jr.; Akgerman, A.; Erkey, C. *Ind. Eng. Chem. Res.* **1996**, *35*,1765-1770.
19. Erkey, C.: *J. Supercritical Fluids* **2000**, *17*, 259-287.
20. Li, J.; Beckman, J. *Ind. Eng. Chem. Res.* **1998**, *37*, 4768-4773.
21. Wang, J. Marshall, W.D. *Anal. Chem.* **1994**, *66*, 1658-1663.
22. Lagalante A.F.; Hansen B.N.; Bruno T.J.; Sievers R.E. *Inorg. Chem.*, **1995**, *34*, 5781-5785.
23. Yazdi, A.V.; Beckman, E.J. *Ind. Eng. Chem. Res.* **1997**, *36*, 2368-2374.
24. Yazdi A.V.; Beckman E.J.: *Ind. Eng. Chem. Res.* **1996**, *35*, 3644-3652.
25. Darr, J.A.; Poliakoff, M. *Chem. Rev.,* **1999**, *99*, 495-541.
26. Glennon J.D. ;Harris S.J. ;Walker A. ;McSweeney C.C.; O´Connell M. *Gold Bulletin* **1999**, *32(2)*.
27. Murphy, J.M.; Erkey, C. *Environmental Science & Technology* **1997**, *31*, 1674-1679.
28. Jarvis, K.E.; Parry, S.J.; Piper, J.M. *Environ. Sci. Technol* **2001**, *35,* 1031-1036.
29. Laintz, K.E.; Wai, C.M. ; Yonker, C.R. ; Smith, R.D. *J. Supercrit. Fluids* **1991**, 4, 194.
30. Fay, R.C.; Piper, T.S. *J. Am Chem. Soc.* **1962**, *84*, 2303-2308
31. Wolf, A.; Doctoral dissertation, "*Determination of Metal Chelate Solubilities in Supercritical Carbon Dioxide*", University Siegen, **1997**.
32. Ammann, A; Diploma thesis, "*Synthesis and Characterization of Symmetrical Dialklydithiocarbamates of Copper, Palladium and Rhodium and Solubility of Copper Chelates in Supercritical Carbon Dioxide*", University Siegen, **1997**.
33. Beer, H.; Doctoral dissertation, "*Synergetic Extraction of Palladium and Rhodium as chelates with Supercritical Carbon Dioxide*", University Siegen, **2002**.

34. Uttich, S.; Doctoral dissertation, *"Extraction of Arsenic Species with Supercritical Carbon Dioxide"*, University Siegen, in progress.
35. Kautz, C.B.; Wagner, B.; Schneider, G.M.: *J. Supercritical Fluids*, **1998**, 43-47.
36. Tuma, D.; Schneider G.M. *"High Pressure Investigations on the Solubility of Disperse Dyes in Near- and Supercritical Fluids: Measurements up to 100 Mpa by a static Method"* The 4th International Symposium on Supercritical Fluids **1997**.
37. Glennon J.D.; Holmes J.D.; oral presentation 223rd ACS Meeting Orlando, 2002, paper 212 J&EC
38. Eastoe J.; Steytler D.C.; Dupont A.; Johnston K.P.; oral presentation 223rd ACS Meeting Orlando, 2002, paper 316 J&EC

Chapter 6

Extracting Gold in Supercritical CO_2: Fluorinated Molecular Baskets and Thiourea Ligands for Au

Jeremy D. Glennon[*], Josephine Treacy, Anne M. O'Keeffe,
Mark O'Connell, Conor C. McSweeney, Andrew Walker,
and Stephen J. Harris

Analytical Chemistry and the Supercritical Fluid Centre, Department of
Chemistry, University College Cork, Cork, Ireland
[*]Corresponding author: email: j.glennon@ucc.ie

Selective ligands for precious and heavy metal complexation
and extraction in sc-CO_2, capable of replacing environmentally
unfriendly processes using solvents or hazardous chemicals, are
being researched in our supercritical fluid centre at UCC. In
particular, the design and synthesis of a series of fluorinated
calixarene ligands, aptly named molecular baskets, has led to
demonstrations of their extractive power for gold(III). Further
research on a series of new 3,5-di(trifluoromethyl)-phenyl
thiourea derivatives proved remarkably successful for the
solubilisation and extraction of gold in sc-CO_2.

67

Gold is a metal that continues to be precious in the market place, in exploration, in medicine and in science and technology *(1)*. Its unique chemical properties ensure that it continues to be researched and applied in many established and in many new fields, including the treatment of arthritis *(2)*, as advanced biocompatible materials in medicine *(3,4)*, in the nanoengineering of optical properties as gold nanoshells *(5)* and in biosensing *(6)*.

This plethora of applications demands efficient methods of extraction and analysis of gold in a variety of matrices. Chromatographic, spectroscopic and atomic absorption spectrometric methods have received particular attention, particularly for application in gold exploration and mining. For example, ion-interaction reversed-phase chromatography *(7,8)* was used to analyse aurocyanide in liquors drawn from the carbon in pulp process and capillary zone electrophoresis (CZE) has been used in the determination of gold (I) and silver (I) cyanide complexes in alkaline cyanide solution *(9)*. Gold is also frequently determined using spot tests and spectrophotometrically, as for example using the chelating reagent 2,3-dichloro-6-(3-carboxy-2-hydrox-1-napthazo)quinoxaline, to yield a yellow brown complex (λ_{max} = 575 nm) *(10)*. However, atomic absorption spectrometry has proven to be an extremely useful technique for the determination of gold in geological materials, especially following an effective gold extraction step.

Meier described a method for the dissolution of gold from geological material into hydrobromic acid and bromine *(11)*. The gold was extracted into methyl isobutyl ketone and determined using an atomic absorption spectrometer equipped with a graphite furnace atomiser. Reddi *et al.* reported on an acid decomposition procedure at room temperature where sample decomposition is effected by treatment with hydrochloric acid and bromine for 24 hr *(12)*. From the resulting bromine extract the noble metal is co-precipitated with tellurium. The precipitated tellurium metal is collected in toluene. The noble metal, together with tellurium, is then re-extracted into *aqua regia* for determination by atomic absorption spectrometry. Several other authors have reported the use of HBr-Br$_2$ and *aqua regia* for the extraction of gold from geological material prior to analysis by atomic absorption *(13-15)*.

Once in solution, more advanced extraction approaches can be used including the use of free or immobilised macrocyclic extractant carriers. Derivatives of 18-crown-6 were used by Yakshin and coworkers *(16)* in the extraction of trace quantities of gold from HCl, while, Fang and Fu *(17)* used benzo-15-crown-5 in the presence of potassium chloride to extract Au^{3+}. Trom *et al.(18)* studied the solvent extraction and transfer through bulk liquid membranes of gold and silver cyanide complexes using dicyclohexano-18-crown-6. The same group investigated the solvent extraction and the transport through a supported liquid membrane of metal cyanide complex salts of gold (I) and silver (I) by macrocyclic extractant carriers *(19)*. Bradshaw and co-workers *(20,21)* have developed silica gel bonded thiamacrocycles, which have shown high selectivity for Au^{3+}. Miller *et al.* *(22)* studied the solvent extraction of gold from alkaline cyanide solution by alkyl phosphorous esters. A method was also developed for the rapid and selective extraction of Au^{3+} with 2-mercapto-benzothiazole into chloroform *(23)*.

Supercritical Fluid Extraction (SFE)

The use of supercritical fluids as alternatives to organic solvents is revolutionising a huge number of important science areas *(24)*. Scientific applications vary from established processes, such as the decaffeination of coffee and the extraction and synthesis of active compounds, to the destruction of toxic waste in supercritical water, the production of nanoparticles and new materials, to novel emerging clean technologies for chemical reactions and extraction.

Supercritical CO_2 has so far been the most widely used because of its convenient critical temperature, cheapness, non-explosive character and non-toxicity. Unmodified sc-CO_2 can be used to extract large organic solute molecules even if they have some polar character; the addition of small amounts of modifiers such as the lower alcohols, extends the use to polar compounds. Above the critical temperature T_C and P_C, increasing the pressure increases the solvating power of the fluid. It is this solvating power that makes supercritical fluids useful in the synthesis and extraction of many important industrial chemicals including natural products, oils, flavors, medicinal compounds and organic pollutants but which with innovative design, can be the medium for selective and efficient environmental processes, for cleaning, for advanced material generation, and for metal extraction and analysis.

Innovative research carried out in this area of green chemistry can alleviate the environmental problems created in many anthropogenic activities, such as the remediation of contaminated soil and waste, and lead to the development of new chemistries for tomorrow's clean technologies. Clean chemistries for precious and heavy metal complexation and extraction by sc-CO_2, capable of replacing environmentally unfriendly processes using solvents or hazardous chemicals, are being researched in our supercritical fluid centre at UCC. Considerable progress has been made on the design and development of new linear and macrocyclic reagents for selective metal extraction. In particular, the synthesis of fluorinated hydroxamic acids has led to demonstrations of their extractive power in supercritical fluid CO_2. In this way the biochelation ability of siderophores found in soil microorganisms has been harnessed for metal extraction from solid samples by SFE *(25)*.

Gold in Supercritical Fluids

Research into elucidating how gold is solubilised and carried, reacts and is extracted under supercritical conditions is of growing scientific importance but has great practical potential in precious metal analysis and recovery. There is an environmental imperative to explore such cleaner technology for efficient gold extraction to avoid environmental damage, such as the extensive environmental contamination that has occurred in the Brazilian Amazon from the mercury-gold amalgamation process.

The published work on the SFE of gold has been limited to date. Wai *et al.* *(26)* have successfully extracted Au^{3+} using bistriazolo-crowns. In the presence of 5% methanol as a modifier to the CO_2 and microlitre quantities of spiked water onto the filter paper containing Au^{3+}, up to 79% was extracted. Otu *(27)* recently reported the desorption of gold from activated carbon using supercritical carbon dioxide. Ion pair solvation of sodium dicyanoaurate $(Na^+...Au(CN)_2^-)$ by tributyl phosphate facilitates the charge neutralisation necessary for the elution of the ionic $Au(CN)_2^-$ by the non-polar supercritical carbon dioxide.

In this chapter, a review of our recent SFE work is provided, covering the design and application of new fluorinated macrocyclic calixarenes and the discovery of the effectiveness of linear flourinated thiourea reagents for gold complexation and extraction in unmodified sc-CO_2.

Experimental

Materials

The thiourea derivative, 1-[3,5-di(trifluoromethyl)phenyl]-2-thiourea (T1) was purchased from Fluorochem Ltd (Derbyshire, UK). A 1000 ppm spectrosol solution of Au^{3+} as an $AuCl_4^-$ solution was obtained from BDH Chemicals Ltd. (Poole, England). All CO_2 gas cylinders were fitted with dip tubes and bought from Irish Oxygen (Cork, Ireland). All extracted samples after SFE were collected in either methyl iso-butyl ketone (MIBK), DMSO (both purchased from BDH) or methanol (Merck, Germany) as indicated.

Synthesis of Ligands

A series of novel fluorinated macrocyclic calixarene and fluorinated thiourea ligands were synthesised at UCC for used in the SFE of metal ions (Figures 1 and 5). Among the fluorinated calixarenes listed, C2-C4 were designed for gold extraction and are of particular relevance in the results presented in this paper. The synthetic methods are either published elsewhere *(28,29)* or in preparation for detailed publication.

Supercritical Fluid Extraction and Flame Atomic Absorption Analysis

Extractions were performed using an Isco SFX supercritical fluid extraction system (Isco Inc., USA, supplied by Jones Chromatography, UK). The SFE system was controlled by the 260D Series Pump controller, allowing programming of modifier addition, pressure automatic refill and continuous flow and consisted of a syringe pump and heated extractor block. A heated variable restrictor was used (part no. 68 395 005) and the flow rate was set at 1.0 ml min^{-1}. Extracted samples were collected in a liquid-trap containing methanol, MIBK or DMSO. The various pressures were set in atmospheres and the temperature of the extractor was set manually. Analysis and detection of extracted samples were carried out using a Pye Unicam SP9 atomic absorption spectrophotometer (FAAS) or the CARY/1E/UV visible spectrophotometer. Unmodified sc-CO_2 (i.e. without the addition of an organic modifier, such as MeOH) was used for all extractions.

Solubility Measurements of Calixarene Ligands in Supercritical CO_2

Solubility measurements of the series of novel fluorinated macrocyclic calixarene ligands were carried out using the Isco SFE system. A weighed amount (*ca.* 80 mg) was placed in an opened ended glass tube (3 x 0.5 cm i.d.), which was plugged with glass wool at both ends and inserted into the extraction cell (2.5 ml) reducing the volume of the cell to 2.2 ml. The sample was statically extracted at 60°C under 200 or 350 atm of sc-CO_2 for 30 min, unless otherwise stated. After this time, the fluid was vented into a collection vial containing 5 ml of MIBK, DMSO or methanol. The sample tube was removed from the cell and weighed. The solubility was calculated from the loss in weight of the sample tube divided by the volume of the extraction cell and given in terms of mmol of sample per litre of CO_2.

Optimisation of SFE of Au^{3+} from cellulose paper

For the Au^{3+} extraction experiments using the fluorinated ligands, 40 μL of Au^{3+} (1000 ppm gold(III) chloride standard, BDH) was spiked onto filter paper (3x1cm). The filter paper was allowed to dry in air for 30 min and was then loaded into the glass tube along with 20 or 30 mg of the ligand as indicated. The glass tube was plugged with glass wool at both ends and also between the ligand and the filter paper. The glass tube was mounted inside a stainless steel extraction vessel, tightened into the heating block and statically extracted using unmodified sc-CO_2 for 30 min at 60°C and at applied pressures between 200 and 400 atm. The extraction cell was then vented into a collection vessel containing 5 ml of methanol, DMSO or MIBK for 15 min (dynamic extraction). For the ligands studied, T1-T4 were soluble in methanol, C2, C3 and C4 were soluble in MIBK and R4 was soluble in DMSO. Having established the optimum pressure for each ligand, the temperature was then varied between 60 and 120°C. Collected extracts were analysed by FAAS or UV-Visible spectrophotometry. The percentage Au^{3+} extracted was determined by direct comparison with collecting solutions spiked with standard Au^{3+} solution.

Finally, the extraction was studied as a function of the amount of water added to the spiked Au^{3+} on the cellulose filter paper. The papers (3x1cm) were spiked with 40 µL Au^{3+} as previously described. The wet filter papers were allowed 30 min to air dry and were each spiked with different amounts of water (0-80 µL). Following the extraction procedure described above, under the optimum temperature and pressure conditions, a static extraction of 30 min followed by 15 min dynamic extraction into a collecting solution of 5 ml methanol, MIBK or DMSO was applied for each sample. The collected solutions after SFE were analysed by FAAS for their gold content.

Results and Discussion

Molecular Baskets in sc-CO₂

Considerable progress has been made at UCC on the design and development of new linear and macrocyclic reagents for metal extraction, based on the chemistries given below. The project work to-date has focussed on organic synthesis, solubility measurements and the study of the efficiency and selectivity of extraction. The first examples of the use of calixarenes in supercritical fluids have been reported by this laboratory (28). The fluorination of calixarenes at the upper rim provides a convenient means to increase the solubility of these molecular baskets in sc-CO₂. These new molecular recognition calixarene reagents have been synthesised to exhibit unique selectivity for metal ions through the use of a suitably sized cavity for complexation.

Solubilities of Calixarene Ligands in sc-CO₂

Solubility measurements were carried out for a number of the above calixarenes, using the method of weight loss, at a temperature of 60°C and at pressures of 200 and 350 atm, including those designed to be gold selective, C2-C4 (measured at 300 atm). The solubility of each ligand increased significantly at the higher pressure studied. The presence of a fluorinated side chain greatly improves the solubility of the calixarenes; this is clearly evident when the solubilities of a range of different fluorinated and non-fluorinated calixarenes are compared. Lower rim functionalisation of the calixarenes to contain chelating groups based on oxygen, nitrogen or sulphur donor atoms lowers the solubility relative to the fluorinated calix[4]arenes R1 and R3 (Table I).

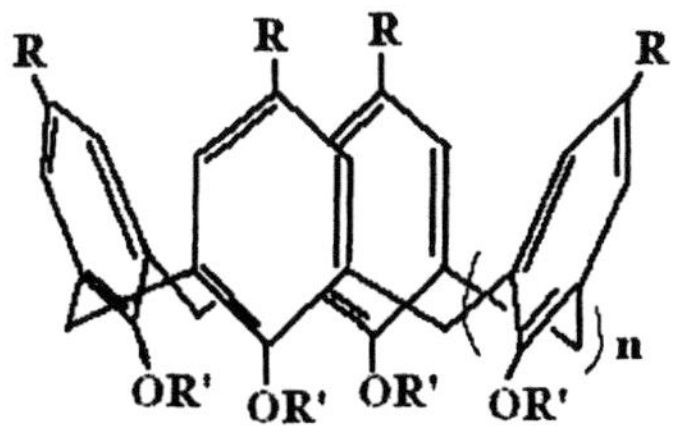

R1^t	R = t-butyl	R' = H	n = 1
R1	R = $(CH_2)_3S(CH_2)_2(CF_2)_7CF_3$	R' = H	n = 1
R2^t	R = t-butyl	R' = H	n = 3
R3^t	R = t-butyl	R' = $CH_2(C=O)OCH_2CH_3$	n = 1
R3	R = $(CH_2)_3S(CH_2)_2(CF_2)_7CF_3$	R' = $CH_2(C=O)OCH_2CH_3$	n = 1
R4^t	R = t-butyl	R' = $CH_2(C=O)NHOH$	n = 1
R4	R = $(CH_2)_3S(CH_2)_2(CF_2)_7CF_3$	R' = $CH_2(C=O)NHOH$	n = 1
R5^t	R = t-butyl	R' = CH_2COOH	n = 1
R5	R = $(CH_2)_3S(CH_2)_2(CF_2)_7CF_3$	R' = CH_2COOH	n = 1
C2:	R = $-(CH_2)_3S(CH_2)_2(CF_2)_7CF_3$;	R' = $-CH_2(C=S)N(C_2H_5)_2$	n = 1
C3:	R = $-(CH_2)_3S(CH_2)_2(CF_2)_7CF_3$;		
	R' = $-CH_2(C=O)N(CH_2CH_2SH)(CHCH_3CH_2CH_3)$		n = 1
C4:	R = $tert$-butyl- ;		
	R' = $-CH_2(C=O)O(CH_2)_2NH(C=S)NH(CH_2)_3S(CH_2)_2(CF_2)_7CF_3$		n = 1

Figure 1: Chemical structures of the macrocyclic calixarenes synthesised and studied in this research work.

Table I. Measured solubilities of selected fluorinated and non-fluorinated calixarenes in sc-CO$_2$

Ligand	60°C/200 atm mmol/L	60°C/350 atm mmol/L
R1	>120	>120
R1^t	0.62	1.12
R2^t	0.50	0.70
R3	>94	>94
R3 ^{t}	0.18	0.27
R4	1.64	3.55
R4^t	0.10	1.01
R5	0.03	0.03
C2	0.12	2.31*
C3	0.13	0.29*
C4	0.74	1.96*

(300 atm)

SFE of Au^{3+} using Macrocyclic Ligands

The SFE of metal ions as the guest in a "*host-guest*" cavity of fluorinated tetrameric calixarenes in unmodified sc-CO_2 has been described previously *(28)*. The selective extraction of Fe(III) in unmodified sc-CO_2 from a range of other metal ions was monitored by ion chromatography (30). The present work focuses on the series of fluorinated calixarenes designed to be Au^{3+} selective in unmodified sc-CO_2. This selectivity was examined using sulphur donor atoms in lower rim functionalisation with thioamide, thiol and thiourea groups, combined with upper rim fluorination. The results obtained from the variation of temperature and pressure are given below in Figures 2 and 3 for this C2-C4 series.

Using a pressure of 400 atm and analysing the collected extracts for traces of Au^{3+} by FAAS, the temperature was varied between 60 and 120°C (Figure 2). No extraction of Au^{3+} was recorded for R4 (collected in DMSO) as expected under any conditions.

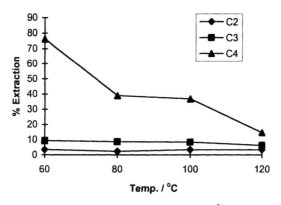

Figure 2. Effect of temperature on the SFE of Au^{3+} from cellulose paper at a pressure of 400 atm using synthesised fluorinated tetrameric calixarene ligands (20 mg ligand, 40 µL Au^{3+}, 30 min static, 15 min dynamic). (Reproduced with permission from reference 29. Copyright 1999)

The percentage gold extraction versus operating pressure is plotted in Figure 3.

Traces of Au^{3+} were extracted with C2 ($\approx$ 17%) and C3 ($\approx$ 9 %) at the optimum conditions, 60°C, 200 atm and 60°C and 400 atm, respectively. Much higher extractions were recorded with the thiourea bearing calixarene, C4. With increased applied pressure, the solubility of C4 in sc-CO_2 increases and the percentage extraction of Au^{3+} increases accordingly. The experiment involving 400 atm was the only pressure where all the C4 ligand was completely solubilised from the glass tube during SFE.

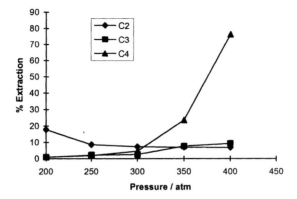

Figure 3. Effect of pressure on the SFE of Au^{3+} using synthesised fluorinated tetrameric calixarenes at their optimum temperatures (40 μL Au^{3+}, 20 mg ligand, 30 min static, 15 min dynamic). (Reproduced with permission from reference 29. Copyright 1999)

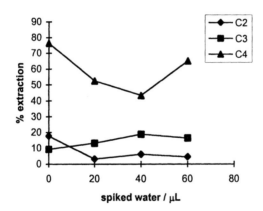

Figure 4. Effect of water on the SFE of Au^{3+} from cellulose paper using synthesised fluorinated calixarene ligands at their optimum temperature and pressure conditions (40 μL Au^{3+}, 20 mg ligand, 30 min static, 15 min dynamic).

The effect of spiked water on the SFE of Au^{3+} is shown in Figure 4. Several authors have studied the beneficial effects of water on the extraction of metal ions in sc-CO_2 (31,32). The addition of water decreased the amount of Au^{3+} extracted for C2 and C4. However, an opposite trend was observed for the ligand C3 where the percentage extraction was seen to increase from ≈ 9 % to ≈ 19%.

SFE of Au^{3+} using Fluorinated Thiourea Derivatives

The performance of the fluorinated calixarene thiourea reagent C4 lead to further investigations of simpler fluorinated thiourea derivatives *(33)* such as shown in Figure 5. Some preliminary findings are reported here that testify to the effectiveness of these ligands and details will be published elsewhere.

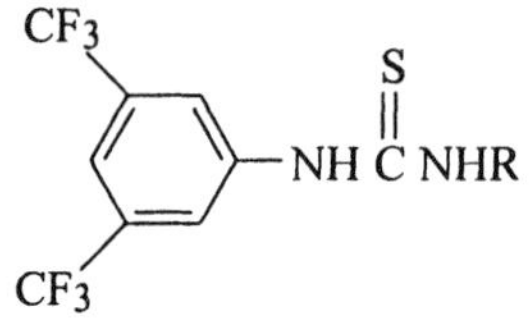

Ligand:	R:
T1	H
T2	CH$_2$CF$_3$
T3	CH$_2$(CF$_2$)$_2$CF$_3$
T4	(CH$_2$)$_2$SH

Figure 5. Chemical structures of the linear thiourea ligands studied in this work.

Initial experiments were performed to determine the optimum conditions for the SFE of Au^{3+} from cellulose paper using different fluorinated thiourea ligands (T1-T4) in unmodified sc-CO$_2$. Like calixarene compounds, the extraction of these ligands is readily monitored by UV-visible spectroscopy. Figure 6 shows the UV spectra recorded for T1 after SFE at 60°C and at pressures between 200 and 400 atm.

With an increase in pressure the solubility of the ligand in sc-CO$_2$ clearly increased. Even under these conditions (temperature 60°C and pressure 400 atm), T1 however, remained very insoluble in sc-CO$_2$ with most of the ligand remaining in the glass tube after SFE and gold extraction efficiencies were very low. However the optimum pressure for the SFE of Au^{3+} using T2, and T3 was significantly lower at 250 atm and 60°C.

These longer chained fluorinated thiourea ligands, in particular T3, have much higher solubility in sc-CO$_2$ at lower operating pressure and are very efficient ligands for the extraction of Au^{3+}. Increasing temperature or the addition of spiked water had marginally detrimental effects on the extraction of gold. Degree and position of fluorination, together with chain length, clearly determine the efficiency of extraction *(34)*.

Conclusions

From a series of fluorinated calixarenes, functionalised at the lower rim to contain oxygen, nitrogen and sulphur donor atoms, a calix[4]arene thiourea derivative proved to be an effective Au^{3+} extractant in unmodified sc-CO$_2$.

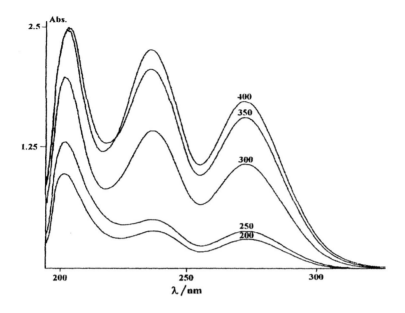

Figure 6. UV spectra recorded in methanol for the extraction of T1 at different pressures in the presence of filter paper spiked with 40 μL Au^{3+} (20 mg ligand, 60°C, 30 min static, 15 min dynamic).

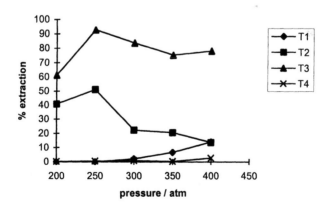

Figure 7. Effect of pressure on the SFE of Au^{3+} from cellulose paper using four selected thiourea reagents (20 mg each, 40 μL Au^{3+}, 60°C, 30 min static, 15 min dynamic).

Simpler fluorinated 3,5-di(trifluoromethyl) phenyl thiourea derivatives, however, displayed higher solubilities in sc-CO_2, at lower operating pressure and proved to be more effective ligands for the extraction of Au^{3+}. Further studies on the optimisation of ligand design and on the applications of these reagents for precious metal extraction in sc-CO_2 are in progress.

Acknowledgements
The authors would like to thank Enterprise Ireland (formerly Forbairt) for Basic Research Grant SC/97/504 and for Strategic Research Grant ST/97/413. Acknowledgement is made to the Donors of The Petroleum Research Fund, administered by the American Chemical Society, for partial travel support to the National ACS meeting in Orlando, Florida.

References

1. *Gold – Progress in Chemistry, Biochemistry and Technology*; Schmidbaur H., Ed.; John Wiley & Sons (Chichester, England) 1999, ISBN 0-471-97369-6.
2. Kothbauer O. Implantation of gold particles into acupuncture points for treatment of painful hip processes in dogs: Three cases. *Wiener Tierarztliche Monatsschrift* **1997**, 84: (2), 47-52.
3. Grabar, K.C.; Deutsch J.E.; Natan M.J. Polymer-suported Gold Colloid Monolayers – A new approach to biocompatible metal-surfaces. *Abstracts of papers of the American Chemical Society* **Apr. 2 1995**, 209:27-Poly, Part 2
4. Ignatius, M.J.; Sawhney, N.; Gupta, A.; Thibadeau, B.M.; Monteiro, O.R.; Brown, I.G. Bioactive surface coatings for nanoscale instruments: Effects on CNS neurons. *Journal of Biomedical Materials Research* **May 1998**, 40:(2) 264-274.
5. Halas, N.J. (Rice University) "Nanoengineering of Optical Properties" lecture summary at *Commercialisation of Nanostructured Materials and Carbon Nanotubes Conferences*, **April 6-11 2000**, Wyndam Miami Beach, Florida, USA.
6. Leech, D., *Chem. Soc. Rev.*, **1994**, 23, 205-213.
7. Haddad, P.R.; N.E. Rochester, N.E. *J. Chromatogr.*, **1988**, 39, 23-36.
8. Otu, E.O.; Robinson, C.W.; Byerley, J.J. *Analyst*, **1993**, 118 1277-1280.
9. Aguilar, M.; Farran, A.; Martinez, M. *J. Chromatogr.*, **1993**, 635, 127-131.
10. Amin, A.S.; Shakra, S.; Abdalla, A.A. *Bull. Chem. Soc. Jpn.*, **1994**, 67, 1286-1289.
11. Meier, A.L. *J. Geochem. Explor.*, **1980**, 13 (1980) 77-85.
12. Reddi, G.S.; Ganesh, S.; C.R.M. Rao, C.R.M.;Ramanan, V. *Anal. Chim. Acta*, **1992**, 260, 131-134.
13. Chattopadhyay, P.; Sahoo, B.N. *Analyst*, **1992**, 117, 1481-1484.
14. Rivoldini, A.; Haile, T. *Atomic Spectroscopy*, **1989**, 10 No.3, May-June.
15. Tewari, R.K.; Gupta, J.R. *Atomic Spectroscopy*, **1993**, 14 No.5 Sep/Oct.
16. Yakshin, V.V.; Vilkova, O.M.; Tsarenko, N.A. *J. Anal. Chem.*, **1994**, 49, 158-161.

17. Fang, S.; L. Fu, L. *Ind. J. Chem.*, 33A (1994) 885-888.
18. Trom, M.; Burgard, M.; El-Bachiri, A. *Analusis*, **1991**, 19, 97-102.
19. Trom, M.; Burgard, M.; Leroy, M.J.F.; Prevost, M. *J. Membrane Sc.*, **1988**, 38, 295-300.
20. Bruening,R.L.; Tarbet, B.J.; Krakowiak, K.E.; Bruening, M.L.; Izatt, R.M.; Bradshaw, J.S. *Anal. Chem.*, **1991**, 63, 1014-1017.
21. Bruening,R.L.; Tarbet, B.J.; Krakowiak, K.E.; Izatt, R.M.; Bradshaw, J.S. *J. Heterocyclic Chem.*, **1990**, 27, 347-349.
22. Miller, J.D.; Wan, R.Y.; Mooiman, M.B.; Sibrell, P.L. *Sep Sc. & Tech.*, **1987**, 22, 487-502.
23. Rajadhyaksha, M.; Turel, S.R. *J. Radional. Nuc. Chem., Letters*, **1985**, 96, 293-300.
24. Clifford, T.; Bartle, K. "Chemistry goes Supercritical" *Chemistry in Britain* **1993** (June) 499-502.
25. Glennon, J. D.; Hutchinson, S.; Walker, A.; Harris, S. J.; McSweeney, C. C. *J. Chromatography A* **1997**, *770*, 85.
26. Wang, S.; S. Elshani, S.; Wai, C.M. *Anal. Chem.*, **1995**, 67, 919-923.
27. Otu, E.O. *Sep. Sci. & Tech.*, **1997**, 32, 1107-1114.
28. Glennon, J.D.; Hutchinson, S.; McSweeney, C.C.; Harris, S.J.; McKervey, M.A. "Molecular Baskets in Supercritical CO_2" *Analytical Chemistry* **1997**, 69(11), 2207-2212.
29. Glennon, J.D.; Harris, S.J.; Walker, A.; McSweeney, C.C.; O'Connell, M. " Carrying Gold in Supercritical CO_2" *Gold Bulletin* **1999**, 32(2), 52-58.
30. O'Connell, M.; O'Mahony, T.; O'Sullivan, M.; Harris, S.J.; Jennings, W.B.; Glennon, J.D. "Monitoring the Selectivity of Supercritical Fluid Extractions of Metal Ions using Ion Chromatography" Advances in Ion Exchange for Industry and Research: Proceedings of the Fifth International Conference and Industrial Exhibition on Ion Exchange Processes (Ion-Ex '98) Wrexham 5-9th July 1998, A. Dyer, A.; P.A. Williams, P.A. Eds.; Royal Society of Chemistry Special Publication No. 239, 1999, 137-143.
31. Lin, Y.; Brauer, R.D.; Laintz, K.E.; Wai, C.M. *Anal. Chem.*, **1993**, 65, 2549-2551.
32. Lin, Y.; Wai, C.M. *Anal. Chem.*, **1994**, 66, 1971-1975.
33. Glennon, J.D.; Harris, S.J. PCT Patent Application No. PCT/IE 00/00018 February 8th 2000.
34. Smart, N.G.; Carleson, T.E.; Elshani, S.; Wang, S.; Wai, C.M., *Ind. Eng. Chem. Res.*, **1997**, 36, 1819-1826.

Chapter 7

Ligand-Assisted Extraction of Metals from Fly Ash with Supercritical CO$_2$: A Comparison with Extraction in Aqueous and Organic Solutions

Christof Kersch[1], Daniela Trambitas[1], Geert F. Woerlee[2], and Geert J. Witkamp[1]

[1]Laboratory for Process Equipment, Leeghwaterstrasse 44, 2628 CA Delft, The Netherlands
[2]FeyeCon D&I B.V., Rijnkade 17a, 1382 GS Weesp, The Netherlands

The extraction efficiencies (EE) for the removal of *Cu, Cd, Pb, Mn*, and *Zn* from fly ash are investigated for supercritical fluid extraction (SFE) with CO$_2$/Cyanex 302, for leaching with different acidic solutions and for solvent extraction with Cyanex 302 in Kerosene. Although initially delayed, after some time the EE obtained by SFE is promoted by water. Leaching at pH=3 proceeds faster and more complete than SFE, although both have a similar expected pH at the solid surface. An empirical fitting model is presented for the aqueous leaching of fly ash. The dissolution-desorption step from the solid phase is probably the limiting step of the overal extraction.

Large volumes of fly ashes are produced by municipal waste incinerators. The amount of produced fly ash is increasing due to an increasing number of incineration plants. In the Netherlands and Germany, together approximately 35 million tons of municipal waste are burned annually, generating about 1.5 million tons of fly ash (1, 2). Fly ashes contain a complex system of crystalline phases and a range of leachable metals (3, 4). This fly ash is contaminated with toxic heavy metals. These (toxic) metals leach out after contact with water, and pollute the groundwater. Therefore, isolated and expensive disposal of the ash is required. It is becoming increasingly important to find ways of utilising this fly ash. For reuse of ash as filler for cement or pavements only a minimal leachability of metals is allowed by national legislation. SFE offers a method to reduce the metal content to such an extent, that leachability is reduced and the demands of legislation are observed.

Supercritical (SC) CO_2 is advantageous for extraction of metals from solid particles such as e.g. fly ash (5), because CO_2 is a benign and cheap solvent for which suitable extractants are available. SFE does not require any expensive drying of the final product. Evaporation of solvent CO_2 by release of pressure results in both a solvent free matrix and a separate metal-extractant complex.

The extraction efficiency is enhanced by addition of small amounts of water to the solid phase before or during SFE. This positive effect has been shown by the group of Wai (6, 7) and by Kersch *et al.* (8). Water leads to dissolution of metal-anions such as oxides *MeO*, eq. 1, and dissociation to metal cations Me^{2+}. The metal cation then further reacts, according to eq. 2, with a monovalent extractant (ligand) anion X:

$$MeO^{(S)} + 2H^{+(L)} \;\rightleftharpoons\; Me^{2+(L)} + H_2O^{(L)} \tag{1}$$

$$2HX^{(G)} + Me^{2+(L)} \;\rightleftharpoons\; MeX_2^{(G)} + 2H^{+(L)} \tag{2}$$

The aim of this work was to study the influence of water on metal extraction from the fly ash, for SFE at a scale of 2 kg solids. The final aim is to provide design parameters for a larger scale SFE unit. We concentrate on the main metals of municipal solid waste incinerator (MSWI) fly ash (*Zn, Pb, Cu, Cd,* and *Mn*, Figure 1). Since the *ZnO* content of the studied ash was high, *Zn* was chosen as model compound. Three types of experiments were performed:

1. SFE from MSWI fly ash (2 kg) with various ash humidities (2wt%. and 38wt.%, see Figure 2)
2. Aqueous leaching of MSWI fly ash under various pH values:
   The presence of CO_2 influences the pH of an aqueous phase, due to formation and dissociation of carbonic acid. The resulting pH is equal to values of about 2.9 at 200 bar and 40°C (9). The dissolution of metals in water determines the extent of leaching and extraction. To study the pH

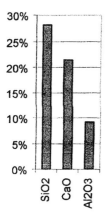
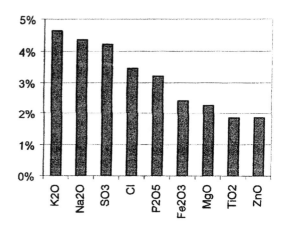
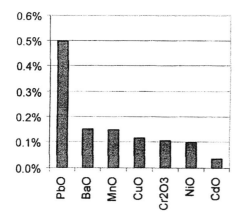

Figure 1. XRF-analysis of the original MSWI fly ash (wt.%) calculated as oxides

effect on that dissolution, aqueous leaching experiments at constant pH=3, pH=4, and pH=5 have been carried out.

3. Solvent extraction of MSWI fly ash with various ash humidities:
    In a third series of experiments, extractant Cyanex 302 (Bis(2,4,4 trimethyl-pentyl)monothiophosphinic acid) was dissolved in Kerosene. These experiments were carried out to (i) study the influence of extractant on the metal ions in solid phase with and without a layer of water, and to (ii) compare the extent of extraction with SFE.

Theory

A stepwise extraction mechanism is shown in Figure 3. The fly ash contains a number of metals, as presented in Figure 1. Exemplarily the extraction of zinc(II) ion (Zn^{2+}) with the acidic chelating agent HX=Cyanex 302 is described.

Dissolution and desorption of ZnO at the solid-liquid phase

This step includes both dissolution of ZnO and then desorption of Zn^{2+}. The pH dependent solubilisation of ZnO with the release of oxide and the consumption of a hydrogen ion is expressed with

$$ZnO^{(S)} + 2H^{+\,(L)} \;\rightleftharpoons\; Zn^{2+(S)^*} + H_2O^{(L)}. \qquad (3)$$

with $^{(S)}$ in the solid phase,
$^{(L)}$ in the liquid phase, and
$^{(S)^*}$ at the solid surface.

The desorption not only depends on the dissolved concentration of Zn^{2+} at the surface of the solid particle $C_{Zn}^{(S)^*}$ [mol m^{-3}] but also on the gradient of that concentration perpendicular to the surface. The desorption is described by

$$J^S = k_S \cdot \left(C_{Zn}^{(S)^*} - C_{Zn}^{(L)^*} \right) \qquad (4)$$

where J^S [mol m^{-2} s^{-1}] is the amount of Zn^{2+} transferred into the liquid phase,
k_S [m s^{-1}] is the overall rate constant, and
$C_{Zn}^{(L)^*}$ [mol m^{-3}] the concentration in the liquid near the solid surface.

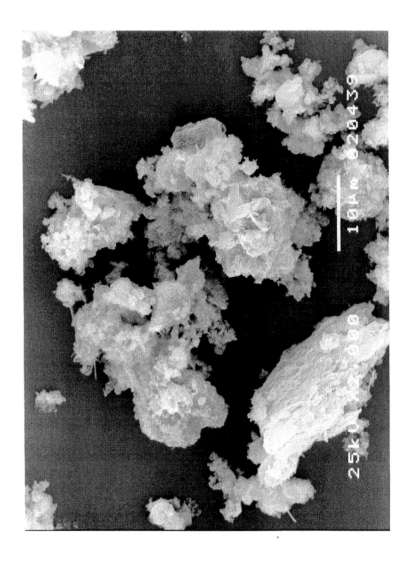

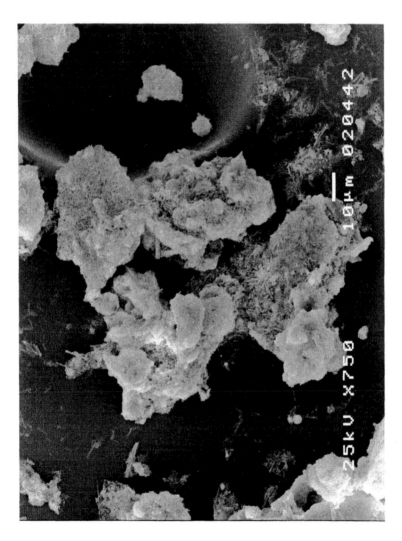

Figure 2. SEM photos of MSWI fly ash (a) 2wt.% humidity (×2 000), and (b) 38wt.% humidity (×750)

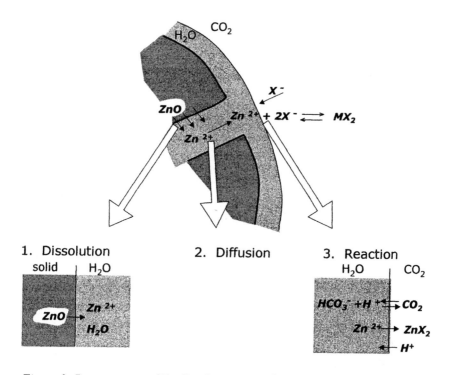

Figure 3: Process steps of Zn dissolution, transfer and extraction from wet fly ash

Diffusion of Zn^{2+} through the aqueous phase

An assumed fickian diffusion of Zn^{2+} through the aqueous phase to the H_2O-CO_2 interface is expressed as,

$$J^D = -D_{Zn^{2+}}^{(L)} \frac{\varepsilon}{\tau} \nabla C_{Zn^{2+}}^{(L)} \qquad (5)$$

where J^D [mol m^{-2} s^{-1}] is the Zn^{2+} flux transferred through the aqueous phase to the CO_2 phase,

$D_{Zn^{2+}}^{(L)}$ [m s^{-1}] is the binary diffusion coefficient,

ε [1] the bed porosity,

τ [1] the pore tortuosity, and

$C_{Zn^{2+}}^{(L)}$ [mol m^{-3}] the concentration of Zn^{2+} in the liquid phase.

The average porosity ε is the ratio of porous volume V_P to total volume apparent volume V_A.

Chemical reaction at liquid-SC CO_2 interface

At the liquid-SC CO_2 interface two reactions takes place,

$$CO_2^{(G)} \rightleftharpoons CO_2^{(L)} \qquad (6)$$

$$H_2O^{(L)} + CO_2^{(L)} \rightleftharpoons H_2CO_3^{(L)} \rightleftharpoons HCO_3^{-\,(L)} + H^{+\,(L)} \qquad (7)$$

Due to the reactions in eq. 6 and 7, the pH of the liquid phase is estimated to be about 3 (9). The zinc cation reacts with the extractant as:

$$2HX^{(G)} + Zn^{2+\,(L)} \underset{\longleftarrow}{\overset{K}{\longrightarrow}} ZnX_2^{(G)} + 2H^{+\,(L)} \qquad (8)$$

where $K = \dfrac{[ZnX_2^{(G)}]\,[H^{+\,(L)}]^2}{[HX^{(G)}]^2\,[Zn^{2+\,(L)}]}$.

From this reaction, only the reaction to the right is considered, resulting in Zn-complex. Whereas the extractant Cyanex 302 is virtually insoluble in the aqueous phase, the Zn^{2+} ion is insoluble in pure SC CO_2. Due to its insolubility in CO_2, the released hydrogen is transferred into the aqueous phase. Since that reaction is fast, its rate does not limit the overall extraction. The reaction of Zn^{2+} into the SC CO_2 phase depends only on a reaction rate constant k' and the concentration of Zn^{2+},

$$r_{apparent} = k' \cdot C^{(L)}_{Zn^{2+}} \qquad (9)$$

where $r_{apparent}$ [mol m^{-2} s^{-1}] is the rate of Zn reaction,

$C^{(L)}_{Zn^{2+}}$ [mol m^{-3}] the concentration of Zn^{2+} in the liquid phase, and

k' [m s^{-1}] rate constant.

Experimental

Reagents

Fly ash was supplied by a municipal waste incinerator plant (AVR) in Rotterdam. For SFE the acid extractant Cyanex 302[*] (Bis(2,4,4 trimethylpentyl)monothiophosphinic acid), kindly donated by Cytec Industries, Canada, was employed as extractant. The pH of the aqueous solution was set with analytical grade nitric acid from J.T. Baker. Solvent extraction was carried out with Kerosene from Chemproha.

Supercritical Fluid Extraction (SFE)

A flow diagram of the process is shown in Figure 4. CO_2 was supplied from a liquid storage vessel (V4), maintained at a pressure of approximately 5 MPa.

---

[*] Cyanex 302 has been shown to be an effective extractant for a variety of metal ions. In combination with organic solvents successful extraction of Cd^{2+} (10) and Pb^{2+} (11) was carried out from aqueous solutions and at pH 3 the extraction efficiency for Cu^{2+}, Zn^{2+}, and Mn^{2+} was 100%, 100%, and 20% respectively (12, 13). With SFE at 200 bar, the extraction efficiency of Cd^{2+}, Cu^{2+}, Pb^{2+}, and Zn^{2+} from sand was about 100%, 90%, 80%, and 75% respectively (14).

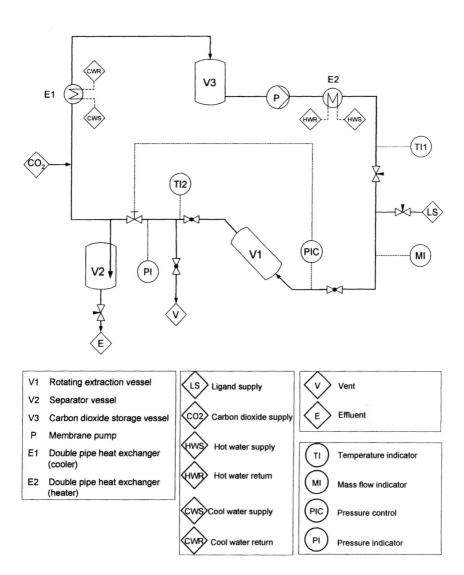

Figure 4. Flow diagram of the supercritical extraction system

That vessel was filled with liquid CO_2 from the recycle stream and fresh CO_2 from storage. CO_2 was pressurised and passed through a heat exchanger (E2) before it entered the preheated 12 litre extraction vessel (V1). After the extraction vessel, the pressure of the CO_2 stream was reduced to 5 MPa via a pressure control valve. This extract was fed into a 17 liter separator (V3) containing 3 litre of 3 M nitric acid to collect the metal-ligand complex. The expanded neat CO_2 was evaporated from the separator, condensed in a heat exchanger (E1) and then returned to the CO_2 storage vessel (V4).

In each of the experiments 2000 g fly ash was placed in the pre-heated (about 40°C) extraction vessel. After pressurisation, warm neat CO_2 was passed through the vessel for about 60 min to heat the fly ash, to attain a constant flow. Metal extraction was commenced by continuously adding the complexing agent into the warm CO_2 flow. The working conditions were 40°C and 20 MPa. The tilted extraction vessel contained a mixing device that maintained both continuous mixing and extraction when the vessel was rotated at (50 rpm). After extraction, a washing step with pure CO_2 removed the remaining complexing agent and metal complexes.

The fly ash that was used for all experiments came from a single homogenous 200 kg batch. This batch showed variation in metal concentration of up to 15% (sample size 0.5 g). To compensate for this variation a homogenous sample (0.5 g) of untreated ash was taken prior to each extraction and analysed. The variation of those samples was then about 5%. An average of the analyses served as a reference for every set of extraction experiments, allowing a more reliable expression of the extraction efficiency. For the studies of wet ashes, homogeneous humidification was desired. The ash was mixed with neutral water (pH=7) prior to SFE, an excess of water was removed after sedimentation, and the ash was dried to a humidity of 38wt.%.

Aqueous leaching

All tools or parts used during the experiments were either cleaned or new to insure analytical metal free conditions. A 500 ml glass beaker was filled with 400 g water and the pH was adjusted to the desired values (pH=3, pH=4, pH=5) at 21(±0.5)°C. The pH electrode (model PHM95, MeterLab) was connected with a dosimeter (model 614+665, Metrohm), and setpoint for the desired pH was chosen. The leaching experiment was commenced by addition of 4.00 g (±0.01) fly ash into the stirred mixture. A magnetic stirrer was used to obtain homogeneity and complete suspension of fly ash. The dosimeter ensured a constant pH by addition of a 6wt.% metal free nitric acid solution. The addition of acid was registered during the experiment to correct for concentration changes. Leaching samples were withdrawn periodically with a PP+PE syringe, about 1cm below the liquid surface. To restrain ash particles, the samples were injected through a filter into PE sample pots and diluted with nitric acid and analysed by ICP-MS. The volumes withdrawn were sufficiently small (1.5 ml) and influences on the extraction were negligible. All experiments were carried out at 20.5-22°C and the temperature change during each test was <1°C.

Solvent Extraction (SE)

Extractant Cyanex 302 (6.7±0.5 g) was placed in a metal free glass beaker (500 ml) and 335g (±0.3) Kerosene were added. Extraction at 21(±0.5)°C commenced by addition of 4.00 g (±0.01) fly ash into the stirred (750 rpm) mixture, using a magnetic stirrer for complete suspension of fly ash. Liquid samples were taken with a PP+PE syringe. To restrain ash particles, the samples were injected through a filter into metal free PE sample pots (0.25 g). After addition of 70wt.% HNO_3 and 70wt.% $HClO_4$ (~5 g and 1 g respectively) and subsequent microwave digestion (for about 20 min), the samples were diluted with nitric acid and analysed by ICP-MS.

Extraction efficiency

For the determination of the extraction efficiency (EE) of SFE experiments the solid fly ash (0.2 g) was mixed with 70wt.% HNO_3, 38wt.% HCl, and 40wt.% HF (5 ml each). The microwave digested (30 min) samples were mixed with 4wt.% H_3BO_3 (15 ml), heated in the microwave (10 min) and diluted with 3wt.% HNO_3, prior to analysis with ICP-MS. The SFE-processed and unprocessed fly ash were compared (always referring to dry initial ash) and the EE was defined as,

$$EE = 1 - \frac{c_{metal}^{ash} \left[ppm \right]}{c_{metal,t=0}^{ash} \left[ppm \right]} . \tag{10}$$

Liquid phases were analysed by ICP-MS for the determination of EE after both aqueous leaching and solvent extraction. Based on these analyses, the amount of extracted metals per mass of dry initial fly ash was related to the initial metal concentration of the applied fly ash as,

$$EE = \frac{m_{metal}^{liquid} / m_{ash,t=0} \left[mg / kg \right]}{c_{metal,t=0}^{solid} \left[mg / kg \right]} . \tag{11}$$

Results and discussion

SFE

Figure 5 shows the extraction efficiency (EE) for six different experiments at $T=40°C$, $p=20$ MPa, with a CO_2 flow of about 9.5 kg/h and concentration of 0.11 mol% Cyanex 302 in CO_2. The water content was 2wt.% and 38wt.%, and time between 3600 s and 22000 s. For both, Zn and Pb, addition of water enhanced the EE. Where after 3600 s almost no extraction of dry ash was obtained, the wet ash resulted in EE of Zn and Pb, of 14±2% and 21±3% respectively. The curves for Zn and Pb in Figure 5 suggest a slight increase of EE with longer time. The extraction level of Cd, Mn, and Cu was fair, in the order of about 50%. However, the decrease of EE in time, for Cd and Mn, was unexpected. Further research is required, looking for co-precipitation for these elements as (bi-) carbonates. Prolonged extraction with SC CO_2 at 40°C lead to decrease in fly ash humidity from initially 2wt.% and 38wt.% to 0.6wt% and 18wt% (after 6 h). The continuous addition of water to the SFE extraction system might be recommendable.

Aqueous leaching

A first leaching test with neutral water (pH=7) and a liquid solid ratio of L/S=10 resulted in a final pH value of about pH=11.7. This effect is due to the basic character of the fly ash, due to its high content of calcium (20wt.%), aluminum (9wt.%), potassium (5wt.%), and sodium (4wt.%) mainly as oxides (Figure 1). After $4·10^3$ s, the obtained leaching level was constant at a very low level for both Zn (0.11±0.01%) and Pb (0.98±0.01%).

In a second series of leaching experiments, the three pH values in the relevant range for water-CO_2 were studied. Throughout the experiments at 21(±0.5)°C, the pH was kept constant at pH=5±0.01, pH=4±0.01, pH=3±0.01 by controlled addition of acid. Figure 6 shows that extraction efficiencies of Zn, Pb, Cd, Cu, and Mn increased dramatically with decreasing pH. After $5·10^4$ s the EE of Zn varied from 15% at pH=5, to 78% at pH=4 and 83% at pH=3, EE of Cd was 20%, 36%, and 51%, and EE of Cu was 10%, 60%, and 78% respectively. The extraction of Zn and Cu at pH=3 seemed to approach almost 100%, but also the curves at pH=4 and pH=5 promise a further increase for longer extraction time. Whereas extraction of Pb was poor at pH=5 and pH=4 (2-7%), a great rise in extraction was obtained at pH=3 to values of 45%. However, after an extraction maximum at $3·10^3$ s, triple reproducible extraction tests at pH=3

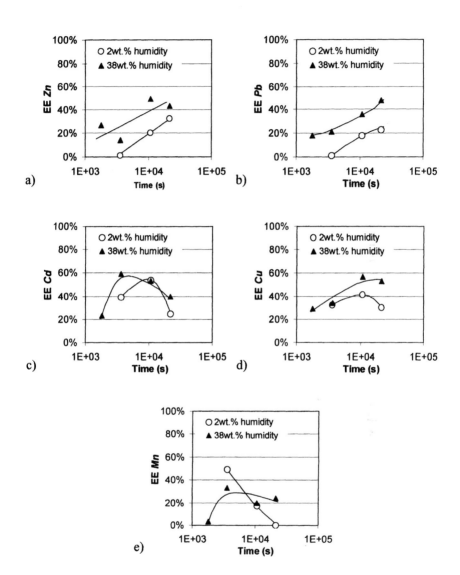

Figure 5: Extraction efficiency (%) of SFE, (total L/S ratio): 1800 s (14), 3600 s (18), 11000 s (36), 22000 s (53), a) Zn, b) Pb, c) Cd, d) Cu, e) Mn

94

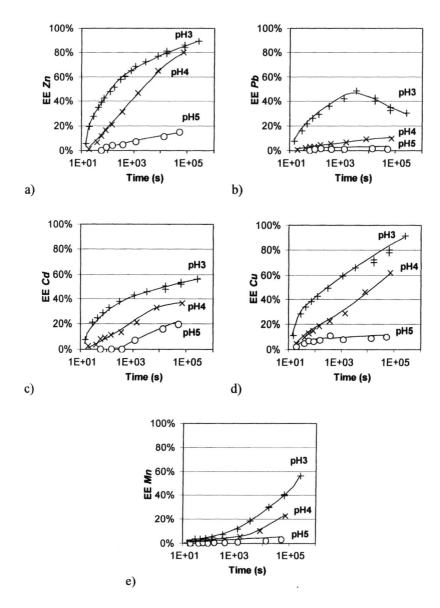

Figure 6: Extraction efficiency (%) of aqueous leaching from MSWI fly ash, a)
Zn ±1%, b) Pb ±2%, c) Cd ±4%, d) Cu ±5%, and e) Mn±2%), L/S=100

resulted unexpectedly in a decrease of EE with time. This might be related to precipitation or reverse reactions, due to an enrichment of other compounds during the batch-process.whereas the extraction rate of *Zn, Pb, Cd,* and *Cu* declined with time, the extraction rate of *Mn* increased over the whole time span (Figure 6e) for pH=4 and pH=3. The sequence of EE at e.g. $5 \cdot 10^4$ s is for pH=3 and pH=4 *Zn>Cu>Cd>Mn>Pb*, is for pH=5 *Cd>Zn>Cu>Mn>Pb*.

The addition rate of the acid is an indication for the reactivity and leaching behaviour of the metals. The semi-log plot in Figure 7 indicates, that lower pH values lead to fast leaching, in agreement with the metal leach plots in Figure 6. Addition of 10 ml HNO_3 dissolved about 0.3 g fly ash corresponding with almost 10% of the applied sample.

Solvent extraction

Figure 8 shows an increasing extraction behaviour with time for all studied metals. The extraction of *Zn* from dry ash initially ($t=30$ s) was about 20%, whereas with increasing humidity to 4.3wt.% and 16wt.% the initial EE minimised to 16% and 0% respectively. After $6 \cdot 10^4$ s, however, the extracted values were similar for all three humidities. The extraction of *Pb* from dry ash commenced with 25% and increased to 70% for dry ash and 80% for wet ashes. In both cases, dry ash resulted in initially similar EE and same or lower EE after some process time. Dry ash gave a faster start. On the longer run, however, dry ash gave no better performances than the humid ash. The extraction from humid ash increased, due to constant removal of metal from the water phase into the solvent phase. Only for *Cd* the humid fly ash lead to higher EE values than the dry ash. During SE with Kerosene, the pH of aqueous phase surrounding the particle can not be adjusted. An equilibrium pH of about 12 in that phase is expected, as in an experiment with neutral water (see previous section 'Aqueous leaching'). Such high pH retards dissolution. Built-up compounds (such as *Na, K, Ca*) in the more alkali system may retard the dissolution of all metals.

Comparison of aqueous leaching, SE and SFE

From the above it can be concluded that for extraction with SC CO_2 or Kerosene, water is not a necessary attribute. With SFE, however, an expected extraction as fast as for leaching at pH=3 was not observed. This might be explained either by formation and/or precipitation of (bi-)carbonates or by decrease of humidity. In contrast with aqueous leaching of *Pb* at p, where unexpectedly the EE decreased with time, the extraction of *Pb* during the continuous SFE kept rising in time. With SE, the initial extraction of *Zn* from dry

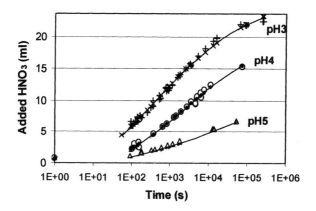

Figure 7: Required addition of 6wt.% HNO₃ (ml) to maintain a constant pH during aqueous leaching from MSWI fly ash at pH=3, pH=4, and pH=5

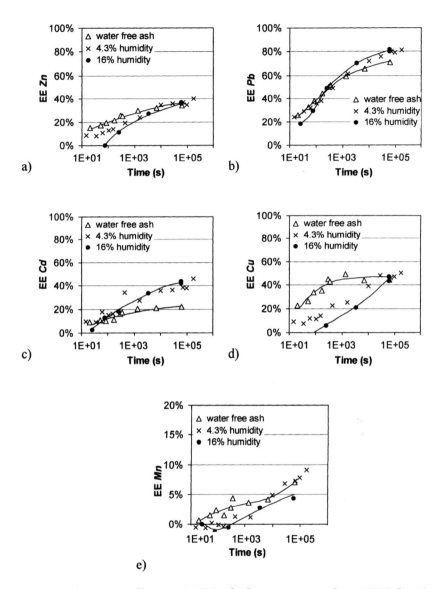

Figure 8: Extraction efficiency (±1%) of solvent extraction from MSWI fly ash with Kerosene and 0.9 mol% Cyanex 302, L/S=84, a) Zn, b) Pb, c) Cd, d) Cu, e) Mn

ash was faster compared to SFE, but it had the same values after $2 \cdot 10^4$ s. For extraction of Zn from wet ash, however, SFE was better. A lower pH of the water phase, caused by dissociation of carbonic acid, seems to have improved that Zn-extraction. SE with either dry or wet ash resulted in more extraction of Pb than SFE. For Cd, Mn, and Cu similar plots were obtained.

The fast and complete aqueous leaching at pH=3, however, shows a tempting possibility to enhance the extraction by either extraction with excess acid or a pre-wetting step and a larger amount of (dispersed) water in SC CO_2.

Empirical model of aqueous leaching

The experimental data of aqueous leaching under constant pH were fitted with Avrami-Erofeev rate law (15) for reactions in the solid state. The reactions of metal dissolution, desorption and transport were combined into one reaction, which is expressed as

$$\left[-\ln\left(\frac{EE}{EE_{t=\infty}} \right) \right]^m = \frac{t}{k_0} \qquad (12)$$

where m [1] is an index of reaction,

k_0 [s] is a rate constant, and

EE and $EE_{t=\infty}$ [1] are extraction efficiencies

The best fits of each set of m, k_0, and $EE_{t=\infty}$ (in Table I) were optained by eq. 12 and combination with the least square of

$$\sigma_{n-1}^2 = \frac{1}{n-1} \sum \frac{\left(x^{fit} - x^{exp} \right)^2}{\left(x^{exp} \cdot s^{rel} \right) + s^{abs}} \qquad (13)$$

where σ^2 [1] is the least square deviation of fitted and experimental data,

n [1] is the number of experimental values,

s^{rel} [1] a relative error of 0.02,

s^{abs} [1] an absolute error of 0.005,

x^{exp} [1] the experimental value of EE, and

x^{fit} [1] the fit value of EE.

For *Zn*, *Pb*, and *Cd* the reaction index m results in values of -2.5, whereas for *Cu* m=-3.3 and for *Mn* a value of m=-5 is best. With decreasing pH value the k_0 increases, expressing the lower leaching rate. The highest k_0 values for *Mn*, express the increasing leaching rate within the first 10^5 s. The plots of measured and calculated *EE* in Figure 9 illustrates the sigmoidal curves of *Zn*, *Pb*, and *Mn* and the fits of the Avrami-Erofeev rate law. The experimental data revealed a decrease in *EE* with increasing pH. This phenomenon is included in the present model by estimation of a maximum *EE* at infinite time ($EE_{t=\infty}$).

The fit of the acid addition in order to maintain a constant pH constant (Table II, Figure 7) results in decreasing reaction indices m such as –3.5 (pH=3), -3.7 (pH=4), and –4.8 (pH=5). These reaction indices are in the order of the indices, obtained for each metal (between –2 and–5, Table I) and combine, thus, the reactions of all metals. The resulting exponential behaviour of the leaching rate with time is yet unknown but possibly due to the complex structure of the studied municipal fly ash. Further research, however, is necessary to assess the validity of these fits.

Rate limiting step for aqueous leaching

The overall rate of *ZnO* aqueous leaching might be controlled by mass transfer in the boundary layer solid-liquid (16). At this layer, a constant low pH in the aqueous phase suggests a saturation of dissolved Zn^{2+}. Kakovskii *et al.* (16) have reported, that at high acid concentrations the rate of *ZnO* dissolution is controlled by diffusion of Zn^{2+} away from the surface. This local saturation appears to influence eq. 4 and suggests a pH dependent dissolution-desorption step. This step was estimated with the Avrami-Erofeev rate law applying the results of the aqueous leaching data.

An estimation of the diffusion time for the solute in the aqueous phase, shows that limitation of diffusion is negligible. The measured values of a total fly ash volume V_A=1.7·10^{-6} m³/g, and a pore volume V_P=1.1·10^{-6} m³/g of initial fly ash, results in a porosity of ε =0.64. The binary diffusion coefficient for Zn^{2+} in the liquid phase is about 5.4·10^{-10} m²/s, estimated with the Stokes-Einstein equation. With an estimated length of a single pore of L_p=20·10^{-6} m, the estimated diffusion time for the solute in the pore is less than 2 s with eq. 14:

$$t = \frac{L_p^{\,2}}{D_{Zn^{2+},eff}^{(L)}} \tag{14}$$

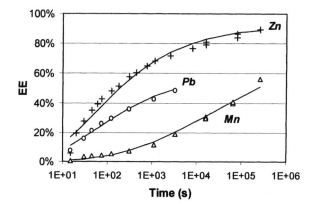

Figure 9: Plots of measured (symbols) and with Avrami-Erofeev calculated (line) extraction efficiency (%) of aq. leaching from MSWI fly ash at pH=3

Table I. Avrami-Erofeev constants for aqueous leaching of metals from MSWI fly ash under constant pH values.

		m	k_0	$EE_{t=\infty}$	σ^2_{n-1}
Zn	pH=3	-2.5	55	0.91	0.16
	pH=4	-2.5	391	0.89	0.03
	pH=5	-2.5	600	0.16	0.02
Pb	pH=3	-2	44	0.54	0.02
	pH=4	-2	108	0.10	0.003
	pH=5	-2	108	0.01	0.003
Cd	pH=3	-2.5	35	0.56	0.07
	pH=4	-2.5	293	0.40	0.04
	pH=5	-2.5	5323	0.31	0.08
Cu	pH=3	-3.3	64	0.90	0.15
	pH=4	-3.3	466	0.71	0.07
	pH=5	-3.3	21	0.11	0.04
Mn	pH=3	-5	35430	1.00	0.03
	pH=4	-5	120000	0.66	0.01
	pH=5	-5	120000	0.11	0.0003

Table II. Avrami-Erofeev constants of overall reaction during aqueous leaching at constant pH (addition of 6wt.% HNO₃)

	m	k_0	$EE_{t=\infty}$	σ^2_{n-1}
pH=3	-3.5	448	27.5	0.73
pH=4	-3.7	2140	22.9	0.97
pH=5	-4.8	5739	12.5	0.24

Rate limiting step for SFE

At the liquid-SC CO_2 interface, a constant, and sufficient high concentration of extractant Cyanex 302 is assumed. Concentration effects of generated metal-complexes are assumed to be negligible, due to diffusion coefficient of solutes in supercritical fluids of about 10^{-8} m²/s (17), which is approximately 2 orders of magnitude faster than in the aqueous phase. A continuous flow of solvent during extraction even reduces surface effects, due to both continuous supply of extractant and continuous removal of metal-extractant complex. Thus, mass transfer is a limiting factor at the liquid-SC CO_2 interface, as studied previously by Tai et al. (18). Further research is required to study a possible impact on the overall extraction of SFE from humid MSWI fly ash.

Conclusion and Recommendation

The SFE can be performed with almost the same results with both dry and humid fly ash. The limiting step of the reaction is most probably the dissolution-desorption step from the solid phase.

For a good extraction of the studied metals from municipal waste fly ash, SFE might be enhanced by a pre-leaching step. Leaching efficiency by aqueous acidic solutions obeys a logarithmic power rate law model.

Acknowledgement

Acknowledgment is made to the Donors of The Petroleum Research Fund, administered by the American Chemical Society, for partial travel support to Orlando for G. Witkamp.

References

1. Puch, K.H.; Hartan, J. In *Proceedings of Int. Workshop on Novel Products from Combustion Residues*; Nugteren, H., Ed; Delft Univ. of Technology, The Netherlands, 2001, pp 43-46.
2. Verwoerd, J., internal communication, AVR-Rotterdam, 2002
3. Fermo, P.; Cariati, F.; Pozzi, A.; Demartin, F.; Tettamanti, M.; Collina, E.; Lasagni, M.; Pitea, D.; Puglisi, O.; Russo, U. *Fresenius J. Anal. Chem.* 1999, *365*, 666-673
4. Towler, M. R.; Stanton, K. T.; Mooney, P.; Hill, R. G.; Moreno, N.; Querol, X. *J. Chem. Technol. Biotechnol.* 2002, *77*, 240-245
5. Glennon, J.D.; O'Connell, M.; Leahy, S.; Mehay, H.; Smith, C.M.M. In *Proceedings of the "REWAS'99: Global Symposium on Recycling, Waste Treatment and Clean Technology"* San Sebastian, Spain, Sept. 5-9 (1999) Volume III, pp 2329-2336
6. Wai, C.M.; Lin, Y.; Brauer, R. *Talanta* 1993, *40 (9)*, 1325-1330
7. Wang, S.; Elshani, S.; Wai C M. *Anal. Chem.* 1995, *67 (5)*, 919-923
8. Kersch, C.; van Roosmalen, M. J. E.; Woerlee G. F.; Witkamp, G. J. *Ind. Eng. Chem. Res* 2000, *39*, 4670-4672
9. Toews, K. L.; Shroll, R. M.; Wai, C. M.; Smart, N. G *Anal. Chem.* 1995, *67(22)*, 4040-4043
10. Almela, A.; Elizalde, M.P. *Hydrometallurgy* 1995, *37*, 47-57
11. Menoyo, B.; Elizalde, M.P.; Almela, A. *Solvent Extr. Ion Exch.* 2001, 19, 677-698
12. Sole, K.C.; Hiskey, J.B. *Hydrometallurgy* 1995, *37*, 129-147
13. Cyanamid Co. *Technical Brochures of Cyanex 301 and Cyanex 302* (1993)
14. Smart, N.G.; Carleson, T.E.; Elshani, S.; Wang, S.; Wai, C.M. *Ind. Eng.Chem. Res.* 1997, *36*, 1819-1826
15. House, J.E. in *Principles of Chemical Kinetics*, WC. Brown Publ.: Times Mirror Higher Education Group, USA, 1997
16. Kakovskii, I.A.; Khalezov, B.D. *Izv. vyssh. ucheb. Zaved., Tsvetn. metall.*, 1977, *2*, 26-31 (Russian text) Chem. Abstr., 1977, *87*, 12172s
17. Paulaitis, M. E.; Krukonis, V. J.; Krunik, R. C.; Reid, R. C. *Rev. Chem. Res.* 1983, *32*, 179
18. Tai, C.Y.; You, G.S.; Chen S.L. *J. of Supercritical Fluids*, 2000, *18*, 201-212

Organics

Chapter 8

Coupled Processing Options for Agricultural Materials Using Supercritical Fluid Carbon Dioxide

Jerry W. King

Supercritical Fluid Facility, Los Alamos National Laboratory,
Chemistry Division, C-ACT Group, P.O. Box 1663, Mail Stop E-537,
Los Alamos, NM 87545

Supercritical carbon dioxide (SC-CO_2) has been documented
numerous times as a benign processing agent, particularly for
the processing of agriculturally-derived substrates and foods.
However SC-CO_2 alone as a medium for conducting
extractions, fractionations, or reactions has certain limitations
which can be overcome by coupling it with other processing
options . Here combinations of multiple fluids, phases, and
processing will be presented that allow a final end result to be
achieved. Several examples will be presented of using
different fluid compositions, including the use of cosolvents
to affect the extraction or enrichment of targeted solutes from
complex natural products. In addition, the coupling of
supercritical fluid extraction (SFE) with supercritical fluid
fractionation (SFF), and/or a reaction conducted in the
presence of supercritical media (SFR) will be cited, using
specific examples that produce useful industrial products.

Introduction

The application of supercritical fluids (SF) and similar media for the processing of agricultural or natural products has traditionally focused on the extraction mode utilizing carbon dioxide in its supercritical ($SC-CO_2$) or liquid (LCO_2) state. Beginning in the mid-1980's, options other than varying the extraction fluid's density in the SFE mode were developed, such as columnar and chromatographic techniques, which facilitated SF-derived extracts or products having more specific composition and properties. This development was followed by the advent of conducting reactions (SFR) in the presence of SFs as documented in the literature (1,2). Further examination of alternative fluids, such as subcritical water have expanded the "natural" fluid base available to the processor of agriculturally- derived products. Therefore it should be possible to process natural product matrices utilizing a series of pressurized fluids such as suggested by the sequence below:

$SC-CO_2$ or LCO_2
↓
$SC-CO2$/Ethanol or H_2O
↓
Pressurized H_2O or H_2O/Ethanol

The above sequence suggests that some degree of selective solvation should be possible, with $SC-CO_2$ or LCO_2 extracting non-polar solutes followed by the enhanced solubilization of more polar moieties via the addition of ethanol to the SF. Processing with pressurized water, i.e., subcritical H_2O (sub-H_2O) expands the range of extractable solutes into the polar solute range with selectivity being controlled by the temperature of extraction or addition of ethanol. Depending on the composition or morphology of the natural product being extracted, there is no reason in theory or practice that the above process could not be done in the reverse order. Therefore by combining such discrete unit processes such as SFE, fractionation (SFF), or SFR in various combinations with a matrix of extraction fluids, a number of coupled processing options can be devised yielding unique products.

A more detailed example of this coupled processing concept is cited below in Table I for the processing of citrus oils using pressurized fluids. Here six discrete unit processes are listed which include standard SFE with $SC-CO_2$, SFF employing stage-wise pressure reduction, SFF as practiced using column-based deterpenation (3), supercritical fluid chromatography (SFC), another variant of SFF called subcritical water deterpenation (4), and utilization of a $SC-CO_2$ or LCO_2 with a permselective membrane described by Towsley et al.

(5). As shown in Table I, combinations of these processes can be coupled to advantage to allow a total processing scheme to be conducted using critical fluids. For example, a combination of processes (1) and (2) in Table I could be combined to yield a more specific composition in the final extract. Unit process 1 if conducted by sequentially increasing the extraction density when coupled with a sequence of let down pressures (unit process 2) can amplify the SFF effect. Likewise, by combining unit process 1 using SC-CO_2 followed by application of unit process 2 utilizing subcritical H_2O to deterpenate the extract from unit process 1, can yield a more specific final product from the starting citrus oil. To obtain a more enriched and/or concentrated product from the latter process, one could add on unit process 6, a supercritical fluid membrane-based separation of the aqueous extract/fractions from unit process 5 as indicated below (Table I).

Table I. Coupled Processing Options for Citrus Oils Using Pressurized Fluids

Process: (1) SFE (SC-CO_2)
          (2) SFF (SC-CO_2) – Pressure Reduction
          (3) SFF (SC-CO_2) – Column Deterpenation
          (4) SFC (SC-CO_2/Cosolvent)
          (5) SFF – (Subcritical H_2O)
          (6) SFM – (Aqueous Extract/SC-CO_2)

Combinations: (1) + (2)
              (1) + (2) + (3)
              (1) + (4)
              (1) + (5)
              (1) + (5) + (6)

Several other combinations of the above unit processes will be discussed shortly. The author and his colleagues have also developed several combinations of processes for the production of oleochemical industrial products (6-9). Two examples of these processes which can link discrete processing steps to advantage are discussed in the next section.

Examples of Coupled Critical Fluid-Based Processes

Recent studies in our laboratory, as well as cost considerations, suggest that there would be a great advantage to employing sub- and supercritical fluid media for multiple processing operations in a sequential fashion in a production plant. This is based on the hypothesis that the capital equipment costs are relatively high to implement a critical fluid-based process, hence multiple applications should distribute the initial costs over an entire production sequence, rather than concentrate the economics on just one unit process.

Toward this end, we have investigated tandem or coupled processes that embodied the use of pressurized fluids, namely carbon dioxide, for both extraction, fractionation and reaction. Related examples to the work described here are coupling supercritical fluid extraction (SFE) with production scale supercritical fluid chromatography (SFC) for the enrichment of high value tocopherols from natural botanical sources (10), or subcritical water hydrolysis of vegetable oils (11) followed by partition into dense carbon dioxide to produce industrially-useful mixtures of fatty acids.

Production of Monoglyceride-Enriched Mixtures

The initial example we wish to cite involves the synthesis of monoglycerides from seed oil botanical sources via glycerolysis in the presence of supercritical carbon dioxide (SC-CO_2) followed by thermal gradient fractionation using SC-CO_2 to produce highly purified monoglycerides of high economic value. Such monoglycerides command not only a premium price but are used extensively as surfactants and food additives (240 million pounds - 1984). They are tradi- tionally made by reacting an excess of glycerol with a vegetable oil in the presence of a metal oxide catalyst at relatively high temperatures (240 -260°C) for a multi-hour period, yielding a composition that is 35-45 wt. % monoglycerides. Such a composition finds utility as a lower grade emulsifier, but in recent years has been supplanted by higher monoglyceride compositions approaching or exceeding 90 wt.%. Such enrichments are usually obtained using molecular or vacumn distillation techniques. Our method utilizes synthesis in the presence of SC-CO_2, first with CO_2 acting as the catalyst, or alternatively a lipase catalyst, to convert the raw vegetable oil to a mixture containing an intermediate level of monoglycerides, followed by further enrichment using a fractionation tower under supercritical conditions. Using this sequential manufacturing approach, we have obtained monoglyceride-containing mixtures from 35 to 95 wt.%, equivalent to those produced by vacumn distillation techniques. The described sequence of processes makes use of only naturally-derived materials; CO_2, vegetable oils, or enzymes in producing the desired end products, and recycles not only CO_2 for further use as a reaction or fractionating medium, but intermediate glycerolysis products for further conversion to monoglycerides using the enzymatic catalyst described below.

Glycerolysis in the Presence of SC-CO$_2$

It is important when developing alternative processing techniques to avoid radical changes in operations, consequently we initially investigated the effect of conducting glycerolysis in the presence of SC-CO$_2$ (*6*) using a traditional batch stirred reactor approach incorporating an excess of glycerol with several diffferent types of vegetable oils. Carbon dioxide has been shown to be an autocatalytic agent for glycerolysis conducted under ambient conditions (*12*), but it was not apparent that this conversion could be conducted in the presence of SC-CO$_2$. Consequently, a ladder of pressures (21 - 62 MPa) and temperatures (150 - 250°C) were used to optimize the glycerolysis process in the presence of SC-CO$_2$. Maximum production of monoglyceride was found to occur in about 3 hours at 250°C and 21 MPa; higher CO_2 pressures actually supressed monoglyceride formation. Also, higher yields of monoglycerides were obtained at 21 MPa in the presence of SC-CO$_2$ than under the ambient conditions noted early. The resultant composition of the product was verified by analytical capillary SFC and found to be somewhat dependent on the starting vegetable oil. Comparison of synthesized monoglyceride mixtures in the range of 35-45 wt.% monoglyceride with an equivalent standard industrial product (Eastman Chemical Co. - Myvatex Mighty Soft Softener) indicated that the product synthesized in the presence of SC-CO$_2$ was much lighter than the standard industrial monoglyceride mixture, an advantage when using such emulsifiers in food compounding applications. This coupled with the zero solvent residue left in the manufactured product makes the SC-CO$_2$ -based method particularly attractive in synthesizing glyceride compositions intended for human food use.

Perhaps of more importance, is that the SC-CO$_2$ - based synthesis avoids the use of any inorganic catalyst that would have to be filtered out of the oil at the conclusion of a traditional glycerolysis reaction. The jettisoned CO_2 is also available for reuse, or for diversion to other manufacturing processes, such as supercritical fractionation. This makes the above reaction conditions attractive for integrating into an all supercritical fluid-based manufacturing process that is environmentally-compatible.

Enzymatic-Based Glycerolysis of Vegetable Oils in the Presence in SC-CO₂

An alternative route for conducting glycerolysis coupled with supercritical fluid fractionation is to use an enzyme-based catalyst in conjunction with SC-CO_2 to produce enriched monoglyceride mixtures. The use of enzymes for synthetic purposes in the presence of supercritical fluids has been demonstrated, often with model systems in the literature, and we have recently applied this approach for the "green" synthesis of fatty acid methyl esters (FAMES) from vegetable oils such as soybean and corn oil (*13*), or to synthesize margarine substitutes from the same substrates via intraesterification (*14*). In most of these systems, including the one to be described, the reaction is conducted by solubilizing the reactants in a flowing stream of supercritical fluid with subsequent transport over a fixed bed of enzyme catalyst. Use of the flow reactor approach provides a large degree of synthetic options to the chemist or engineer since a number of experimental variables can be easily altered and the enzyme continously reused over multiple cycles. For this process, as well as the other examples cited above, a Novzyme 435 lipase supported on polyacrylamide resin has proven very facile for multiple types of esterifications as noted by King and Turner (*15*).

Initial glycerolysis experiments were run in the pressure range of 21 - 35 MPa and at temperatures between 40 - 70°C. The optimal pressure and temperature consistent with maintenance of enzymatic activity was found to be 27.6 MPa and 70°C. The vegetable oil feedstock and glycerol were each pumped into the flowing SC-CO_2 stream and the reactants mixed in a tee followed by equilibration in a mixing coil. Passage over the supported enzyme bed in the reactor produced the desired glyceride compositions, solvent-free, having improved color properties over those derived with metallic catalysts. It was found that the Novozym 435 did not lose activity over a 72 hr. period and produced glycerides having a random fatty acid distribution from a parent oil (*7*).

The advantage of this enzymatic/supercritical fluid mode of synthesis, besides its ecological compatibility, lies in the ability to "custom" design glyceride containing mixtures by varying the reaction conditions. First, it was observed that the resultant monoglyceride yield was dependant on the flow rate of the supercritical fluid containing the reactants throught the reaction vesssel. At low flow rates, compositions having a monoglyceride content between 80-90 wt.% could be produced, while a 20-fold increase in fluid flow rate produced glyceride mixtures that contained between 40-50 wt.% monoglyceride. Secondly, the water content of the reaction mixture was found to have an obvious effect on the monoglyceride content of the resultant glyceride mixture. Hence, if glycerol cotaining a 0.7% wt. water content was used as one of the starting reactants, then a glyceride mixture containing 84 wt.% monoglyceride could be produced. Likewise, using glycerol with a 4.2 wt.% water content produced a 67% monoglyceride-containing mixture.

Based on reported solubilities of glycols and alcohols in SC-CO_2 , it was apparent that synthesis was being conducted in a multi-phase region, where one of the components (CO_2) was in its supercritical state. Despite the lack of phase homogeneity, the glycerolysis could be successfully run in the mixed phase system. A similar observation has also been reported by Gunnlaugsdottir, et al. (16), who like ourselves, observed that a small amount of enzymatic catalyst resulted in a very high production rate of the targeted end products. Other glycols besides glycerol can also be reacted with vegetable oil substrates; e.g., 1,2-propanediol with soybean oil yields predominately monoester, which finds use in the lubricant market. This observation coupled with the above cited impact of moisture and critical fluid flow rate on the yield of monoglyceride in the product mix, offers great flexibility to the synthetic chemist or engineer for producing emulsifiers of specific composition.

Based on the above research, a patent entitled, "Monoglyceride Produced Via Enzymatic Glycerolysis of Oils in Supercritical Carbon Dioxide" was filed and granted by the U.S. patent office, as U.S Patent 5, 747, 305 (17). This particular synthesis option is one of the two synthetic options that can be coupled with the supercritical fluid fractionation step described below.

Supercritical Fluid Fractionation of Monoglyceride-Containing Synthetic Mixtures

The use of SC-CO_2 as a reaction medium also facilitates movement and introduction of product mixtures to another stage of the production process, namely supercritical fluid fractionation of the glyceride mixtures in a high pressure column held under a thermal gradient to affect a density gradient. This method as a stand alone technique is not new, and has been used for the purification of fish oil esters (18). However coupling it with the synthetic methods just described offers an additional manufacturing dimension, including further incorporation of SC-CO_2 as a processing aid, and ultimately superior or equivalent products to those available commercially.

The technique is described in two publications (8, 18). Briefly, the CO_2 source is delivered to the column after passing through a preheater which converts it to its supercritical state, where it passes upward through a series of increasing temperature zones of 65, 75, 85 and 95°C. This decreases the density of the SC-CO_2 as it passes upwards through the column which is packed with a conventional distillation type of packing. The glyceride feedstock (from the previously described reaction sequences) to be enriched in monoglyceride is then introduced at the top of the first thermally-heated zone so as to allow equilibration along the column. Fractionation then commences with the introduction of CO_2 flow to yield a purified monoglyceride fraction after leaving the last heated zone through an expansion valve. It should be noted that although CO_2 density and hence solvation power is lost as the fluid travels up the column into zones of increasingly higher temperature, that the vapor pressure, particularly of the more volatile components in the synthetic product reaction mixtures (i.e., the monoglycerides), facilitates their enrichment at the top of the column. This can result in glyceride mixtures that can exceed 95 wt.% in monoglyceride content.

Utilization of the fractionating tower approach when coupled with the aforementioned supercritical synthesis adds considerable additional versatility in the production of glyceride mixtures of varying composition. There is a trade off between fractionating capability and throughput, i.e., monoglyceride compositions exceeding 90 wt.% occur at lower CO_2 densities or pressures, while greater throughput is achieved at higher CO_2 densities at the expense of monoglyceride content. However, this feature does allow the production of designer emulsifiers based on monoglyceride content varying from 60 - 90 wt.% monoglyceride using SC-CO_2 densities up to 0.75 g/ml. One attractive outcome of operating the fractionating tower at peak efficiency is the ability to match or exceed the high monoglyceride content products that are achieved via the conventional vacumn distillation method. This has been verified by analytical capillary SFC and visually (their color and physical state) by inspection of the products. It is also possible by operating the fractionating column with internal reflux, to not only obtain highly purified monoglyceride fractions, but to also obtain a 90 wt.% diglyceride fraction after approximately 60% of the original charge of synthesized product has been processed on the fractionating column. However, it is most likely that the di- and triglyceride-enriched bottom fraction would be preferably transported back to the synthesis stage for further conversion to monoglyceride since it is the more economically-viable product ($2.46/lb for the 90 wt.% monoglyceride product).

Production of Fatty Alcohol Mixtures

Fatty alcohols and their derivatives are important in many industrial processes where they are used as raw materials for surfactants and lubricants. A fatty alcohol is, in general defined as a monohydric aliphatic alcohol with six or more carbon atoms. The annual production of fatty alcohols is over 1 million metric tons. Commercially fatty alcohols are produced by one of three processes the Ziegler process, the Oxo process or by a high pressure hydrogenation of fatty acids or esters. The latter process is the only one process that uses renewable natural fats/oils whereas the two first processes utilize petrochemical feedstocks. Depending on their application fatty alcohols are divided into subgroups. Thus fatty alcohols having eleven or more carbon atoms are usually called detergent range alcohols because they are used in the detergent industry mainly as sulfate or ethoxy sulfate derivatives. Fatty alcohols with less than eleven carbon atoms are called plasticizer range alcohols, and they are used as plasticizers and lubricants mainly in the form of ester derivatives.

Here a coupled process is described which involves the synthesis of saturated fatty alcohol mixtures from seed oil botanical sources via two discrete synthetic steps, conducted in the presence of supercritical carbon dioxide (SC-CO_2) or propane (SC-C_3H_8), followed by exhaustive hydrogenation in binary mixtures of hydrogen with the above two supercritical fluids. A number of single-step chemical reactions have been successfully conducted in supercritical fluids, such as esterifications and glycerolysis, however multiple step synthesis in critical fluids are rare. The described process is very environmentally attractive since it uses benign catalysts, recycles the gaseous reaction media and co-products, and utilizes a naturally-occurring starting substrate.

Optimization of the Transesterification Step

Successful transesterification of a naturally-derived oil such as soybean oil requires optimization of several parameters to achieve high yields of methyl esters. The use of an enzyme to catalyze the reaction required screening of several candidate lipases, among which, Novozym SP 435, a lipase derived from *Candida antarctica*, proved successful. Experiments were run to optimize the lifetime of the lipase, and a temperature of 50°C and pressure of 17 MPa proved consistent with the long term use of enzyme (please note that the enzyme could be reactivated even after utilization for over 20 consecutive runs). Addition of the methanol to the SC-CO_2 had to be adjusted to 0.8 mole fraction of fluid flow through the supported enzyme bed in order to maintain 100% enzyme activity under the above conditions. Lower flow rates of the methanol into the SC-CO_2 lowered the efficacy of the lipase, while higher flow rates of methanol also inhibited the reaction. Likewise, the volume percentage of water in the flow system must be kept below 0.05% to achieve greater than a 99% conversion to methyl esters (*13*) and avoid the competing hydrolysis reaction. It was found that the initial hydration level of the enzyme coupled with the water-carrying capacity of the SC-CO_2 facilitated multiple runs using the above reported lipase. The observed conversions were ascertained using capillary supercritical fluid chromatography (SFC), analysis which showed total conversion to the methyl ester moieties; and gas chromatography (GC) which confirmed the resultant methyl ester mixture expected for soybean oil as a starting substrate, by comparison to a BF_3/methylated soybean oil.

Optimization of the Hydrogenation Step

Full hydrogenation of methyl esters formed from vegetable oil substrates has been traditionally achieved through the use of a supported copper chromite-based catalyst. For our hydrogenations, we utilized a non-chromium hydrogenation catalyst, T-4489, from United Catalysts, Inc. The hydrogenation stage of the binary reaction sequence was initially studied separately in order to optimize the hydrogenation step. A device was constructed for generating binary fluid mixtures of hydrogen (H_2) with carbon dioxide, and later propane. This consisted of using digital flow controllers dispensing the proper amounts of each gas into a stirred autoclave, with the resultant mixture then being brought to the various desired hydrogenation pressures using a gas booster pump.

Optimization experiments were initially done using a 2^{5-1} factorial design, which included four center points, resulting in the need for 20 total experiments. The chosen experimental parameters encompassed a pressure range from 15-25 MPa, a temperature range from 210-250°C, and mole fractions of hydrogen in supercritical CO_2 and propane (C_3H_8) from 0.10-0.25. Reaction time in the supported catalyst bed and feed rates of methyl esters to the reactor were also studied in both of the above binary fluid mixtures. It was found from response surface plots, that two variables, namely the reaction temperature and mole fraction of hydrogen in the compressed fluid mixture, were critical to achieving high yields of hydrogenated product.

Using a reaction pressure of 25 MPa, the saturated alcohol mixture yield was over 97% at a 0.25 mole fraction hydrogen in SC-CO_2 (and slightly lower in SC-C_3H_8) at a temperature of 250°C. Analysis of the response surface graph for the H_2/C_3H_8 system indicated that a high alcohol conversion could be accomplished using lower mole fractions of hydrogen in propane than for the corresponding H_2/SC-CO_2 system, however the final product quality was poorer due to the appearance of more n-alkane by-product. At high mole fractions of H_2, the H_2/SC-CO_2 system is superior to the H_2/SC-C_3H_8 system, in terms of product conversion, i.e, less alkane by-product. However, at low mole fractions of H_2, the H_2/SC-C_3H_8 system gave a higher yield than the H_2/SC-CO_2 system. Also, the H_2/SC-C_3H_8 system has a five-fold greater mass throughput than the H_2/SC-CO_2 system. This would make the use of propane seem more attractive, however there are always more n-alkane by-products produced when conducting the hydrogenation step using propane, and its incorporation introduces a potential flammability problem into the synthesis, despite the cited advantages of a propane-based hydrogenation system (*19*).

Coupling the Tranesterification and Hydrogenation Stages

Utilizing the above two optimized reactions systems, with SC-CO$_2$ as a common solubilization and reaction agent, a coupled reaction system was constructed. The all flow reactor system required pumps for the carbon dioxide, vegetable oil, and methanol, respectively for the initial transesterification stage. Conversion to the methyl esters was achieved at a >98% level before transfer to the hydrogenation reactor.

Conversion of the methyl ester substrate dissolved in the SC-CO$_2$ is very rapid due to the demonstrated superior mass transfer kinetics which occur in supercritical fluid media relative to rates in condensed liquid media. Calculations of reaction times in the hydrogenation catalyst bed were between 4-9 seconds based on pulse injection experiments. Conversions for the second stage of the overall synthesis were found to average 96.5 %, yielding mixtures consisting of 90% steryl alcohol, approximately 8% palmityl alcohol, and trace levels of unreacted fatty acid methyl esters and n-alkanes. Upon depressurization of synthesized product, the methanol phase separates from the solid alcohol mixture and can be used as feed to the initial tranesterification stage. Attrition and lifetime of the hydrogenation catalyst was minimized and extended by using the methyl ester substrate, which also minimizes corrosion to the entire system. Initial charges of hydrogenation catalyst could be used for over two months without requiring a change in the catalyst charge.

The above overall presented process incorporates several features of "green" processing. These are as follows:

1. The use of environmentally-compatible CO$_2$ as solvent and reaction medium.
2. Utilization of a natural enzyme derived from *Candida antarctica* as a lipase during the transesterification stage.
3. Use of a chromium-free catalyst during the hydrogenation sequence.
4. Incorporation of a natural, renewable resource (vegetable oil) as a starting substrate.
5. Recycling of the product methanol to feed the transesterification stage of the synthesis.

Other benefits of the described process are high yields for both stages of the synthesis using the above agents, long catalyst lifetimes under the stated conditions, and rapid reaction conditions; particularly in the hydrogenation stage of the sequenced process. Although not demonstrated, the potential for further fractionation of the resultant saturated fatty alcohol mixture using SC-CO_2 via the use of the fractionating tower described in a previous section exists, and will be demonstrated for other lipophilic mixtures in the next section.

Development of New Critical Fluid-Based Coupled Processes

Plant sterols (phytosterols) are complex alcohols constituted by C_{28} or C_{29} terols, differing structurally from cholesterol (C_{27} by the addition of an extra methyl or ethyl group on the eight-carbon side chain of cholesterol. Approximately 40–50 different known plant sterols occur naturally in several forms: in the free form, as fatty acid esters, as ferulic or p-coumaric esters, and as steryl glycosides, which may also be esterified with a free fatty acid (FFA). In edible oils and human diets beta-sitosterol, campesterol, stigmasterol, and brassicasterol are the major plant sterols. Phytosterols usually constitute less than half of the dietary sterol intake of humans in the United States, the remainder being dietary cholesterol (20). Phytosterols are present in low concentrations as secondary substances, but their cholesterol-lowering effects have been known since the 1950s (21). Thus, recovering phytosterols and similar high-value components is important not only from a nutritional perspective but also from a commercial point of view to add value to processing agricultural crops.

Phytosterols are also used as starting materials in the synthesis of steroids for pharmaceutical purposes, as emulsifiers in the cosmetics and food industries, and as a starting material in pesticide manufacturing; they also find individual applications in the field of liquid crystals as used in the optics industry (22). Recently, plant sterols and plant stanols (hydrogenated forms of the respective sterols) have been incorporated into margarines and vegetable oil spreads. These food products have been shown to lower total and LDL cholesterol levels by 10 to 15% in individuals with high blood cholesterol levels (23,24). These same cholesterol-lowering compounds also have been incorporated into breakfast cereals, cereal bars, and soy beverages (25). Recent clinical studies have demonstrated the cholesterol-lowering properties of free and esterified sitostanol (24).

One approach to increase the phytosterol ester content of vegetable oils is via refining rather than isolating them from the by-products and then adding them back to the oil (26). Such a processing scheme simplifies the enrichment process and improves the economic feasibility of the production. Dunford and King (27) were able to increase the total phytosterol ester content of rice bran oil (RBO) and corn fiber oil from <5% to over 19% utilizing the described SFF process. In general, the economic feasibility of industrial operations is higher for continuous processes when compared to batch or semicontinuous processes. Also countercurrent operations tend to be more efficient due to the larger driving force for mass transfer between solvent and solute. Thus, the objective of this new study was to examine the potential of a continuous countercurrent SC-CO$_2$ fractionation process for enrichment of phytosterols in vegetable oils. In the above processes, initially free fatty acids (FFA) are removed and then a phytosterol-enriched oil fraction is obtained via a second fractionation process. This particular study focused on the retention of phytosterol esters in the rice bran oil during the continuous countercurrent deacidification SFF process.

Corn bran obtained as a by-product from the dry-milling of corn and yields an oil that contains the above mentioned phytosterols (28). However, these ferulate phytosterol esters (FPE) are present at very low levels (1.5 wt%) in the predominately triacylglycerol (TAG)-based oil. Therefore, enrichment of these moieties is desired since they can be used as nutraceuticals, commanding a high value in the functional foods market ($18–20/kg) (29).

Previous reports (30, 31) have appeared on the use of supercritical fluid extraction (SFE) coupled with supercritical fluid fractionation (SFF) for the enrichment of these FPE. Carbon dioxide (CO$_2$) and ethanol (EtOH), as a cosolvent, were utilized to fractionate and enrich the FPE from 1.25 to 14.5 wt% in corn bran oil employing a sorbent bed. However, this prior research was performed on an analytical scale. In the present study, SFF technology of corn bran oil has been scaled up using SFE/supercritical fluid chromatography (SFE/SFC). The oil is removed from the corn bran by utilizing supercritical carbon dioxide (SC-CO$_2$), and then the extract is fractionated by on-line SFC to obtain a fraction enriched in FPE.

Extraction of the berry substrates, such as elderberry or black raspberry, with sub-H$_2$O offers another discrete process that can be coupled with SFE using SC-CO$_2$ or perhaps a SFM option to enrich the aqueous extracts. Extractions of anthocyanins are frequently done with ethanol or aqueous ethanolic solvents, and must be done with care due to light-, heat-, and air-sensitivity of anthocyanins. Extraction using sub-H$_2$O is largely dependent on altering the extraction temperature of the fluid above its normal boiling point while under pressure, thereby changing the dielectric constant of water and hence the solvation power of the fluid [32]. For example, by adjusting temperature and pressure, the dielectric constant of the water at 20°C (~80) can be changed to a value of 48 at 100°C. This is close to the dielectric constant values for furfural (42), glycerol (47) and acetonitrile (38) at 20°C, or methanol (37.5) at 0°C. Hence, sub-H$_2$O offers an extraction medium that is difficult to match using GRAS (Generally-Regarded-As-Safe) organic solvents and somewhat unique in its extraction characteristics. Evidence of the use of sub-

H$_2$O in the literature for natural products is provided for the extraction of kava-kava (*33*), rosemary (*34*), and savory or peppermint (*35*).

Experimental

Three distinct processes were experimentally studied: a coupled process for deacidifiying and enriching the phytosterol content of rice bran oil (RBO) by continuous countercurrent columnar fractionation, a scale up of a coupled supercritical fluid extraction (SFE)/ supercritical fluid chromatography (SFC) process for the enrichment of phytosterol in corn bran oil, and a unit process involving the subcritical water extraction of berry substrates. The experimental aspects of the first two processes are described in the literature (36, 37), and will not be repeated here. Research is currently underway to couple the described process below with other unit processes involving both subcritical water and supercritical carbon dioxide.

Subcritical Water Extraction of Anthocyanins (ANC) from Berries

The experimental apparatus used to conduct sub-H$_2$O extraction on berry substrates is shown in Figure 1. It consists of a modified Applied Separations Inc. (Allentown, PA) Spe-ed pumping unit feeding water from a reservoir into a extraction vessel (cell) contained in a thermo-regulated oven (Model 3710A, ATS, Inc., Butler, PA). The extraction cell was a 316 SS, 1" o.d., 9/16"i.d., approximately 55 mL in volume.

As shown in Figure 1, the water is pumped through an equilibration coil contained in the oven to bring it into its subcritical state at temperatures above its normal boiling point under pressure, and then passed through the extraction cell before exiting the oven into a cooling bath reservoir (Model 801, Polyscience, Inc., Nile, IL). Back pressure was maintained on the system with the aid of a micrometering valve which also allowed adjustment of the water flow rate. Aqueous extracts were collected after exiting the micrometering valve.

The first thermocouple in Figure 1 was connected to the temperature controller (Part No. CN4800, Omega Engineering, Stanford, CT) which regulated the oven temperature while the other thermocouples were connected to a digital meter to obtain an accurate reading of the water temperature, both before and after the extraction cell.

118

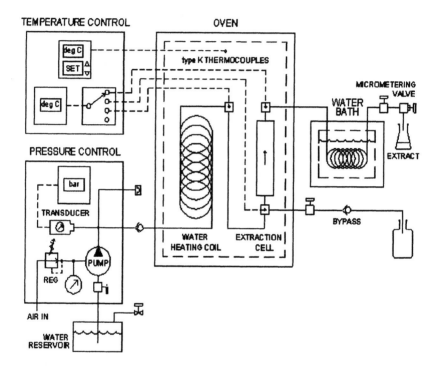

Figure 1. Subcritical water extraction system for extracting anthocyanins from berry substrates.

Extraction Procedure for Berry Substrates

Samples of various fruit berries and their by-products (pomaces) were placed in the extraction cell and the oven heated to temperatures between 110-160°C. Both deionized and Milli-Q-purified neat water as well as acidified water (0.01% HCl, pH ~ 2.3) were fed at a rate of 24 mL/min at a constant pressure of 4.0 MPa. This pressure was well in excess of that required to prevent the formation of steam within the extraction cell. Incremental samples were obtained every 60-80g of aqueous solution expelled from the extractor over a 40 min time interval, however extracts were not taken until the cell was at the desired extraction temperature and pressure.

Color was monitored visually to an approximate equivalent of 20 ppm of cyanin-3-glucoside (a specific anthocyanin). Extract samples were analyzed by the HPLC procedure described by Skrede et al. (*38*). The efficiency of the sub-H_2O extraction was compared to results obtained using a 70% ethanolic extract. The control sample was extracted with 70% ethanol in water for 40 min using sonication and washed with excess ethanol to remove any remaining color from the berry substrate. Because of the extreme sensitivity of ANCs to light, heat, and oxygen; all samples were immediately prepared after extraction for injection into the HPLC as described above.

RESULTS AND DISCUSSION

Results from the columnar fractionation of rice bran oil are initially discussed in this section. For this case, one fractionation column was used to obtain the reported results, however these results may be amplified by using an even longer column with more fractionating power, or two columns operating in sequence (SFF-SFF) to accomplish both deacidification and further enrichment of the phytosterol components in rice bran oil (RBO). The reported results for the coupling of SFE-SFC have a similar purpose, namely the enrichment of phytosterol components from corn bran oil on a preparative scale. Finally, initial results on what is envisioned to be an initial stage in the multi-unit processing scheme for berry substrates, namely the subcritical water (sub-H_2O) extraction of ANCs, above the boiling point of water, are reported.

Results from the Columnar Fractionation of Phystosterols

Fractionation experiments were carried out in a continuous countercurrent mode of operation. Initially the column was filled with CO_2 and allowed to equilibrate at the desired temperature and pressure. Then CO_2 and oil were allowed to flow and fraction collection initiated. Carbon dioxide entered the system from the bottom of the column, right above the raffinate section. In this particular study, oil was delivered into the system from the top of the column so as to allow full countercurrent contact of SC-CO_2 with the feed material. Solute-laden SC-CO_2 then proceeded upward in the column and the resultant extract was collected from the top of the column. RBO oil components, which were not solubilized significantly in the SC-CO_2 accumulating in the raffinate section of the column. The raffinate reservoir was drained in 15-min intervals to avoid overflow of the raffinate fraction into the fractionating section of the column. During a typical SFF experiment, steady state conditions were reached in the column within the first 3 hours of operation. Steady state operation of the column was ascertained by monitoring the weight and composition of the extract fraction collected in 30-minute intervals.

The fractionation experiments were carried out under isobaric and isothermal conditions over the pressure and temperature range of 138–275 bar (13.8-27.5 MPa) and 45–80°C, respectively. Carbon dioxide and oil flow rates were 2 L/min and 0.7 ml/min, respectively, as measured at ambient conditions. After the completion of the experiment the column was depressurized and residual oil drained off at the end of each run. The column was cleaned between runs at a pressure of 34.5 MPa and a temperature of 90°C by flowing CO_2 for more than 6 hours.

Crude RBO was used as starting material for the fractionation experiments. Table II shows the composition of the starting material. It should be noted that FFA composition of the crude RBO is higher (~5%, w/w) than that of the other vegetable oils such as soybean and corn oil (~1–2%, w/w) due to the presence of an active lipase in the rice plant. Hence, high phytosterol and FFA content RBO is an excellent model system applicable to this study. Table III also shows a typical raffinate composition resulting from the continuous columnar SFF process . Note that the oryzanol content of the resultant extract was increased three-fold when half of the FFA were removed from the RBO feed. The phytosterol fatty acid ester composition was also found to be higher than in the feed material, although the StE enrichment was not as significant as that found for oryzanol.

The solute loading of the SC-CO_2 increased with increasing pressure and decreasing temperature. This can be explained by the higher density of SC-CO_2 at higher pressures and lower temperatures, hence the higher solvent power of SC-CO_2 under these conditions. Therefore processing at high pressures and low temperatures requires less solvent (SC-CO_2) and reduces the processing time.

Table II. Composition of the crude rice bran oil and a typical SFF raffinate fraction obtained at 13.8 MPa and 80°C.

Component	Crude RBO (% w/w)	RBO SFF Raffinate Fraction (% w/w)
Free Fatty Acid (FFA)	5 +/- 0.5	2.5+/- 0.5
Free Phytosterols (St)	0.70 +/- 0.05	0.50 +/- 0.03
Free Fatty Esters of Phystosterols (StE)	2.6 +/- 0.3	2.9 +/- 0.4
Ferulic Acid Esters of Phytosterols (FE)	1.5 +/- 0.3	4.9 +/- 0.04

However, examination of the extract compositions showed that the FFA content of the extracts was lowest at the highest pressure and lowest temperature studied as shown in Figure 2, indicating that SFF fractionation under these conditions is not suitable for efficient FFA removal from the crude oil. This is in part due to the large amount of TAG lost in the extract fraction during high pressure and low temperature processing. For example, there is a higher TAG content in the extracts at a higher pressure and lower temperature (i.e. 60% w/w TAG at 275 bar and 45°C as compared to <10% w/w TAG at 138 bar and 80°C) due to the higher SC-CO_2 density and increased volatility of TAGs. These results are similar to the data obtained from the semicontinuous process reported by Dunford & King (27). These results confirm that the deacidification process should be carried out at lower pressures and high temperatures to expedite FFA removal commensurate with lower TAG loss in the extract.

Coupled SFE/SFC Results

Previous SFF studies (30,31) using the SFE/SFC approach were performed on an analytical scale, and were designed to emulate a preparative-scale fractionation process. In this study, solute fractionation was accomplished in two steps. The first step, utilizing neat CO_2, removed the majority of the TAG and the phytosterol fatty acyl esters. The second elution step was designed for FPE enrichment and was achieved with ethanol-modified CO_2.

The initial fractionation experiments were performed utilizing corn bran oil and varying amounts of the amino-propyl bonded silica to check for the possibility of FPE breakthrough. These studies were necessary so that the

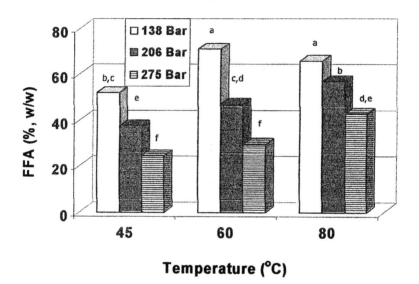

Figure 2. Effect of pressure and temperature on the FFA composition of the columnar SFF extract, i.e., bars with the same letter are not significantly different at the P>0.05 level, e.g, at 138 bar, the wt. % free fatty acid (FFA) content is the same at 60 and 80°C(a), but not the same at 45°C (b,c), etc. (Reproduced from reference 36.)

separation column sorbent bed could be optimized for the preparative-scale SFC of the corn bran extract. It was determined that a 3:1 ratio of sorbent:oil was the minimum ratio for the FPE not to break through on the sorbent column. This finding was scaled up to the preparative-scale SFE/SFC system, but the sorbent:oil ratio was increased to 4:1 for this process. Thus, FPE are retained on the sorbent bed during the neat CO_2 step, but they elute with the introduction of ethanol modifier into the SC-CO_2 mobile phase.

Preparative Scale SFE

Before preparative-scale SFE/SFC trials were undertaken, it was necessary to conduct experiments for both the SFE and SFC stages in order to optimize the processes. The SFE runs yielded an average amount of extract equal to 5.85 g. This equated to an average yield of 3.49 wt% with a relative standard deviation (RSD) of 1.9%. The oil content of the corn bran was also determined in triplicate by the AOCS Official Method Ac 3-44, which uses petroleum ether as the extraction solvent in a Butt-type extraction apparatus. The organic solvent extraction yielded an average of 3.50 wt% with an RSD of 2.0%, in excellent agreement with the SFE result.

However, the SFE was time consuming because of the low solubility (~1 wt%) of TAG in SC-CO_2 at the cited pressure and temperature. A CO_2 volume of 1200 L (STP) was needed for the SFE at a flowrate of 5 L/min, requiring 240 min for the extraction. To operate in an efficient manner, it was determined to stop the SFE stage after 600 L of CO_2 had been used, since SFE at this point yielded ~96% of the available oil. It is this extraction product that was then transferred to the sorbent column for the SFC stage of the SFE/SFC procedure.

Preparative Scale SFC

The sorbent/sorbate ratio of 4:1 was adhered to for these optimization experiments, and preparative-scale SFC was accomplished in three steps followed by a sorbent bed reconditioning as described in the experimental section. The first SFC step removed the majority of the TAG and the phytosterol fatty acyl esters. The second step was designed for maximum FPE enrichment, and the third fraction was run to elute any remaining corn bran extract from the sorbent bed, preventing extract carryover to subsequent runs. Column reconditioning purged the column of any residual ethanol and corn bran oil components and was a necessary step so that the chromatographic sorbent column could be used multiple times for the SFC step.

The cumulative mass collected in the fractions from the SFC runs yielded an average of 4.96 g, which represented an 82.7 wt% recovery of the starting charge of corn bran oil. This is in contrast to earlier research on the analytical-scale SFF of corn bran oil, which exhibited nearly quantitative mass recovery (30). However, this result is not atypical in preparative scale SFC as evidenced by prior investigators (10,39,40). For example, in the first two studies (10,39) involving the SFC of tocopherols, only partial recovery of the tocopherols (76 to 87%) was obtained from silica gel.

Coupled Preparative-Scale SFE/SFC

Data from the preparative scale SFE/SFC experiments using corn bran are shown in Table III. The cumulative mass of the four fractions averaged 5.75 g, which is practically identical to the previously stated mass recovery of 5.85 g obtained during the preparatory-scale SFE studies. The SFE/SFC mass recovery data is more typical than the lower recovered masses noted during the preparatory-scale SFC optimization studies.

Table III. % Composition of Corn Bran Components After SFE/SFC[a,b]

Fraction	Mass (grams)	TAG		FS		FPE	
1	4.9 (3.9)	93.6	(0.6)	0.27	(8.1)	0	
2	0.79 (5.0)	6.3	(9.3)	6.1	(8.5)	12.9	(3.5)
3	0.03 (10.1)	76	(10.4)	2.1	(11.9)	2.7	(8.6)

[a] n = 4

[b] ( ) = % Relative Standard Deviation

The first collected SFE/SFC fractions had an average mass recovery of 4.93 g, which represents 85.7% of the total extract. HPLC analyses showed that TG made up approximately 93.6% of these fractions. This finding corroborated the analytical-scale SFF studies using the 4:1 sorbent/sorbate ratio. In those studies, the first fractions averaged 84.7% of the total extract and TG constituted 94.3% of the fraction. The second fraction had an average mass recovery of essentially 0.8 g, representing 13.7% of the total extract. FPE comprised almost 13% of the fraction. Thus, the FPE were enriched 10-fold from the initial corn bran oil content of 1.25%. Free sterols also showed a slight enrichment in this fraction, constituting better than 6% of the total mass. This shows a 4.5-fold enrichment of free sterols, which constituted 1.3% of the original corn bran oil. Fraction 3 had an average mass recovery of 0.03 g, equaling 0.5% of the total extract, and consisted mainly of TG (76%). Free sterols and FPE were also present at 2.1 and 2.7%, respectively. The sorbent column reconditioning steps yielded an average mass of 0.002 g, equaling 0.03% of the total extract. As in our earlier analytical-scale corn bran SFF study (*30*), extract carryover from one run to the next did not seem to be problematic.

In summary, this study demonstrates a two stage coupled process of combining SFE with SFC on a preparative-scale to enrich and fractionate high-value nutraceutical components. By using the above described process, one can extract the oil from the corn bran, fractionate the majority of the oil away from the FPE, and further enrich the FPE. This process provides an alternative to conventional phytosterol extraction, which requires specialized equipment such as fractional or molecular distillation units and their attendant high energy requirements, and in addition, an environmentally-benign process using only CO_2 and ethanol.

Subcritical Water Extraction of Berries

Results for the sub-H_2O extraction of berries are presented below in Tables IV and V for the acidified sub-H_2O extraction of ANCs from berry pomaces, stems, and seeds at 120°C. The results in Table IV indicate that the volume of sub-H_2O required to carry out an equivalent extraction of dried elderberry seeds is much less than when using ethanol as the extraction solvent. Extraction of only 90% of the available ANCs from the same substrate (the 90%+Sub-H_2O results) yields a much more concentrated aqueous extract, but takes only 15 min versus the 40 min extraction times associated with the other results. Not only does this reduce the extraction time, but more than half the volume of the required solvent.

The above trends are also substantiated by the results shown in Table V for the extraction of black raspberry pomace by sub-H_2O. The pomace is the substance left over after the removal of the juice from black raspberries. Here, extraction with ethanol yields an approximately equivalent result to that obtained on dried elderberry seeds (Table IV). The results for extraction with sub-H_2O and 90%+ Sub-H_2O yield less total ANC than in the case of the whole dried elderberry seeds, but this is due to the reduced levels of ANCs found in all pomaces after the juice is expressed.

Table IV. Subcritical Water Extraction of Elderberry Seeds (Dry)

Solvent	mg ANC/g-seed	g- ANC/g-solvent	Ratio
Ethanol	4.76	142	33:1
Sub-H_2O	4.34	213	21:1
90% + Sub-H_2O	4.17	1853	7:1

ANC = Anthocyanin

Ratio = solvent (fluid)/substrate

Table V. Subcritical Water Extraction of Black Raspberry Pomace.

Solvent	mg ANC/g-seed	g-ANC/g-solvent	Ratio
Ethanol	4.79	141	35:1
Sub-H_2O	3.85	137	28:1
90% + Sub-H_2O	3.50	237	15:1

ANC = Anthocyanin

Ratio = solvent (fluid)/substrate

Similar encouraging results have been achieved with moist elderberry seeds, elderberry stems, and blueberry pomaces. It should be noted that the dried and moist elderberry substrates contained between 7.4 – 9.3 % moisture, while the raspberry and blueberry pomaces contained approximately 65% moisture. The above recoveries of ANCs at an extraction temperature of 120°C might seem somewhat surprising considering their inherent thermal instability, however calculations of the superficial velocity of sub-H_2O through the extraction cell are very rapid (~0.1 cm/sec), facilitating rapid mass transport of the target solutes (ANCs) from the substrate. One additional benefit of the "hot" water extraction process is the in-situ sterilization of resultant product, thereby potentially avoiding the need for thermal retorting of the final product.

Average percentages of ANCs in the final aqueous extract ranged from 8-10% for the extraction of berry seeds/stems to 2.5-4.5% from the pomaces. Although the tintorial strength of such extracts is high, it would be desirable to further concentrate these extracts for applications in the nutraceutical or functional food areas. This potentially could be accomplished by coupling a SFM process step after sub-H_2O to yield a SFE-SFM coupled process. It should be noted that the use of SFE with SC-CO_2 (neat and with cosolvents) has been reported in the literature for extracting both the oil and enriched polyphenolic fractions from grapes (41-*43)*. Such results suggest that by combing sequential extractions using SC-CO$_2$ and sub-H$_2$O, that an array of useful natural product extracts could be obtained, as noted by author previously (*44*).

Conclusions

Several total critical fluid coupled processes have been developed by combining discrete SFE, SFF, and SFR steps with SC-CO$_2$ and other pressurized fluids. Using this approach, a variety of useful target products can be developed that are not accessible when using one critical fluid-based unit process alone. As noted previously, another advantage of using multiple critical fluid processing steps is that it can help offset the capitalization costs that are required in constructing a high pressure processing plant, i.e., permitting more universal application of the SC-CO$_2$ or alternative fluid delivery and recycle system. Such combining of critical fluid-based processes offers many advantages, including the use of environmentally-compatible processing agents and extracts/products that are free of toxic solvents.

References

1. *Chemical Synthesis Using Supercritical Fluids;* Jessop, P.G.; Leitner, W., Eds.; VCH-Wiley, Weinheim, 1995.
2. King, J.W. In *Lipid Biotechnology;* Kuo, T.M.; Gardner, H.W., Eds.; Marcel Dekker, New York, NY, 2002, pp. 663-687.
3. Reverchon, E., *J. Supercrit. Fluids*, 1997, 10, 1-37.
4. Clifford, A.A.; Basile, A,; Jimenez-Carmona, M.M.; Al-Saidi, S.H.R. *Proceedings of the 6th Meeting on Supercritical Fluids;* Nottingham, UK, Institut National Polytechnique de Lorraine, Vand, France,1999, pp. 485-490.
5. Towsley, R.W.; Turpin, J.; Sims, M.; Robinson, J.; McGovern, W. *Proceedings of the 6th Meeting on Supercritical Fluids;* Nottingham, UK, Institut National Polytechnique de Lorraine, Vand, France, 1999, pp. 579-583.
6. Temelli, F.; King, J.W.; List, G.R., *J.Am. Oil Chem. Soc.*, 1996, 73, 699-706.
7. Jackson, M.A.; King, J.W., *J. Am. Oil Chem. Soc.*, 1997, 74, 103-106.

128

8. King, J.W.; Holliday, R.L.; Sahle-Demessie, E.; Eller, F.J.; Taylor, S.L. *Proceedings of the 4th International Symposium on Supercritical Fluids*, Sendai, Japan, 1997, Vol. C, pp. 833-838.
9. Andersson, M.B.O.; King, J.W.; Blomberg, L.G. *Green Chem.*, 2000, 2, 230-234.
10. King, J.W.; Favati, F.; Taylor, S.L. *Sep. Sci. Tech..*, 1996, 31, 1843-1857.
11. King, J.W.; Holliday, R.L.; List, G.R. *Green Chem.* 2000, 1, 261-264.
12. Kochhar, R.K.; Bhatnagar, R.K. Indian Patent 71, 979, 1962.
13. Jackson, M.A.; King, J.W. *J. Am Oil Chem. Soc.*, 1996, 73, 353-356.
14. Jackson, M.L.; King, J.W.; List, G.R.; Neff, W.E. *J. Am. Oil Chem. Soc.* 1997, 74, 635-639.
15. King, J.W; Turner, C. *Lipid Tech. Newsletter* 2001, 13(5), 109-113.
16. Gunnlaugsdottir, H.; Sivik, B. *J. Am. Oil. Chem. Soc.* 1997, 74, 1491
17. Jackson, M.A. U.S. Patent 5, 747, 305, 1997.
18. King, J.W.; Sahle-Demessie, E.; Temelli, F.; Teel, J.A. *J. Supercrit. Fluids* 1997, 10, 127-137.
19. van den Hark, S.; Harrod, M.; Moller, P. *J. Am. Oil Chem. Soc.* 1999, 76, 1363-1370.
20. Ravi Subbiah, M.T. *Mayo Clin. Proc.* 1971, 46, 549–559.
21. Peterson, D.W.; Robbins, R.; Shneour, E.A.; Myers, W.D. *Proc. Soc. Exptl. Biol. Med.* 1951, 78,143–147.
22. Daguet, D.; Coïc, J.-P. *Oleagineaux Corps Gras, Lipides* 1999, 6, 25–28.
23. Miettinen, T.A.; Gylling, H. In New Technologies for Healthy Foods and Nutraceuticals, Yalpani, M., Ed.; ATL Press, Inc., Shrewsbury, MA, 1997, pp. 71–83.
24. Miettinen, T.A.; Puska, P.; Gylling, H.; VanHanen, H.; Vartiainen, V. *N. Eng. J. Med.* 1995, 333, 1308-1312.
25. Yankah, V.V.; Jones, P.J.H. *INFORM* 2001 12, 1011-1014.
26. Dunford, N.T.; King, J.W. *J. Food Sci.* 2000, 65, 1395-1399.
27. Dunford, N.T.; King, J.W. *J. Am. Oil Chem. Soc.* 2001, 78, 121-125.
28. Moreau, R.A.; Powell, M.J.; Hicks, K.B. *J. Agric. Food Chem.* 1996, 44, 2149-2154.
29. Hicks, K.B. *Genetic Eng. News* 1998, 18, 1-4.
30. Taylor, S.L.; King, J.W. *J. Chromatogr. Sci* 2000, 38, 91– 94.
31. Taylor, S.L.; King, J.W. *J. Am. Oil Chem. Soc.* 2000, 77, 687–688.
32. G. Akerof. *J .Am. Chem. Soc.*, 1932, 54, 4125-4139.
33. Kubatova, A.; Miller, D.J.; Hawthorne, S.B. *J. Chromatogr. A*, 2001, 923, 187-194.
34. Basile, A.; Jimenez-Carmona, M.M.; Clifford, A.A. *J. Agric. Food Chem.*, 1998, 46, 5205-5209.

35. Kubatova, A., Lagadec, A.J.M.; Miller, D.J.; Hawthorne, S.B. *Flavor Fragrance J.*, **2001**, 16, 64-73.
36. Dunford, N.T.; Teel, J.A.; King, J.W. *Food Res. Int.*, **2003**, 36, 175-181.
37. Taylor, S.L.; King, J.W. *J. Am. Oil Chem. Soc.*, **2002**, 79, 1133-1136.
38. Skrede, G.; Wrolstad, R.E.; Durst, R.W. *J. Food Sci.*, **2000**, 65, 357-364.
39. Shishikura, A.; Fujimoto, K; Kaneda, T.; Arai, K.; Saito, S. *J. Jpn. Oil Chem. Soc*, **1998**, 37, 8–12.
40. Zhao, S.-Q.; Shi, T.-P.; Wang, R.-A.; Yang, G.-H. *Xibei Daxue Xuebao, Ziran Kexueban*, **2001**, 31,229–231.
41. Palma, M.; Taylor, L.T. *J Chromatogr. A*., **1999**, 849, 117-124.
42. Murga, R.; Sanz, M.T.; Beltran, S.; Cabezas, J.L. *J. Supercrt Fluids*, **2002**, 23, 113-121.
43. Sovova, H.; Kucera, J.; Jez, J. *Chem. Eng. Sci.*, **1994**, 49, 415-420.
44. King, J.W. *Food Sci. Techn. Today*, **2000**, 14, 186-191.

Chapter 9

Supercritical Fluid Extraction of Bioactive Components from St. John's Wort (*Hypericum perforatum* L.) and *Ginkgo biloba*

Mari Mannila[1], Qingyong Lang[1], Chien M. Wai[1,*], Yanyan Cui[2], and Catharina Y. W. Ang[2,*]

[1]Department of Chemistry, University of Idaho, Moscow, ID 83844
[2]Division of Chemistry, National Center for Toxicological Research, Food and Drug Administration, 3900 NCTR Road, HFT-230, Jefferson, AR 72079

Supercritical fluid extraction (SFE) offers an attractive alternative to solvent-based methods for extraction and manufacturing of herbal products. In addition to being a "green" solvent, supercritical CO_2 can also be selective for extraction and separation of active ingredients from certain herbs. The main active constituents in St. John's wort are hyperforin and adhyperforin, which can be effectively extracted with neat CO_2 under mild conditions (30 °C and 80 atm). Furthermore, extraction of hyperforin and adhyperforin from St. John's wort with neat CO_2 is selective, resulting in a fairly enriched and stable extract of these compounds. A successful supercritical CO_2 extraction of terpene trilactones (bilobalide and ginkgolides A, B and C) from *Ginkgo biloba* leaves can be achieved at an elevated temperature (100 °C) and high pressure (350 atm; 0.72 g/ml density) with a small amount of ethanol/acetic acid (9:1) as a modifier. SFE of bilobalide and ginkgolides under these conditions resulted in an approximately 10% higher yield compared with traditional solvent based extractions.

1. Introduction

The use of herbal medicines and food supplements has gained increased popularity in recent years. The total vitamin, mineral, and herbal supplement market in 1999 was $14.1 billion in the USA (*1*). According to a report in 2000, the top purchased herbal products in the USA were gingko, garlic and glucosamine, followed by echinacea and St. John's wort (*1*). Bioactivity studies of these herbal extracts showed beneficial health effects. The extract of *Ginkgo biloba* was found to improve memory, having antioxidant activity and improving blood circulatory disorders (*2*). St. John's wort (*Hypericum perforatum L.*) is used as an antidepressant, antioxidant, and as a remedy for several psychological and neuralgic disorders (*3-5*).

Although herbal products are widely used today, there is a lack of complete knowledge of the active constituents in many herb extracts, their biochemical effects and possible interactions with other medicines. Commercially available herbal products usually have complex compositions and show significant variations in concentrations of active ingredients (*6,7*). It is known that medicinally active compounds often have positive health effects at a low dose but may be toxic at a high dose. Thus, large variations in quality of herbal products could create health problems for the consumers. Therefore, developing a good knowledge of active compounds and their methods of analysis are crucial for safe use and quality control of herbal remedies. For analytical purposes, various organic solvents including methanol can be used for initial extraction of the herbs.

Herb extracts intended for human consumption are generally prepared using organic solvents, such as alcohol or acetone. Extraction and processing of herbal products with a nontoxic and "green" solvent is highly desirable for consumer protection. Conventional solvent extraction methods have several inherent drawbacks including lack of selectivity and generation of liquid wastes. Supercritical fluid extraction (SFE) has gained acceptance in recent years as an alternative to conventional solvent extraction for separation of organic compounds in many analytical and industrial processes. Carbon dioxide is widely used in SFE because of its moderate critical constants ($T_c = 31$ °C and $P_c = 73$ atm), nontoxic nature, and ready availability in a relatively pure form. Thus, developing supercritical CO_2 extraction methods for production of herbal products appear attractive with respect to economy, consumer protection, and environmental point of view.

SFE has been applied to herb and natural product studies for years (*8-10*). This chapter describes procedures of optimization of SFE parameters for the separation of active constituents from two popular herbs: St John's wort and *Ginkgo biloba*. The active compounds in St. John's wort are believed to be hypericins (*11-13*), hyperforins (*14-16*), and some flavonoids (*17*). Hyperforin, a major constituent in St. John's wort, is a lipophilic phloroglucinol (Figure 1A). Adhyperforin, a homologue of hyperforin, is believed to have similar medicinal activities to hyperforin. These compounds are usually extracted from dried stem, leaves and flowers of St. John's wort using organic solvents such as methanol (*18-19*), acetone-ethanol (*20*), or hexane (*20, 21*). Earlier reports have shown that hyperforin and adhyperforin could be specifically extracted by supercritical CO_2 (*22*). Larger-scale extraction methods with stabilizers or improved processing procedures were designed for production of St. John's wort extracts with high hyperforin content (*23*). However, no detailed information was available regarding the SFE conditions. Detection of hyperforin and adhyperforin has been performed most often by reverse phase HPLC combined with a diode array or ultraviolet detector. If the goal is to detect all the known active ingredients in St. John's wort, a long analytical time (up to 90 min) with a gradient elution system is required (*18, 20*). If only hyperforin and adhyperforin are to be detected, a rapid HPLC procedure (about 15 min) with an isocratic solvent system can be used for quantitative analysis (*19, 24*).

The main active ingredients in *Ginkgo biloba* are believed to be bilobalide and ginkgolides, which are terpene trilactones having t-butyl groups (Figure 1B). Initially these compounds were isolated from the roots of *Ginkgo biloba* tree (*25*), but today the leaves of *Ginkgo biloba* are the main source of ginkgolides. Commonly used extraction solvents for separation of ginkgolides are water (*26*), acetone (*27*), and mixtures of water and ethanol (*28*).

Several analytical methods including HPLC (*29, 30*), GC/MS (*26*), CE (*31*) (capillary elctrophoresis), and SFC (supercritical fluid chromatography) (*28*) have been developed for detection of ginkgolides (*32*). Among these methods, SFC with ELSD (evaporation light scattering detector) as an alternative analytical method showed some unique advantages, such as speed and decreased consumption of organic solvents. Furthermore, this method was less demanding in terms of sample preparation. However, the disadvantages are also obvious. For instance, higher requirements in experience, instrument maintenance and relative standard deviation (RSD) values are greater than those of GC-FID (*28*). In addition to this, SFC with ELSD detector is not as easily available in many analytical laboratories as HPLC or GC. Most recently, Lang et al. reported a simple detection method based on gas chromatography (GC) connected to a flame ionization detector (FID) (*8*). In this procedure bilobalide and ginkgolides were first separated from the extract by liquid-liquid extraction followed by derivatization with bis(trimethylsilyl)trifluoroacetamide (BSTFA, 99%) and trimethylsilyl chloride (TMCS, 1%), and finally analyzed by GC-FID.

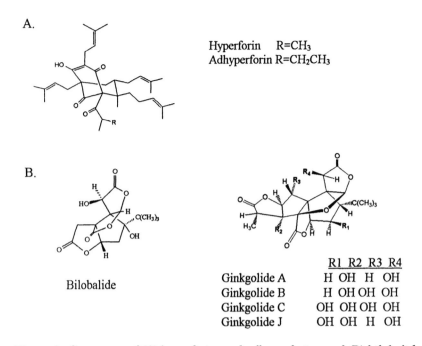

A.

Hyperforin   R=CH₃
Adhyperforin R=CH₂CH₃

B.

Bilobalide

	R1	R2	R3	R4
Ginkgolide A	H	OH	H	OH
Ginkgolide B	H	OH	OH	OH
Ginkgolide C	OH	OH	OH	OH
Ginkgolide J	OH	OH	H	OH

Figure 1. Structures of (A) hyperforin and adhyperforin, and (B) bilobalide and ginkgolides.

2. SFE of Hyperforin in St. John's Wort

2.1 Modifier

Recent reports from our research groups have demonstrated that neat supercritical CO_2 is selective for the extraction of hyperforin and adhyperforin in St. John's wort (*24, 33*) (Figure 2). Modifiers do not increase the SFE recoveries of these phloroglucinols, but tend to cause partial extraction of polar compounds in St. John's wort. Thus, extraction with neat CO_2 is the best choice for selective removal of hyperforin and adhyperforin from St. John's wort.

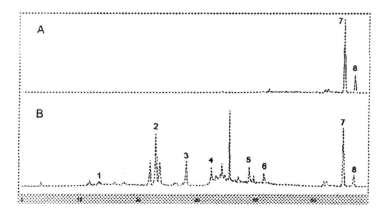

Figure 2. HPLC chromatograms of extract of St. John's wort by (A) SFE with neat CO_2 (40 °C, 100 atm), and (B) ultrasonic extraction with methanol. Peak 1: chlorogenic acid; 2; rutin; 3: quercitrin; 4:quercetin; 5: pseudohypericin; 6: hypericin; 7: hyperforin; 8: adhyperforin.
(Reproduced with permission from reference 32. Copyright 2002 The Royal Society of Chemistry.)

2.2 Effects of temperature and pressure (density)

Density is often an important parameter for achieving high extraction efficiency in the optimization of SFE conditions. Density of a supercritical fluid can be calculated from temperature and pressure of the fluid phase using computer models available in the literature. According to our experiments, the recoveries of hyperforin and adhyperforin from St. John's wort are good if the density of the fluid phase is kept at 0.65 g/mL or higher in the temperature range 30 – 90 °C and pressure range 80 – 300 atm. The density of a supercritical fluid decreases with increasing temperature at a given pressure. For example, at 100 atm pressure the density decreases from 0.85 g/mL to 0.41 g/mL when temperature rises from room temperature (22 °C) to 50 °C. This density decrease at elevated temperatures explains the decrease in extraction efficiency of hyperforin and adhyperforin with increasing temperature at 100 atm as shown in Figure 3A. On the other hand, at 300 atm the density of the fluid phase remains high (0.71 g/mL) even at 90 °C. Consequently, the extraction efficiency of hyperforin stays nearly constant at 300 atm in the temperatures 30 - 90°C. At 100 atm, a high extraction efficiency can be achieved at 40 °C due to a higher density under this condition (Figure 3B). Thus, very mild SFE conditions can successfully extract hyperforin and adhyperforin from St. Johns's wort.

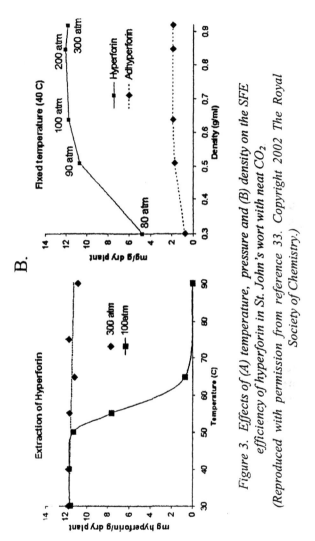

Figure 3. *Effects of (A) temperature, pressure and (B) density on the SFE efficiency of hyperforin in St. John's wort with neat CO_2 (Reproduced with permission from reference 33. Copyright 2002 The Royal Society of Chemistry.)*

Temperature increase (at a fixed density of 0.65 g/mL) does not improve the extraction of hyperforin and adhyperforin in St. John's wort. Maximum extraction was achieved at very mild conditions such as 30°C and 80 atm, even though these conditions are slightly below the critical point of the CO_2. Because the change from liquid to supercritical fluid is known to extend over a small temperature range (several degrees), the property of liquid CO_2 near the critical point is probably close to that of the supercritical state. Using mild conditions for extraction of natural products have several advantages including technical convenience, reduced co-extraction of undesirable products, and minimizing the risk of thermal decomposition of the desired compounds.

2.3 Extraction Time

A relatively short extraction time (total of 20 min) is sufficient to complete the SFE of hyperforin and adhyperforin under optimized temperature and pressure conditions (*33*). Depressurization of the sample matrix during the extraction process could improve the over-all extraction efficiency. Most probably depressurization alters or breaks down the plant cells so that more solutes become extractable by the fluid. In extraction of 300 mg St. John's wort at 40°C and 100 atm, a 5 min static extraction followed by two 7 min (total 20 ml liquid CO_2) dynamic extractions (with depressurization between dynamic steps) was sufficient to complete the extraction. Without depressurization (one 14 min dynamic extraction), about 90% of the maximum extraction could be achieved.

2.4 SFE compared to solvent extraction methods

The efficiency of removing hyperforin and adhyperforin from St John's wort by SF-CO_2 was studied by interlaboratory comparison between our two research groups. The SFE results were also compared to the ultrasound-aided extraction in methanol. The SFE results of hyperforin obtained by the two groups were equal (about 5 mg/g dry wieght), although different SFE conditions were used: 1) 50 °C and 375 atm (density > 0.9 g/ml), and 2) 30 °C and 80 atm (0.65 g/ml). This is reasonable, since limiting factor for efficient extraction was found to be a fluid density of ≥ 0.65 g/ml, that was reached with both SFE methods. Also the reproducibility of the SFE results was comparatively good; both research groups obtained relative standard deviations of 10-13% for hyperforin and adhyperforin.

Compared to the methanol extraction, the SFE results were equal with one group, whereas another laboratory obtained higher efficiencies with methanol extraction. The SFE results were 70% of the higher methanol extraction result. Both groups used the same methanol extraction method (90 min in ultrasound bath, after which the sample was centrifuged). It is possible that one ultrasound bath (PC620, Branson Cleaning Equipment Co., Shelton, CT) was more effective than the other (FS30, Fisher Scientific, Suwanee, GA) and thus hyperforin and adhyperforin were separated more efficiently by the former. Due to the higher efficiency of methanol extraction, SFE can not be used for the quantitative analysis of hyperforin and adhyperforin in St. John's wort.

The selectivity of supercritical CO_2 extraction of these phloroglucinols is much better than most of the organic solvents (33). The peak area of hyperforin and adhyperforin in the crude extract obtained from the SFE process was 85% (69% hyperforin and 16% adhyperforin) of the all peaks detectable by HPLC/UV. Only hexane shows selectivity similar to that of CO_2, but the use of SFE is beneficial due to its non-toxic nature. Furthermore, the stability of hyperforin was found to be poor in non-polar organic solvents such as hexane (34). Concentrating St. John's wort extract in hexane was shown to be a main source for peroxide formation that led to the oxidation and decomposition of hyperforin (21). SFE extracted compounds can be collected in a small amount of suitable solvent (such as ethanol or methanol) and thus there is no need for sample concentration and evaporation of solvents. This can be a very crucial advantage in isolation of phyto-nutraceuticals, because these compounds are often very air- and light-sensitive and thus can easily decompose during sample treatment.

Hyperforin and adhyperforin have been reported to undergo decomposition almost completely (over 90%) in methanol/acetonitrile (1:9) solution when they were exposed to light and air for 2 hr after mixed solid-phase purification (19). Different research groups have achieved different results of the stability of hyperforin in methanol solution. According to one report, 96% of hyperforin was decomposed in 5 min when methanol extract of St. John's wort was exposed to direct daylight (35). On the other hand, only 40% decomposition was observed after 30 days when purified hyperforin was stored in daylight in methanol solution (34). Decomposition of hyperforin also depends on pH: storage in methanol with strongly basic conditions (pH 12) caused complete decomposition of hyperforin during 30 days in daylight (34). Thus, decomposition of hyperforin can be minimized by 1) protecting extract from light and air, 2) using polar storage solvents, 3) using nonbasic pH, and 4) using minimal sample treatment after extraction. When the crude SFE extract was dissolved in methanol, no decomposition of hyperforin and adhyperforin was observed even after 80 hr of exposure to direct daylight and air (33). During SFE, hyperforin can be trapped in an empty vial and dissolved in a small amount of polar solvent, so there is no need for evaporation of solvent after extraction. Furthermore, in the crude SFE extract, some impurities may protect hyperforin from oxidation, and probably the SFE procedure may inhibit peroxide formation. Because the isolation of hyperforin and adhyperforin from SFE extract can be achieved with minimum sample treatment and, consequently, with reduced exposure to air, the decomposition of the hyperforin can be minimized.

2.5 Additional studies of hyperforin and adhyperforin in SFE extract

Because SFE of St. John's wort with neat CO_2 gives a comparatively high concentration and stable mixture of hyperforin and adhyperforin, the crude SFE extract can be used directly for certain studies to improve the knowledge of these compounds. For example, crude SFE extract has been used successfully to study the tautomerism of the hyperforin (*33*) and in animal studies for investigation of bioactivities (*36-38*). Hyperforin is a colorless, oily, and easily decomposed compound in its pure form, and thus the pure form is difficult to isolate. Orth et al. has reported a preparative-HPLC isolation procedure of hyperforin from solvent extract (*21*). The crude SFE extract of St. John's wort using our method may offer increased recovery in the isolation of these compounds in pure form using preparative HPLC. The SFE crude extract which contained approximately 30 to 40% of hyperforin and adhyperforin on a dry weight basis can be purified to different degrees of purity for different purposes. For example, in one of our studies the crude extract in methanol was partitioned with hexane saturated with methanol to remove co-extracted nonpolar substances. The hexane layer was discarded and the methanol layer was evaporated under vacuum and the residue was dissolved in hexane for subsequent clean-up through a silica gel column. Impurities were washed out by hexane and hyperforins were eluted by hexane:ethyl acetate (9:1). The eluted solution was mixed with 10-20 mL methanol and evaporated to near dryness under vacuum. The residue contained about 90-93% hyperforins. It can be dissolved in ethanol for further use in pharmacokinetic and *in vivo* drug-herb interaction studies. For the metabolic study in our laboratory, the residue was further purified by analytical HPLC and resulted in isolated hyperforin and adhyperforin fractions at 98% purity. The purified hyperforin was used in *in-vitro* rat liver microsome systems designed to elucidate biotransformation mechanism of hyperforin and the major metabolic pathway. More than 10 metabolites were found by HPLC/PDA method and 4 major metabolites were isolated and identified by LC/MS and NMR (*39*).

3. SFE of Ginkgolides from *Ginkgo biloba*

3.1 Modifier

Supercritical CO_2 is a poor extraction media for intermediately polar to polar compounds and thus a polar organic modifier is needed to improve the extraction of bilobalide and gingkolides from *Gingko biloba*. Without a modifier, no ginkgolides were detected from the SFE extract. Methanol is the most often used modifier in SFE, but it is not suitable in extraction of food supplements due to its toxicity. Thus, a less toxic mixture of solvents, ethanol and acetic acid (9:1), was selected as a modifier to enhance the extraction of bilobalide and ginkgolides in supercritical CO_2. According to our experiments, a small amount of modifier (0.5 ml per gram of leaves) could significantly improve the SFE efficiency of the terpene trilactones. The highest SFE efficiency was achieved with two times of SFE with total of 5 mL modifer (3 + 2 mL) per gram of ginkgo leaf powder at 100 °C and 350 atm. In fact, at the beginning of the SFE process, there were two phases (extra modifier liquid and supercritical fluid) existing inside the SFE vessel. This is necessary in order to extract the more polar ginkgolide C as well as to prevent tubing clogging. Extraction of ginkgolides using boiling ethanol requires more than 40 mL of the solvent per gram of the leaf powder to achieve a similar efficiency.

3.2 Analysis of ginkgolides

After SFE, the extract was diluted with ethanol in a 25-mL volumetric flask, and 5 mL of the solution was used for analysis. The solvent was first removed by evaporation in a warm water bath, and then the ginkgolides were dissolved in 5 mL of 5% NaH_2PO_4 aqueous solution. In order to save time, the aqueous solution may be centrifuged to settle down the insoluble residue. Precisely 1 mL of the clear supernatant was transferred into a small sample vial, and 3 mL of ethyl acetate/tetrahydrofuran (70:30 v/v) and 25 µg of squalane (in hexane) as an internal standard were added into the vial for liquid-liquid extraction separation. The vial was shaken for 1 min either using a mechanical shaker or manually. After phase separation, about 1 – 2 mL of the organic solution was transferred to a dry and clean sample vial for derivatization. The solvent was blown down to dryness with a gentle stream of nitrogen gas in a hood and 300 µL of DMF and 300 µL of BSTFA/TMSC (99:1) were added to the vial. The capped vial was placed in a GC oven at 120 °C for 45 min for derivatization. The standards of ginkgolides were treated in the same way. After cooling down to room temperature, 1 µL of each sample was injected into a GC for quantification.

For GC analysis, helium was used as the carrier gas and the head pressure was set at 18 psi. The injector was set at 280 °C and the detector at 300 °C. The oven was programmed to have an initial temperature of 200 °C for 1 min, and was then ramped to 280 °C at a rate of 10 °C/min and held at the final temperature for 5 min. The details of the analytical procedure are given in the literature (8).

3.3 Effect of temperature and pressure (density)

Increasing the extraction temperature has been reported to improve the SFE efficiency for many organic compounds (40), including certain natural products from plant matrices. According to our studies, increase in extraction temperature (at 350 atm) improved the extraction of the terpene trilactones (Figure 4). Higher temperatures can lower the viscosity and increase the diffusivity of the fluid. Thus, the fluid can penetrate into the leaf sample and reach the extractable compounds located deeper inside the matrix. Furthermore, a higher temperature tends to increase the vapor pressure of solutes and thus facilitates the transfer of the solutes from the matrix to the fluid phase. The maximum extraction efficiency for bilobalide and ginkgolides was achieved at a temperature around 100 °C. At temperatures greater than 120 °C, the extraction efficiency was found to decrease. This might be related to the lower fluid density or due to decomposition of the solutes. Thus, 100 °C appeared to be a suitable extraction temperature for ginkgolides from ginkgo leaves.

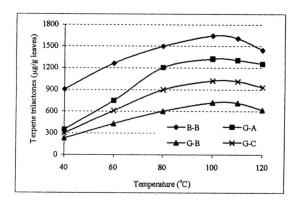

Figure 4. Temperature effects on the SFE results of terpene trilactones.

Raising pressure at a fixed temperature improves the extraction efficiency due to an increased fluid density. In the extraction of bilobalide and ginkgolides, the maximum extraction efficiency was achieved when the density was at least 0.72 g/mL. Because the extraction of bilobalide and ginkgolides is improved at elevated temperature, high pressures are required to obtain the minimum needed density. At the optimal temperature (100 °C), a density of 0.72 g/mL was obtained at 350 atm pressure. With the optimized conditions (ethanol/acetic acid as a modifier, 100 °C and 350 atm), quantitative extraction of terpene trilactones from 1g sample of *Ginkgo biloba* was achieved with 20 min static and 20 min dynamic extraction times (two times of SFE). Under these conditions, the total amount of B-B, G-A, G-B and G-C extracted was about 4.7 mg per gram of the dry ginkgo leaves used in this study.

3.4 SFE compared to the solvent extraction methods

The SFE technique was compared to the boiling liquid methods performed with the following solvents: 1) acetone, 2) ethanol, 3) methanol, 4) water, and 5) water/methanol (70:30 v/v). The SFE method showed the highest extraction efficiency and reproducibility (RSD<5%, n = 3) for removing terpene trilactones from the ginkgo leaves in comparison with the solvent extraction methods. The total efficiencies for recovery of bilobalide and gingkolides (A, B, and C) with the solvent based methods were 60-91% relative to the SFE results (Table I).

Table I. Comparison of extraction results (μg/g dry leaves) using different extraction methods. Results of SFE are the mean values of triplicate extractions and all others are the averages of duplicate extractions.

Method	B-B	G-A	G-B	G-C	Total	%*
Acetone	1074	864	361	514	2813	60
Ethanol	1579	1326	625	771	4301	91
Methanol	1399	1204	617	769	3989	85
Water	1576	1344	629	745	4294	91
W/M (70:30)	1008	926	482	686	3102	66
SFE	1645	1324	721	1020	4710	100

*Percentages relative to SFE results.

For the boiling solvent methods, the relatively lower efficiency and reproducibility probably were caused by several factors including sample loss during the boiling, transfer and filtration steps, plus undesired reactions during the boiling process including decomposition (degradation) and oxidation. For some volatile solvents such as acetone (boiling point 56 °C), insufficient solvation strength due to high volatility of solvents could be another factor contributing to the low efficiency and poor reproducibility. SFE process is performed in a closed CO_2 system that should minimize oxidation and other undesirable reactions during extraction. In addition, SFE does not require multiple sample-handling steps and thus it should also minimize sample loss during extraction. In the boiling solvent experiments, more than 40 mL of each solvent were used for the extraction of one gram of the leaf sample. The amount of ethanol used in the boiling solvent method is about an order of magnitude higher than that used in the SFE process. Removing large quantities of solvents from the ginkgo extracts is both time and energy consuming. Furthermore, boiling solvent extraction is a non-discriminative dissolution process and often ends up with large amounts of other compounds in the extract solution such as chlorophyll.

4. Summary

This chapter summarizes recent developments in our laboratories regarding supercritical fluid extraction of natural products from popular herbs St. John's wort and *Ginkgo biloba*. Supercritical CO_2 is highly selective compared with conventional boiling solvent methods for extracting hyperforin and adhyperforin from St. John's wort. In the case of extracting bilobalide and ginkgolides from *Ginkgo biloba*, ethanol/acetic acid (9:1) modified CO_2 is required to achieve a maximum extraction. The amount of modifier required in this SFE process is an order of magnitude less than that of boiling ethanol to achieve the same level of extraction. We have patented a "Pressurized Water Extraction" technique which is more suitable for the extraction and analysis of terpene trilactones (*41*). The advantage for SFE is that it is not only good for removing bilobalide and ginkgolides from ginkgo leaves but also for removing flavonoids (*42, 43*), which is another important group of active ingredient in ginkgo extract.

Because SFE can provide much higher selectivity than conventional extraction methods, usually the SFE processes produce far less impurities co-extracted with the active compounds. Pure active compounds can be easily isolated from the SFE produced herbal extracts using HPLC for medicinal studies. This green extraction technique provides an efficient way of obtaining active compounds from herbs with minimum waste production and appears attractive for manufacturing high quality herbal products and for isolation and identification of active natural products from herbs and plants in general.

5. References

1. Vitamins, minerals, herbs and supplements: A year in rewiev, year 1999, A National Study Conducted by The Hartman Group, The Hartman Group, Inc., Bellevue, WA, 2000.
2. Glisons, J.; Crawford, R.; Street, S. *The Nurse Practitioner* **1999**, 24(6), 28, 31, 35-36.

3. Linde, K.: Ramirez, G.; Mulrow, C. D.; Pauls, A.; Weidenhammer, W.; Melchart, D. *Brittish Medical Journal* **1996**, 313, 253-258.
4. Bilia, A. R.; Gallori, S.; Vincieri, F.F. *Life Sci.* **2002**, 70, 3077-3096.
5. Vickery, A. R.; *Economic botany* **1981**, 289-295.
6. Consumers Union, Consumer Reports **1999**, 64, 44-48.
7. Lang, Q.; Yak, H. K.; Wai, C. M. *Talanta* **2001**, 54, 673-680.
8. Lang, Q.; Wai, C. M. *Talanta* **2001**, 53, 771-782.
9. Reverchon, E. *J. Supercritical Fluids* **1997**, 10, 1-37.
10. Modey, W. K., Mulholland, D. A., Raynor, M. W., *Phytochemical Analysis* **1996**, 7, 1-15.
11. Lavie, G.; Mazur, Y.; Lavie, D.; Meruelo, D. *Medicinal Research reviews* **1995**, 15, 111-119.
12. Weber, N.D.; Murray, B.K.; North, J.A.; Wood, S.G. *Antiviral Chemistry & Chemotherapy* **1994**, 5(2), 83-90.
13. Meruelo, D.; Lavie, G.; and Lavie, D. *Proc. Natl. Acad. Sci.* USA **1988**, 85, 5230-5234.
14. Singer, A.; Wonnemann, M.; Muller, W. E. *Pharmacol. Exp. Ther.* **1999**, 290(3), 1363-1368.
15. Chatterjee, S.S.; Bhattacharya, S.K.; Wonnemann, M.; Singer A.; Muller, W.E. *Life Sci.* **1998**, 63(6), 499-510.
16. Laakmann, G.; Dienel, A.; Kieser, M. *Phytomedicine*, **1998**, 5(6), 435-442.
17. Butterweck, W.; Jurgenliemk, G.; Nahrsted, A.; Winterhoff, H. *Planta Medica* **2000**, 66(1), 3-6.
18. Brolis, M.; Gabetta, B.; Fuzzatti, R.; Pace, R.; Panzeri, F.; Peterlongo, F. *J. Chromatogr. A* **1998**, 825, 9-16.
19. Gray, D. G.; Rottinghaus, G. E.; Garrett, H. E. G.; Pallardy, S. G. *J. AOAC International*, **2000**, 83(4), 944-949.
20. Liu, F. F.; Ang C. Y. W.; Springer, D. *Agric. Food Chem.* **2000**, 48, 3364-3371.
21. Orth, H. C. J.; Rentel, C.; Schmidt, P. C. *J. Pharm. Pharmacol.* **1999**, 51, 193-200.
22. Chatterjee, S. S.; Nöldern, M.; Koch, E.; Erdelmeier, C. *Pharmacops.* **1998**, 31(S), 7-15.
23. Erdelmeier, C.; Grethlein, E.; Lang, F.; Oschmann, R.; Strumpt, K. H. US Patent 6,280,736, 2001.
24. Cui, Y.; Ang, Y. W. C. *J. Agric. Food Chem.* **2002**, 50, 2755-2759.
25. Hostettmann, K.; Lea, P.J. *Biologically active Natural Products*, Claredon Press, Oxford, 1997, p. 119.
26. Okabe, K.; Yamada, K.; Takada, S. *J. Chem. Soc.*, **1967**, 21, 2201-2206.
27. Chauret, N.; Carrier, J.; Mancini, M.; Neufeld, R.; Weber, M.; Archambault, J. *J. Chromatograp.* **1991**, 588, 281-287.
28. Strode III, J.T.B.; Taylor, L. T.; van Beek, T. A.; *J. Chromatograp. A* **1996**, 738, 115-122.

144

29. van Beek, T. A.; Scheeren, H. A.; Rantio, T.; Melger, W. Ch.; Lelyveld, G.P. *J. Chromatograp.* **1991**, 543, 375-387.
30. Lobstaein-Guth, A.; Briancon-Scheid, F.; Anton, R. *J. Chromatograp.* **1983**, 267, 431-438.
31. Oehrle, S.A. *J. Liquid Chromatograp.* **1995**, 18, 2855-2859.
32. van Beek, T. A. *J. Chromatograp. A* **2002**, 967, 21-55.
33. Mannila, M.; Kim, H.; Isaacson, C.; Wai, C. M. *Green Chem.* **2002**, 4, 331-336.
34. Orth, H. C. J.; Schmidt, P. C. *Pharm. Ind.* **2000**, 62, 60-63.
35. Potaraud, A.; Lobstein, A.; Girardin, P.; Weniger, B. *Phytochem. Anal.* **2001**, 12, 335-362.
36. Bhattacharya, S. K.; Chakrabarti, A.; Chatterjee, S. S. *Pharmacopsychiat.* **1998**, 31(S), 22-29.
37. Perfumi, M.; Panocka, I.; Ciccocioppo, R.; Vitali, D.; Froldi, R.; Massi, M. *Alcohol & Alcoholism.* **2001**, 36(6), 199-206.
38. Dimpfel, W.; Schober, F.; Mannel, M. *Pharmacopsychiat.* **1998**, 31(S), 30-35.
39. Cui, Y.; Ang, C. Y. W.; Leakey, J.; Hu, L.; Berger, R.; Heinze, T. M. NCTR/FDA, Jefferson, AR, USA (Unpublished data).
40. Yang, Y.; Ghairabeh, A.; Hawthorne, S.B.; Miller, D. J. *Anal. Chem.* **1995**, 67, 641-646.
41. Wai, C. M., Lang, Q., Pressurized Water Extraction, US patent, 2002, pending.
42. Ciu, K-L.; Cheng, Y-C.; Jun-Hao, C.; Chen, J-H.; Chang, C. J.; Yang, P-W. *J. Supercritical Fluids* **2002**, 24(1), 77-90.
43. Yang, C.; Xu, Y-R.; Yao, W-X. *J. Agric. Food Chem,* **2002**, 50(4), 846-850.

Chapter 10

Hot Water Extraction Followed by Solid-Phase Microextraction of Active Ingredients in Rosemary: An Organic Solvent-Free Analytical Technique

Youxin Gan and Yu Yang[*]

Department of Chemistry, East Carolina University, Greenville, NC 27858

An organic solvent-free technique was developed in this study to determine the concentrations of the active ingredients (pinene, camphene, limonene, camphor, citronellol, and carvacrol) of rosemary in the water phase after cooking at 100 °C. The water extract obtained from the hot water extraction was then used as the sample for solid-phase microextraction (SPME) followed by GC analysis. Polydimethylsiloxane fiber was employed in SPME. SPME was optimized by evaluating the carryover effect and effects of water volume and sorption time on SPME efficiency. The concentrations of six active components in rosemary in the water phase after cooking were determined using this green analytical technique. Up to 15% of the active ingredients in rosemary were found in the water phase after a 15-min hot water extraction at 100 °C. This organic solvent-free coupling technique could have a good potential in pharmaceutical and food analysis.

INTRODUCTION

Tracing the history of food and pharmaceuticals, plants have served human beings for thousands of years. Chinese people have a long history of preparing and using herbs for the treatment of all kinds of diseases. Traditional Chinese medicines are prepared with dried herbs, which are cooked with boiling water in order to extract the active ingredients into the soup. Patients take the soup according to the prescription. This cooking process is actually a so-called hot water extraction. In order to make better application of medicinal herbs, people extract the active ingredients from the raw plants to make pills, tablets, juice, and injecting solutions. Extraction methods applied to natural products generally include hot water extraction, steam distillation, organic solvent extraction, supercritical fluid extraction, and microwave-assisted extraction [1-6].

Rosemary (*Rosemarinus officinalis*) is one of the widely used natural products in everyday life. It is generally used as a spice for cooking, but it is also an important herb. The leaves and tops of rosemary can be used as herbs for tonic, diaphoretic, stomachic, and antirheumatic purposes [6]. The major active ingredients in rosemary have been studied and identified [7,8], and they include pinene, camphene, limonene, camphor, citronellol and carvacrol. The concentrations of these active ingredients in raw rosemary are well understood, but the fraction of the active ingredients in the water phase after hot water extraction (cooking) process is less studied. The water phase is normally treated by solid phase extraction (SPE) or liquid-liquid extraction for quantitation. Therefore, organic solvents are involved in the analysis of the active ingredients in the water phase.

Solid-phase microextraction (SPME) uses a very small amount of organic phase coated on the needle of a specially designed syringe to extract analytes from water samples or the headspace [9-11]. The SPME process involves two steps: partitioning of analytes between the coating and the sample matrix, followed by desorption of the concentrated extracts into an analytical instrument. In the first step, the coated fiber is exposed to the sample or its headspace, which causes the target analytes to partition from the sample matrix into the coating. The fiber bearing extracted analytes is then transferred to an instrument for desorption, whereupon separation and quantification of extracts can take place. Since SPME can only be applied to gas or liquid samples, hot (or subcritical) water extraction can convert solid samples into water samples so that SPME can be used for solid matrices. Therefore, the coupling of SPME with subcritical water extraction has also been studied [12-15].

The goal of this research was to develop an organic solvent-free technique to determine the concentrations of the active ingredients of rosemary in the water phase after the hot water extraction at 100 °C. Hot water extraction was used to simulate the cooking process, and the water extracts from the hot water extraction were then used as samples for solid-phase microextraction followed by GC analysis. Therefore, no organic solvents were involved in the entire extraction/analysis process. Please note that the emphasis of this technique is not to exhaustively extract the active ingredients from herbs but to determine how much of them can be found in the water phase after cooking, the traditional way of preparing Chinese medicinal herbs.

EXPERIMENTAL

Samples, Chemicals, and Reagents

The rosemary used in this project was directly purchased from a local grocery store. The raw rosemary was grounded using ceramic mortar and pestle. Methylene chloride and acetone used for solvent extraction were of analytical grade and purchased from Fisher Scientific (Pittsburgh, PA). Pinene, camphene, limonene, camphor, citronella, and carvacrol were obtained from Aldrich (Milwaukee, WI).

Hot Water Extraction of Rosemary

Approximately 0.5 g of grounded rosemary was weighed into a 20-mL glass vial as a single sample for hot water extraction at 100 °C. Approximately 8 mL of distilled water was added into the rosemary vial. The loaded vial was put on a hot plate to perform the hot water extraction. The vial was loosely sealed with a rubber cap. The water extraction time was 15 min. After the hot water extraction at 100 °C, the glass vial was removed from the hot plate and cooled down to room temperature. Clear aqueous solution was obtained by filtration using small filter paper. This water sample was then ready for solid-phase microextraction.

Solid-Phase Microextraction

Before applying solid-phase microextraction to the water extracts obtained from hot water extraction of rosemary, optimization of experiments were conducted to determine the optimum working conditions for SPME, including water volume, SPME time, and carry over effect. Polydimethylsiloxane fiber (PDMS, 100 μm, Supelco, Bellefonte, PA) was used for performing solid-phase microextraction.

Three different extraction volumes (2, 4, and 8 mL) were tested with SPME extraction time set to 15 min. The solid-phase microextraction time studied was 15, 30, and 60 min while the water volume was set to 4 mL. In order to study the fiber carryover effect in GC analysis, the total thermal desorption time (at 250 °C) was set to 10 min. Two sets of experiments were performed: 1-min thermal desorption followed by 9-min desorption, and 5-min desorption followed by 5-min desorption.

The water extractant of rosemary was separated from the rosemary residue. The PDMS fiber was dipped into the glass vial containing the water extracts.

SPME was performed for 15 min. In order to determine the concentration of the target analytes in the water phase, calibration curves were prepared using the same SPME procedure. All SPME extractions were performed at room temperature while the water samples were stirred during the SPME extraction process.

Sonication Extraction of Rosemary

In order to determine the total concentration of the target analytes in rosemary, solvent extraction of rosemary was also performed using a Fisher FS5 sonication bath. A solvent mixture of methylene chloride and acetone (50:50, v%) was chosen for solvent extraction. For the determination of the total concentration of analytes in rosemary, approximately 0.3-0.4 g of grounded rosemary was weighed into a glass vial, and 3 mL of methylene chloride and acetone mixture was then added into the vial. The solution was thoroughly mixed before setting into a water bath for sonication. Sonication extractions were performed for 12 hours. After the sonication extraction, the upper layer of the clear solution was injected into GC for analysis.

Gas Chromatographic Analysis

A Hewlett-Packard (Wilmington, DE) 6890 GC was used in this project. A HP-35MS capillary column (30 m × 0.25 mm id, 0.25 μm film thickness) was used for separation, while a flame ionization detector (FID) was employed for detection. Both injector and FID temperatures were held at 250 °C. The initial oven temperature was 40 °C. After 5-min holding time, the oven temperature was increased at 8 °C/min to a final temperature of 300 °C, with 15-min holding time. Splitless injection was employed for SPME injections while split mode was used for most solvent injections.

RESULTS AND DISCUSSION

SPME Carryover Effect

Unlike any other traditional GC injection methods (e.g., solvent injection using a traditional syringe), carryover effect plays an important role in SPME/GC analysis. If the thermal desorption time is not properly chosen, the compounds remaining on the SPME fiber lead to poor accuracy and precision in SPME/GC analysis.

Two experiments were performed. First, the SPME fiber was thermally desorbed in the hot GC injector (250 °C) for 1 min, then GC analysis was performed. After the GC analysis for the 1-min desorption, the fiber was replaced in the GC injector for another 9-min desorption then GC analysis was again performed. Another experiment was done in a similar way but with 5-min desoprtion/GC analysis followed by a second 5-min desorption/analysis. The total thermal desorption time was 10 min for both experiments. As shown in Table 1, the 5-min desorption yielded higher percentage of desorption, therefore, 5-min desorption was used as desorption time for the remainder of this work.

Table 1. Influence of Thermal Desorption Time on Carryover Effect of SPME Fiber

Compound	% Desorbed (%RSD[a])			
	1 min		5 min	
Pinene	99.34	(<1)	99.85	(<1)
Camphene	98.57	(<1)	99.28	(<1)
Limonene	99.44	(<1)	99.83	(<1)
Camphor	99.58	(<1)	99.53	(<1)
Citronellol	99.39	(<1)	99.90	(<1)
Carvacrol	94.12	(3)	99.58	(<1)

a. Triplicate measurements.

Effect of Water Volume on SPME Efficiency

The effect of water volume on SPME efficiency was evaluated using three different volumes: 2, 4, and 8 mL, while the concentration of analytes remained the same. As shown in Table 2, the peak area ratio (the peak area of analyte divided by the peak area of the internal standard) increased about 10 times when the volume was increased from 2 to 4 mL. However, when water volume further increased from 4 to 8 mL, only slight enhancement of peak area ratios was found. Since the volume of water extracts after the hot water extraction of rosemary was ~5 mL, the 4 mL of water volume was chosen in this work.

SPME Sorption Time

The effect of SPME sorption (extraction) time on SPME efficiency was investigated by using three time intervals: 15, 30, and 60 min. Table 3 lists the peak area ratios as a function of the sorption time. The SPME sorption time had no significant effect on SPME efficiency. In order to save analysis time, the 15-min sorption time was chosen for the remainder of this work. Since a typical GC run lasts approximately 20 min, shorter SPME extraction time is also more practical because both SPME and GC analysis can be performed simultaneously.

Table 2. Effect of Water Volume on SPME Efficiency

Compound	Peak Area Ratio[a] (%RSD[b])					
	2 mL		4 mL		8 mL	
Pinene	0.55	(11)	5.6	(7)	7.66	(49)
Camphene	0.72	(6)	7.1	(10)	9.54	(45)
Limonene	0.72	(8)	7.1	(2)	7.97	(39)
Camphor	0.10	(33)	0.94	(26)	1.2	(54)
Carvacrol	0.040	(36)	0.40	(27)	0.57	(59)

a: The peak area of the analyte divided by the peak area of the internal standard. b: Triplicate measurements.

Table 3. Effect of Sorption Time on SPME Efficiency

Compound	Peak Area Ratio (%RSD[a])		
	15 min	30 min	60 min
Pinene	29 (60)	37 (55)	28 (51)
Camphene	6.0 (45)	8.4 (38)	7.5 (35)
Limonene	8.8 (63)	11 (42)	7.8 (48)
Camphor	5.8 (10)	5.3 (7)	5.9 (25)
Citronellol	3.0 (15)	3.7 (11)	4.0 (19)
Carvacrol	2.0 (22)	2.4 (9)	2.5 (15)

a: Triplicate measurements.

Total Concentration of the Active Ingredients in Rosemary

For purpose of comparison, the total concentrations of the target analytes in raw rosemary were determined with sonication extraction and GC analysis. Figures 1 and 2 demonstrate the chromatograms obtained for the standard solution and the rosemary extract using a mixture of acetone and methylene chloride, respectively. The concentrations obtained by sonication extraction using the mixture of acetone and methylene chloride are given in Table 4.

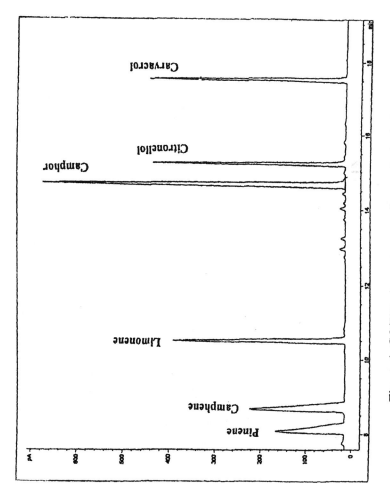

Figure 1. GC/FID chromatogram of a standard solution in methylene chloride and acetone.

152

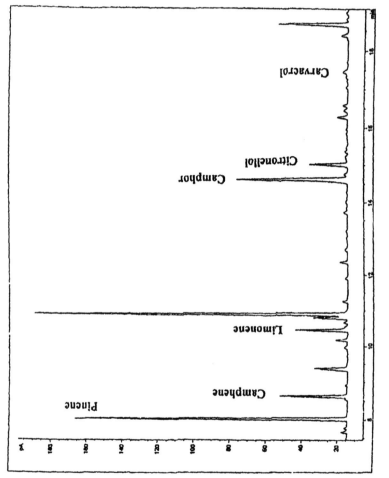

Figure 2. GC/FID chromatogram of rosemary extract after sonication extraction using a mixture of methylene chloride and acetone.

Table 4. Concentrations of Target Analytes in Raw Rosemary Determined by
Sonication Extraction Followed by GC Analysis

Compound	Concentration (mg/g)	(%RSD[a])
Pinene	2.9	(1)
Camphene	0.97	(2)
Limonene	0.47	(1)
Camphor	2.0	(1)
Citronellol	0.12	(1)
Carvacrol	0.20	(3)

a: Triplicate measurements.

Hot Water Extraction Coupled With SPME

Raw rosemary was extracted using hot water at 100 °C for 15 min. After the
hot water extraction, the water phase was cooled and removed from the rosemary
residue. The 100-μm PDMS fiber was dipped into 4 mL of the water sample.
The SPME was performed for 15 min. Following the SPME process, the SPME
fiber was inserted into the GC injector, and the extracted analytes were thermally
desorbed at 250 °C for 5 min. Figure 3 shows the SPME/GC chromatogram of
the standard solution in deionized water, while the chromatogram obtained by
SPME/GC/FID for rosemary water extract is shown in Figure 4. The
concentrations of the target analytes extracted into the aqueous phase (15 min of
hot water extraction at 100 °C) are summarized in Table 5. The percentage of
the mass of target analytes extracted into the aqueous phase compared to the
total mass of analytes in raw rosemary is listed in the last column in Table 5. Up
to 15% of analytes were found in the water phase after the hot water extraction
of rosemary.

Table 5. Concentrations of Target Analytes in the Water Extract after Hot Water
Extraction of Rosemary and SPME/GC Analysis

Compound	Concentration in Water Extract (mg/g) (% RSD[a])		% Extracted into the Aqueous Phase
Pinene	0.040	(8)	1.4
Camphene	0.029	(6)	3.0
Limonene	0.015	(4)	3.2
Camphor	0.14	(9)	7.2
Citronellol	0.018	(3)	15
Carvacrol	0.016	(5)	8.1

a: Triplicate measurements.

154

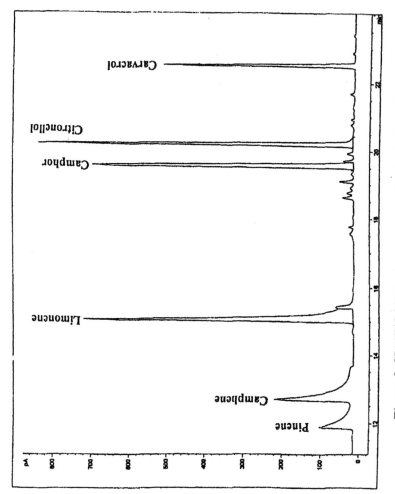

Figure 3. SPME/GC/FID chromatogram of a standard solution in deionized water.

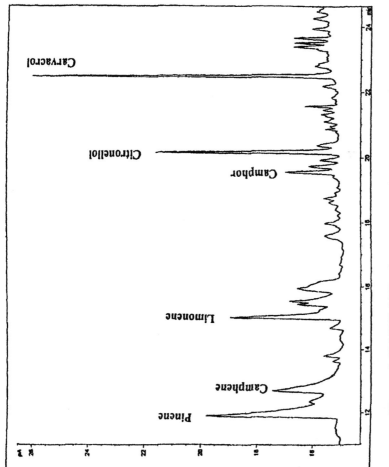

Figure 4. SPME/GC/FID chromatogram of a water extract after the hot water extraction of rosemary at 100 °C.

156

CONCLUSION

In this study, an organic solvent-free technique for the extraction, separation, and analysis of the active ingredients from rosemary was developed. The active ingredients in raw rosemary were extracted into hot water at 100 °C. The water sample was then used for solid-phase microextraction. Polydimethylsiloxane fiber with a coating thickness of 100 μm was employed for solid-phase microextraction. The concentrations of six active components including pinene, camphene, limonene, camphor, citronellol, and carvacrol in the water extract of rosemary were determined using this technique. Up to 15% of the active ingredients in rosemary were found in the water phase after a 15-min hot water extraction at 100 °C. This organic solvent-free coupling technique completely eliminates the use of organic solvents in the entire procedure and could have great potentials in pharmaceutical and food analysis.

REFERENCES

1. Ma, X.; Yu, X.; Zheng, Z.; Mao, J. *Chromatographia*, **1991**, *32*, 40-44.
2. Guo, Z.; Jin, Q.; Fan, G.; Duan, Y.; Qin, C.; Wen, M. *Analytica Chimica Acta*, **2001**, *436*, 41-47.
3. Fang, Q; Yeung, H.W.; Leung, H.W.; Huie, C.W. *J. Chromatogr. A*, **2000**, *904*, 47-55.
4. Ling, Y.C.; Teng, H.C.; Cartwright, C. *J. Chromatogr. A*, **1999**, *835*, 145-157.
5. Li, K.-L.; Sheu S.-J. *Analytica Chimica Acta*, **1995**, *313*, 113-120.
6. Gan, Y. "Development of an organic solvent-free analytical technique for natural products and studies of the sorption mechanism in solid-phase microextraction," *MS thesis*, East Carolina University, Greenville, NC, July **2000**.
7. Tena, M.T.; Valcarcel, M.; Hidalgo, P.J.; Ubera, J.L. *Anal. Chem.* **1997**, *69*, 521-526.
8. Basile, A.; Jimenez-Carmona, M.M.; Clifford, A.A. *J. Agricultural and Food Chemistry*, **1998**, *46*, 5205-5209.
9. Arthur, C.L.; Pawliszyn, J. *Anal. Chem.* **1990**, *62*, 2145-2148.
10. Zhang, Z.; Yang, M.J.; Pawliszyn, J. *Anal. Chem.* **1994**, *66*, 844A-853A.
11. Zhang, Z.; Pawliszyn, J. *Anal. Chem.* **1995**, *67*, 34-43.
12. Hageman, K.J.; Mazeas, L.; Grabanski, C.B.; Miller, D.J.; Hawthorne, S.B. *Anal. Chem.* **1996**, *68*, 3892-3898.
13. Daimon, H.; Rittinghaus, S.; Pawliszyn, J. *Anal. Communi.* **1996**, *44*, 421-424.
14. Hawthorne, S.B.; Grabanski, C.B.; Miller, D.J. *J. Chromatogr. A*, **1998**, *814*, 151-160.
15. Krappe, M.; Hawthorne, S.B.; Wenclawiak, B.W. *Fresenius' J. Anal. Chem.* **1999**, *364*, 625-630.

Chapter 11

Cleanup of Disperse Dye Contaminated Water by Supercritical Carbon Dioxide Extraction

Jya-Jyun Yu[1] and Kong-Hwa Chiu[2]

[1]Department of Environmental Engineering and Science, Feng-Chia University, Taichung, Taiwan, Republic of China
[2]Division of Natural Science, National Science Council, Taipei, Taiwan, Republic of China

Abstract

Hydrophobic disperse dyes including color index Red-60, Red-73, Blue-60, and Blue-79 present in the textile wastewater can be effectively removed by supercritical carbon dioxide (SC-CO_2) extraction. Near quantitative removal of the dye from the aqueous solution can be achieved by SC-CO_2 modified with 10% of ethanol at 85℃ and 300 atm for 30 minutes of static extraction followed by 60 minutes of dynamic flushing of SC-CO_2. The extraction efficiency increases dramatically with increasing both temperature and pressure. After SC-CO_2 extraction, the American Dye Manufacturer Institute (ADMI) true color value of the mother dyebath was dropped significantly below the regulatory level. The extracted dyes were collected in a small amount of aqueous solution in the presence of acetone, which performed as a photo-sensitizer. The collected dyes were illuminated by UV and were degraded completely within a short time. The combinations of SC-CO_2 extraction of disperse dyes from aqueous solution and subsequent degradation by photo-sensitization of the extracted dyes provide an innovative purification technique for treating textile wastewater.

Introduction

Synthetic dyes are used extensively by a variety of industries including textile dyeing and paper printing. In the dyeing and finishing processes of textile manufacturing plants, a considerable amount of wastewater containing residual dyes is generated. A severe threat may be imposed on the ecosystem if the colored wastewater is released into the environment without appropriate treatment. As the discharge regulations are becoming more stringent, cleanup of dye contaminated water stream is one of the most important jobs in pollution control and has become a major focus of policy debate in this country. Many commercial disperse dyes are designed to be chemically stable under typical usage conditions. The dye effluent poses treatment problems mainly owing to its high color and low biodegradability. The hydrophobic disperse dyes are too recalcitrant to be degraded by indigenous microorganisms which are present in the aquifer matrix. Adsorption and chemical coagulation employing granular activated carbons, lime, alum or iron salts are quite effective for color removal (1). Drawbacks of these techniques include a secondary waste generated by transferring dyes from aqueous solution to adsorbents, which still creates a waste disposal problem (2,3). Recently, photochemical degradation has been employed for treating textile water by using ultraviolet catalyzed wet-oxidation to decompose the dye molecules. However, commercial disperse dyes are scarcely soluble in water and are designed to resist photo-degradation, therefore, direct photolysis of dyeing wastewater can be difficult (4,5).

Supercritical fluid extraction (SFE) of organic compounds from environmental samples has been the subject of many studies (6-8). The high diffusivity, low viscosity, and T-P dependence of solvent strength are some attractive characteristics that make supercritical fluids excellent candidates for removing organic chemicals from solid matrices. Supercritical carbon dioxide ($SC-CO_2$) offers considerable potential as a replacement solvent in the laboratory due to its moderate critical constants ($Tc=31\ ℃$, $Pc=73atm$), non-toxic, non-flammable, sufficient solvating power, and availability in pure form. In the past decade, much attention of SFE applications in environmental studies was focused on solid matrices. Using SFE for environmental remediation was first reported by Hawthorne and Miller (9). In this study, the extraction efficiency of poly-aromatic hydrocarbons (PAHs) from fly ash, marine sediment, and urban dust were investigated with three different supercritical fluids: CO_2, nitrous oxide (N_2O), and ethane (C_2H_6). Modified N_2O yielded the highest recoveries for all PAHs. A variety of organic pollutants such as chlorinated phenols, N-heterocycles, pesticides, and PAHs from real environmental samples using $SC-CO_2$ have been studied (10). The USEPA draft method 3560/8440 for the analysis of total petroleum hydrocarbons (TPH) in soil using SFE has also been published (*11*). Removal of metals from both solid and aqueous matrices by SC-CO$_2$ extraction is an interesting subject in the past decade. Several comprehensive studies have reported that materials contaminated with metal oxides could be cleaned-up through SC-CO$_2$ extraction containing a suitable chelating reagent (*12-16*). One attractive feature of the SFE technique is its ability to extract organic chemicals from environmental samples. Many previous studies in SFE were generally performed with solid matrices. There is little information in the literature regarding SFE of organic compounds from aqueous solutions. In fact, matrix effects should be more complicated for SFE of solid samples relative to that of aqueous phase. The strong analyte-matrix interactions that frequently occur in the environmental solid matrices have prompted many studies to evaluate extraction parameters for solid samples. Recently, Yu (*17*) has reported that organophosphate pesticides present in the water can be effectively removed by neat SC-CO$_2$ extraction. Lee *et al.* (*18*) have demonstrated that disperse dyes were able to dissolve in the SC-CO$_2$. Therefore, it is reasonable to assume that chemically stable organic compounds such as disperse dyes in aqueous solution should be extractable by SC-CO$_2$. This study was conducted to investigate the dissolution of hydrophobic disperse dyes in supercritical CO$_2$ and the extraction efficiency of dyes from aqueous solution using supercritical CO$_2$ as the extraction medium. The American Dye Manufacturer Institute (ADMI) Tri-stimulus Filter method was used to evaluate the effectiveness of SFE in terms of removing disperse dyes from water. The supercritical CO$_2$ extraction of disperse dyes from aqueous solution holds potential for the textile industry and could become positioned as the favored technology used in dye houses to alleviate the impact of dye wastewater on our environment.

Materials and Methods

Reagents

Four commercial disperse dyes including Red-60, Red-73, Blue-60, and Blue-79 were supplied by BASF Taiwan Ltd. The chemical characteristics of the four disperse dyes are listed in Table I. Dye solutions with concentrations ranging from 50 to 250 mg/L where prepared for SFE experiments. Acetonitrile, ethanol, acetone and other chemicals used in this study were purchased from Aldrich. SCF-grade CO$_2$ with a purity of 99.9995% was provided by Scott Specialty Gases.

Instruments

All extractions were performed using a SFE system (Isco, USA). The SFE system was controlled by two 260D series pump controller which allowed programming of pressure and continuous flow of SC-CO$_2$. A stainless steel supercritical fluid extractor (High Pressure Company, Erie, PA) with internal volume of 200 mL was used for aqueous phase extraction. The extraction vessel was equipped with a water jacket which the water bath can circulate through it and the extraction temperature was controlled by a thermocouple. A high-pressure liquid chromatography (HPLC, Lab Alliance, series II pump) equipped with a variable-wavelength UV detector (Isco V4) and a 3.9 × 300 mm C$_{18}$ - reversed phase column (Waters Novapack) was used for the analysis of disperse dyes (sample loop volume of 20 μL). The mobile phase employed was 50% H$_2$O and 50% CH$_3$CN (by volume), which was pumped through the column at a flow rate of 1.5 mL/min.

Table I. Chemical characteristics of 150mL of four disperse dye solutions at the concentration of 250 mg/L

C.I. code	Chemical structure	ADMI	COD * (mg/L)	pH
Disperse Red 60		1113	304	6.1
Disperse Red 73		1031	352	6.3
Disperse Blue 60		447	336	6.0
Disperse Blue 79		1170	288	5.9

*Chemical oxygen demand（COD）

Extraction of disperse dyes

Liquid CO_2 was charged into the syringe pumps through a 1/16-inch (i.d.) stainless steel tube and compressed to the desired pressure. The CO_2 stream was counter-currently passed through a preheated stainless steel extraction vessel which was filled with 150 mL of 100 mg/L of dye. Each sample was extracted under a static condition (extraction vessel pressurized with SC-CO_2 having no flow through the vessel) for 30 min, followed by 40 min of dynamic (SC-CO_2 flow through the vessel) extraction. These extraction times were sufficient to complete the extraction of the dyes according to our results. In co-solvent SFE system, because of the non-toxic characteristics of ethanol, the ethanol (10% v/v) modified CO_2 was used as the fluid phase. A 25 cm × 300μm i.d. fused silica tubing was used as a restrictor which allowed a flow rate of about 2 mL/min of SC-CO_2 during the dynamic extraction. Extracts were collected by inserting the restrictor through a silicon stopper into 20 mL of acetone contained in a 100 mL glass tube. It should be noted that SFE of aqueous samples is known to plug fused silica restrictors (*19*), therefore, the restrictors should be heated to prevent plugging from occurring. After extraction and complete depressurization, the dye content of the original sample solution was analyzed by HPLC. The percent extraction was determined by direct comparison with the dye concentration in the aqueous solution before and after the extraction. Percent color reduction was measured by the American Dye Manufacturer Institute (ADMI) Tri-stimulus filter method.

Photosensitization of disperse dyes

The stock solution of disperse dye was prepared by dissolution of dye in a mixed 1:2 (v/v) acetone/H_2O solution and the pH of the resulting solution was adjusted to 9. The UV irradiation of disperse dyes was carried out in a 50mL cylindrical quartz bottle. The dyebath was placed in a photo-reactor equipped with a UV lamp and the dye solution was stirred to maintain homogeneity. The light sources were ten 254 nm phosphor-coated low-pressure mercury lamps with a total output power of 500 watts. The light intensity at 5 cm distance from the lamps was about 2.25×10^{16} photons/sec/cm^2.

Results and Discussion

Extraction of disperse dyes by supercritical CO_2

The recoveries of the spiked dyes from dyebath (150 mL, 250 mg/L) by neat supercritical CO_2 extraction at 85°C and 300 atm were generally low (40-61%). Table II summarizes the percent recoveries of each disperse dye using a 250 mg/L dyebath against various SFE pressure under neat SC-CO_2 and 5% and 10%

ethanol modified (by volume) supercritical CO_2, at an extraction temperature of 85°C. It was observed that the overall extraction efficiency appears to increase with higher quantities of ethanol. The efficiency of extracting the spiked disperse dyes from aqueous solution was significantly improved when ethanol modified CO_2 was used as the solvent. For example, the percent recoveries (Table II) of the spiked disperse dyes Red-60, Red-73, Blue-60, and Blue-79 from water were in the range of 56-85% using 5% ethanol modified SC-CO_2 compared with 40-61% recoveries observed with neat SC-CO_2.

Due to the low solubility of disperse dyes in supercritical CO_2, the addition of a co-solvent, such as ethanol is crucial to achieve extraction of disperse dyes from aqueous solution. Without sufficient ethanol modifier, the supercritical CO_2 extraction could not generate high solvating power which leads to the high

Table II. Percent recovery of Red-60, Red-73, Blue-60, and Blue-79 from a spiked dye solution with 0 %, 5%, and 10 % ethanol modified by SC-CO_2 extraction.

	% Recovery of disperse dyes											
	Red-60			Red-73			Blue-60			Blue-79		
Pressure (atm)	Ethanol modified			Ethanol modified			Ethanol modified			Ethanol modified		
	0 %	5 %	10%	0 %	5 %	10%	0 %	5 %	10%	0 %	5 %	10%
75	21	32	40	26	29	45	18	22	27	18	32	44
100	23	35	43	27	32	52	19	24	31	19	34	45
150	29	46	56	32	35	70	21	29	50	24	39	50
200	37	57	81	40	51	90	28	38	81	32	53	65
250	42	71	97	50	59	99	35	49	95	41	73	94
300	52	82	100	61	72	100	40	56	100	51	85	100

Experiment conditions : 150 mL and 250 mg/L of dyebath, Extraction temperature was 85°C, %RSD for triplicate extractions was 2-3%

removal efficiency of disperse dyes from the dyebath. It was observed that the overall extraction efficiency appears to increase with high quantities of ethanol modified SC-CO$_2$. However, a percent increase in ethanol will not result in a proportional increase in the recoveries of the spiked dyes. Our study found that 10% ethanol modified CO$_2$ was suggested as the optimal SFE operation consideration. Quantitative recoveries of four spiked dyes from dye solution (150mL, 100mg/L) were observed by 10% ethanol modified CO$_2$ extraction at 85 °C and 300 atm. The percent recoveries of the spiked disperse dyes Red-60, Red-73, Blue-60, and Blue-79 from the dyebath (150mL, 100mg/L) by 10% ethanol modified SC-CO$_2$ extraction at 85°C and at various pressure were shown in Figure 1 (a)-(d) by graphical illustration. It is also known that pressure and temperature are the most important parameters in SFE and they have both theoretical and practical implications for the extraction process. As shown in the Figure 1, increasing temperature from 45°C to 85°C at pressures above 250 atm, the recoveries of the dyes generally increase from around 40% to >92%. For example, by setting the extraction pressure of SC-CO$_2$ at 260 atm, the extraction efficiency of disperse Red-60 at 45, 65, and 85°C was found to be 45.6, 71.2, and 98.6%, respectively (Figure 1 (a)). At a lower extraction temperature such as 45°C the extraction efficiency of dyes from dye solution was generally poor. High SFE recoveries were observed at higher pressure for all of the four disperse dyes investigated. For example, at the extraction temperature of 85°C, the percent recoveries of the spiked disperse Blue-60 at 75, 150, 200, 250, and 300 atm were found to be 40.8, 50.1, 79.5, 92.2, and 99.9%, respectively (Figure 1 (c)). The enhanced SFE efficiencies at higher temperatures and pressure were also observed for all the four types of disperse dyes. It is likely that the overall extraction could have been more efficient at elevated pressures and temperatures. In this study, it was found that overall extraction efficiencies of the four disperse dyes from aqueous solution were reaching 99.9% by using 10% ethanol modified CO$_2$ at a pressure of 300 atm and a temperature of 85°C.

Langenfeld (20,21) has reported that at the extraction temperature of 50°C, raising the extraction pressure of SC-CO$_2$ from 355 to 659 bar had little effect on extraction of PCBs and PAHs from solid samples. These authors concluded from their results that temperature was more important than pressure for achieving high extraction efficiencies when the interactions between pollutant molecules and sample matrices were strong. This conclusion, however, is not necessarily true for the extraction process that is carried out in a more homogeneous sample matrix such as aqueous phase. Based on our results, the extraction efficiency of the disperse dyes from water was greatly enhanced by increasing both the pressure and temperature. The reproducibility of SC-CO$_2$ extraction of disperse dyes was evaluated by percent relative standard deviation (%RSD). The %RSD of these data is about 2-3% for triplicate measurements.

164

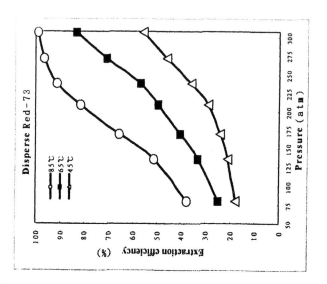

(b)

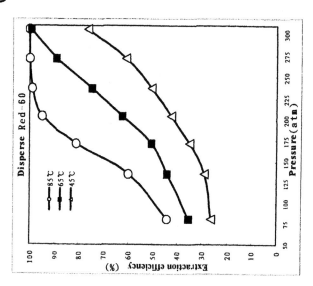

(a)

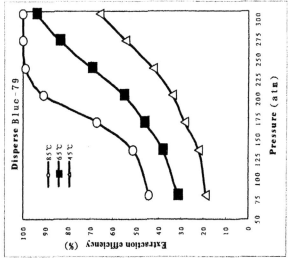

(d)

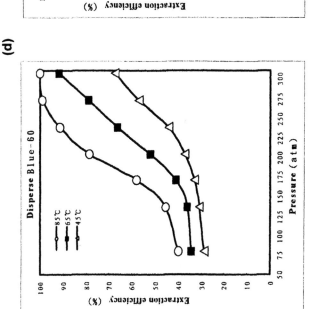

(c)

Figure 1. *Extraction profiles of (a) Red-60, (b) Red-73, (c) Blue-60, and (d) Blue-79 from a spiked solution at various temperatures and pressure. 150 mL and 100mg/L of dyebath, 10% ethanol modified SC-CO₂, Restrictor flow:2mL/min, %RSD for 3 extractions was 2-3% for each extraction profile.*

Because the extraction efficiency was determined by the direct comparison of dye concentration in the spiked dyebath before and after the extraction, the higher SFE recoveries (e.g. efficiency >99%) should have relative standard deviations <1%. For the purpose of this study, >99% of recovery is sufficient to illustrate the effectiveness of the SFE technique. According to our experiments, no decomposition or breakdown of these disperse dyes was observed during SFE at the specified experimental conditions described above. The restrictor flow rates of SC-CO_2 often dominate the success of SFE, and can be varied to provide information on the dynamics of the extraction process. It is known that if the flow of supercritical fluid is sufficient to sweep the cell void volume, the effectiveness of the extraction is enhanced. In fact, changing the flow rate is a simple way to determine the extraction efficiency (7). In this study, no obvious difference in extraction efficiency was observed at the SC-CO_2 flow rate of 2.0, and 5.0 mL/min. It is also noted that SFE of samples with high concentrations of water tends to plug fused silica restrictors (19). Therefore, a restrictor temperature controller was used in our experiments to avoid restrictor plugging.

Decolorization of disperse dyes by supercritical CO_2 extraction

Currently, the proposed ADMI of color characterization is the best available and acceptable method for evaluation the water quality of textile effluent before it can be discharged into the water stream. In our country, the ADMI regulatory value for textile effluent was 400. Therefore, the dye-contaminated water must be treated in order to meet the regulatory requirement. For the SC-CO_2 extraction, the color removal was very efficient. Figure 2 (a) and (b) give the results of ADMI reduction of the SFE trial of four disperse dyes by 5% and 10% of ethanol modified SC-CO_2 versus extraction pressure. For example, the ADMI for 150 mL of 250 mg/L of Blue-79 dye solution was initially at 1200. During the SFE trial, the ADMI dropped rapidly below the regulated level. To test the efficiency of extracting disperse dyes from real environmental samples, a textile wastewater sample was collected from the effluent of a dye house. This textile wastewater was contaminated with disperse Blue-79 and other organic pollutants. The concentration of Blue-79 in the original dyebath was measured to be 188 mg/L. The contaminated dye solution (150 mL) was placed in a 200 mL extraction vessel and extracted with 10% ethanol modified SC-CO_2 at 85°C and 300 atm for 30 min statically followed by 40 min of dynamic flushing. The extracted dyes were collected in 20 mL of acetone and the extraction efficiency was determined by HPLC analysis. The total percent removal of disperse Blue-79 from dye solution was greater than 99.9%. After SFE, the analysis results revealed that the dye concentrations and ADMI in the sample solution were reduced below the regulation level. Apparently, the disperse Blue-79 in the original dyebath was essentially all removed by supercritical CO_2 extraction.

Treatment of extracted disperse dyes by photo-sensitization.

Conventional water purification methods such as biological treatment often deal effectively with synthetic organic compounds such as hydrophilic dyes. Unfortunately, with hydrophobic disperse dyes, have rather poor treatment efficiencies were reported using the biological process and contradictory results were obtained via chemical oxidation methods such as ozonation and Fenton's reaction. In order to cleanup water contaminated with these recalcitrant organic compounds, adsorption and chemical coagulation using ferrous iron or lime have been developed and deployed. However, the safe and inexpensive disposal of the settled sludge is becoming difficult. Recently, photochemical degradation has become more important and the goal of these methods is to minimize the pollutants leaving neither chemical sludge nor toxic residues in the treatment effluent. Although disperse dyes are designed to resist photo-degradation, Chu *et al.*, (*4*) have demonstrated that the rate of UV photo-decomposition can be enhanced by using acetone as a sensitizer which could effectively decompose disperse dyes from aqueous solution.

Experimental trials were performed to investigate the degradation efficiency of disperse dyes by photo-sensitization. The degradation reaction for each test lasted up to 40 min and the residues of disperse dye was determined by HPLC. After the photo-sensitization, degradation of disperse dyes in the spiked solution was more than 95% depending on the dyes. The main objective of water treatment is to regenerate the wastewater to clean water. Although photo-sensitization is effective and could be used in treating hydrophobic disperse dye wastewater, a technical drawback of this method is the need to add large quantities of photo-sensitizer such as acetone to the original dyebath. Addition of chemicals to water could cause secondary contamination of that treated water and may rule out the adoption of these methods for water purification. The main advantage of this SFE method is that the hydrophobic disperse dyes in water can be removed by $SC-CO_2$ and the extracted dyes can be collected in a small amount of solution for further suitable wastewater treating techniques. Our results have demonstrated the feasibility of treating disperse dyes contaminated wastewater by using the $SC-CO_2$ extraction of hydrophobic dyes and subsequent degradation by photosensitization.

Conclusions

Supercritical CO_2 can be effectively applied to the extraction of four hydrophobic disperse dyes from simulated dyebath. Quantitative removal of spiked dyes from aqueous solutions can be achieved by using 10% ethanol

(a)

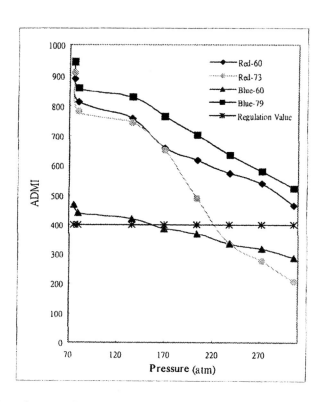

Figure 2. *The reduction of ADMI for Red-60, Red-73, Blue-60, and Blue-79 by*
(a) 5% ethanol modified SC-CO₂ and (b) 10% ethanol modified
SC-CO₂ extraction under various pressure.
Experiment conditions: 150 mL and 250 mg/L of dyebath, Extraction
temperature was 85 ℃

(b)

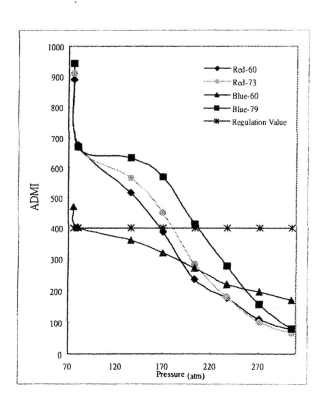

Figure 2. *Continued.*

modified SC-CO$_2$ at 85°C and 300 atm. An extraction protocol was developed for real contaminated samples. Analysis results indicated that the disperse dyes in dyebath were essentially all removed by this SFE method. The proposed method is fast, environmentally safe and requires no solvent. The extracted disperse dyes can be collected in a limited volume of acetone for further degradation by irradiation with 254 nm UV light. The optimal conditions for photosensitization of the disperse dyes was found at a pH of 9 and in a mixed solution of 1:2 (v/v) acetone/water ratio. A combination of this SC-CO$_2$ extraction of disperse dyes together with degradation by photosensitization may provide an innovative technique for treating textile wastewater.

Acknowledgements

The author gratefully acknowledges the National Science Council of Taiwan (ROC), for the financial support under the Grant No. NSC 89-2211-E-035-030. Acknowledgement is made to the Donors of The Petroleum Fund, administered by the American Chemical Society, for travel support to attend the symposium on Separations and Processes Using Supercritical Carbon Dioxide at the ACS meeting in Orlando in April 2002.

References

1. Arslan, I. Treatability of a simulated disperse dyebath by ferrous iron, ozonation, and ferrous iron-catalyzed ozonation. *J. of Hazardous Materials* **2001**, *B85*, 229-241.
2. Hsu, Y. C.; Yen, C. H.; Huang, H. C. Multistage treatment of high strength dye wastewater by coagulation and ozonation. *J. Chem. Technol. Biotechnol.* **1998**, *71*, 71-76.
3. Kuo, W. G. Decolorizing dye wastewater with Fenton's reagent. *Water Res.* **1992**, *26*, 881-886.
4. Chu, W.; Tsui, S. M. Photo-sensitization of diazo disperse dye in aqueous acetone. *Chemosphere* **1999**, *39(10)*, 1667-1677.
5. Yang, Y.; Wyatt, D.T.; Bahorsky, M. Decolorization of dyes using UV/H$_2$O$_2$ photochemical oxidation. *Text. Chem. Color,* **1998**, *30*, 27-35.
6. Burford, M. D.; Hawthorne, S. B.; Miller, D. J. Extraction rates of spiked versus native PAHs from heterogeneous environmental samples using supercritical-fluid extraction and sonication in methylene-chloride. *Anal. Chem.* **1993**, *65*, 1497-1505.
7. Hawthorne, S. B.; Miller, D. J.; Burford, M. D.; Langenfeld, J. J.;

Eckert-Tilotta, S. E.; Louie, P. K. Factors controlling quantitative supercritical-fluid extraction of environmental-samples. *J. of Chromatogr.* **1993**, *642*, 301-317.

8. Maio, G.; von Holst, C.; Wenclawiak, B. W.; Darskus, R. Supercritical fluid extraction of some chlorinated benzenes and cyclohexanes from soil: optimization with fractional factorial design and simplex. *Anal. Chem.* **1994**, *69*, 601-606.

9. Hawthorne, S. B.; Miller, D. J. Extraction and recovery of polycyclic aromatic hydrocarbons from environmental solids using supercritical fluid. *Anal. Chem.* **1987**, *59*, 1705-1708.

10. Hawthorne, S. B.; Miller, D. J. Direct comparison of soxhlet and low-temperature and high-temperature supercritical CO_2 extraction efficiencies of organics from environmental solids. *Anal. Chem.* **1994**, *65*, 4005-4012.

11. Lopez-Avila, V.; Benedicto, J.; Dodhiwala, N. S.; Young, R.; Beckert, W. F. Development of an off-line SFE method for petroleum-hydrocarbons in soils. *J. of Chromatogr. Sci.* **1992**, *30*, 335-343.

12. Wai, C. M.; Wang, S.; Liu, Y.; Lopez-Avila, V.; Beckert, W.F. Evaluation of dithiocarbamates and β-diketones as chelating agents in supercritical fluid extraction of Cd, Pb, and Hg from solid samples. *Talanta* **1996**, *43*, 2083-2091.

13. Ashraf-Khorassani, M.; Combs, M. T.; Taylor, L. T. Supercritical fluid extraction of metal ions and metal chelates from different environments. (Review). *J. Chromatogr. A*, **1997**, *774*, 37-49.

14. Ager, P.; Marshall, W. D. Mobilisation/purging of copper, chromium, and arsenic ions from aqueous media into supercritical carbon dioxide. *Spectrochim. Acta Part B*, **1998**, *53*, 881-891.

15. Foy, G. P.; Pacey, G. E. Specific extraction of chromium(VI) using supercritical fluid extraction. *Talanta* **2000**, *51*, 339-347.

16. Tomioka, O.; Enokida, Y.; Yamamoto. I. Cleaning of materials contaminated with metal oxides through supercritical fluid extraction with CO_2 containing TBP. *Progress in Nuclear Energy* **2000**, *37(1-4)*, 417-422.

17. Yu, J. J. Removal of organophosphate pesticides from wastewater by supercritical carbon dioxide extraction. *Water Res.* **2002**, *36*, 1095-1101.

18. Lee, M. J.; Lin, H. M.; Liu, C. Y.; Cheng, C. H.; Chen, Y. T. Solubilities of disperse dyes of blue-79, red-153, and yellow-119 in supercritical carbon dioxide. *J. Supercrit. Fluids* **2001**, *21*, 1-9.

19. Burford, M. D.; Hawthorne, S. B.; Miller, D. J. Comparison of methods to prevent restictor plugging during off-line supercritical extraction. *J. of Chromatogr.* **1992**, *609*, 321-332.

20. Langenfeld, J. J.; Hawthorne, S. B.; Miller, D. J.; Pawliszyn, J. Effects of temperature and pressure on supercritical fluid extraction efficiencies of polychlorinated biphenyls. *Anal. Chem.* **1994**, *65*, 338-344.

21. Langenfeld, J. J.; Hawthorne, S. B.; Miller, D. J.; Pawliszyn, J. Effects of temperature and pressure on supercritical fluid extraction efficiencies of polycyclic aromatic-hydrocarbons. *Anal. Chem.* **1994**, *66*, 909-916

Chapter 12

Approaches to Soil Remediation with Green Procedures

Q. Wu, T. Yuan, and W. D. Marshall*

Department of Food Science and Agricultural Chemistry,
Macdonald Campus of McGill University, 21, 111 Lakeshore Road,
Ste-Anne-de-Bellevue, Québec H9X 3V9, Canada

At elevated temperature, a reactor column (25 x1 cm) of zero-valent metal (or bimetallic mixture) mediated efficient dechlorinations so that 20-30 mg/min of polychlorinated biphenyl compounds (PCBs) or pentachlorophenol (PCP) was converted to biphenyl or to phenol respectively. The one other principal product was chloride that accumulated on the metal surfaces. The hydroxylic solvent used to dissolve the substrate must have represented the hydrogen source for the reaction. GC-MS monitoring of the effluent indicated ~99% dechlorination. Moreover, sequential 10-30 minute fractions of reactor effluent collected during 14h, indicated that the reaction was efficient and repeatable provided that the column was washed with 30 mL methanol/water at 2.5-3 h intervals. For soil cleaning, a 3% (v/v) surfactant emulsion mobilized PCBs with procedures designed to mimic either *ex-situ* washing or *in-situ* flushing. The mobilized PCBs were then recovered from the emulsion by back-extraction into $scCO_2$ and dechlorinated *on-line*. The principle shortcoming of the overall processing was that an appreciable fraction of the surfactant was lost to the $scCO_2$ during back-extraction.

Many large cities are faced with soil contamination problems inherited from past industrial activities. There are some 3,000 contaminated "Brownfields" sites in Canada (1) and some 450,000 such sites in the United States (2). A number of these urban sites within the Island of Montreal, have mixed contamination problems. Hazardous (mixed) contamination can be found in soils and groundwater where it commonly corresponds to a mixture of inorganic and organic contaminants. Lands that contain such contaminants, in varying concentrations, cannot be redeveloped unless they have been decontaminated.

We were anxious to evaluate the use of supercritical carbon dioxide ($scCO_2$) to remediate an urban soil. A two-stage approach was prompted by our efforts to make the process continuous. $ScCO_2$ extraction of particulate media is inherently a batch process. Pressure within the extractor is maintained with a capillary restrictor (in our case a 50 μm diameter silica tube) that is prone to fouling. The non-polar organic fraction of an aqueous (surfactant) extract can be removed continuously (either *on site* or *off site*) with $scCO_2$ in a counter current liquid-liquid like process. The heavy metal contaminants remain with the water fraction for subsequent treatment. The organics fraction is the subject of this short review on $scCO_2$ processing.

The process of extraction can transfer toxicants from one medium to another and can concentrate them efficiently but it does not detoxify the target compounds *per se*. The process of detoxification must involve making them less acutely toxic or less available to intact organisms or to isolated biological processes. For organic toxicants, two strategies can be pursued. The toxiforic functional groups on the contaminant can be altered chemically or the toxicants themselves can be sorbed/fixed to an inert surface so as to make them less bioavailable to contacting organisms. It was envisaged that the overall process of decontamination of natural media (soils/sediments) would become more appealing if extractions of organochlorine (OC) contaminants from polluted media were to be coupled with an *on line* detoxification sequence. Generally, OC compounds have been considered to be environmentally recalcitrant due principally to their relatively non-polar nature (that decreases aqueous solubility and hinders physical dispersal processes) and the general lack of functional groups (that hinders metabolic transformations). Toxicities of OC compounds can be decreased appreciably by reductive dechlorination to form relatively innocuous hydrocarbon and chloride ion. It was further anticipated that once mobilized into $scCO_2$, dechlorination might be accelerated further by contact of the OC solutes with zero-valent (ZV) metal.

Bimetallic Mixtures with Zero-valent Iron

The operation of a short reactor column (Figure 1) designed to be interposed within a flowing stream of $scCO_2$ has been optimized for dechlorinations (3). The column assembly (25 x 1cm) was filled with ZV metal particles - iron (Fe^0)

or preferably a bimetallic mixture [Ag° or Pd° at 2 or 0.2% (w/w) loading on Fe°]. In operation, a stream of substrate feedstock (~0.1 mL/min), dissolved in a hydroxylic organic solvent, was merged with a stream of $scCO_2$ (~1.5 mL/min) and delivered to the heated (350-450 °C) column. The mean efficiencies of PCB dechlorinations (the average of six successive 10-min cumulative trappings of reactor effluent for two reactor columns mounted in series) are summarized in Table 1 (3). The efficiency of dechlorination varied as the composition of the feedstock solvent, as the identity of the ZV-metal and as the operating temperature of the reactor. Somewhat surprisingly, water was detrimental to the dechlorination efficiency. Bimetallic mixtures of ZV metals ($Ag°/Fe°$, $Ag°/Ni°$ or $Pd°/Fe°$) proved to be even more efficient at mediating polychlorinated biphenyl (PCB) dechlorinations (Table 2). It was also observed that acetic anhydride (Ac_2O) in the feedstock solvent could be replaced by a methyl ketone

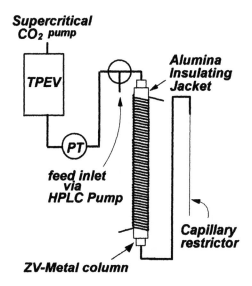

Figure 1. A dechlorination assembly consisting of source of scCO2, a temperature and pressure equilibration vessel (TPEV), a pressure transducer (PT), a three-way substrate inlet tee, a reactor column filled with zero-valent (ZV) metal particles (jacketed with a heated alumina tube) and a terminating capillary restrictor. (Adapted with permission from reference 3. Copyright 2000, Royal Chemical Society.)

Table 1. Variations in the mean [a] percent dechlorination (± 1 RSD) for 2 successive reactor columns (i.) containing various zero-valent metals and (ii.) with different solvent compositions to transfer a 20% (V/V) Aroclor 1242 to the reactor.

ZV-Metal	Feedstock Composition (v/v) in the presence of 20% (v/v) Aroclor 1242	Mean % Dechlorination for two columns
Cu	5% H_2O, 75% DME	36 ±11
Cu	70% Ac_2O, 10% DME	86 ±8
Fe	5% H_2O, 75% DME	36 ±30
Fe	70% Ac_2O, 10% DME	94 ± 4
Ni	5% H_2O, 75% DME	68 ± 8
Ni	70% Ac_2O, 10% DME	91 ± 6
Zn [b.]	5% H_2O, 75% DME	67 ±11
Zn [b.]	70% Ac_2O, 10% DME	90 ± 8

[a] Mean percent dechlorination for six successive traps of reactor effluent.
[b] Performed at 400 °C or [c] 200 °C / 31 MPa
Source: Adapted from reference 3, Copyright 2000, Royal Chemical Society

(either methylisobutyl ketone or heptan-2-one) with only a modest decrease in dechlorinating efficiency. The principal reaction product in the reactor effluent was biphenyl with lesser quantities of benzoic acid esters and traces of monochloro- and dichlorobiphenyl were also detected (3). In companion trials, it was also established that at 300 °C, recovery of organically-bound chlorine in the reactor eluate was virtually quantitative and the conversion of PCB residues in the reactor effluent to biphenyl [by reaction with $K_2PdCl_6/Mg°$ (4)] was also virtually quantitative (99 ± 8%) but was reduced to 84 ± 13% at 400 °C suggestive of further reaction.

That dechlorinations were accelerated by the ZV-metal particles was demonstrated by replacing the metal particles with silica (acid washed sea sand). For the silica column operated under dechlorinating conditions that had been optimal for ZV metal (400 °C, 31.0 MPa), the recovery of organically bound chlorine from the eluate was virtually quantitative. In further trials, acetone-hexane extract of a sandy loam soil (spiked with 600 ppm Aroclor 1254) was fed to the reactor at 0.1 mL/min and dechlorinated efficiently. Chlorinated residues were not detected in the reactor effluent by GC-MS. More importantly, soil co-extractives in the PCB solution did not seem to affect the course or the efficiency of the reaction perceptibly.

Table 2. Variations in the mean [a] percent dechlorination (± 1 RSD) for 2 columns maintained at 400 °C / 31.0 MPa (i.) containing different bimetallic mixtures and (ii.) with varying acetic anhydride – dimethoxyethane mixtures to transfer Aroclor 1242 to the reactor.

ZV-Bimetallic mixture	% (v/v) PCB content	Feedstock Composition (v/v) in the presence of Arochlor 1242	Mean % Dechlorination [a]
Ag/Fe	10	50% Ac$_2$O, 40% DME	98.3 ±1.2
	20	70% Ac$_2$O, 10% DME	97.0 ±2.6
	20	80% Ac$_2$O	88.9 ±4.5
Ag/Ni	10	50% Ac$_2$O, 40% DME	97.8 ±1.1
	20	70% Ac$_2$O, 10% DME	95.8 ±2.1
	20	80% Ac$_2$O	81.1 ±3.9
Pd/Fe	10	50% Ac$_2$O, 40% DME	97.8 ±2.1
	20	70% Ac$_2$O, 10% DME	96.1 ±3.1
	20	80% Ac$_2$O	84.3 ±4.6

[a] Mean ± 1 relative standard deviation (RSD) of six successive 10-min traps of reactor effluent.

Source: Adapted from Reference 3. Copyright 2000 Royal Chemical Society

Pentachlorophenol (PCP)

In subsequent studies (5), a continuous stream (10-20 mg/min) of pentachlorophenol (PCP) was also dechlorinated virtually quantitatively as it was passed through a single heated column of silver-iron (Ag^o/Fe^o) bimetallic mixture. Again substrate dechlorination, although not complete, was very efficient. In this case, the course of the reaction was monitored by gas chromatographic (GC) analysis and by argentometric titration of organically bound chlorine in successive cumulative 10-min traps of reactor eluate (mean % dechlorination of 6 successive traps, 95 ± 0.9). Again, to evaluate the contribution of thermally induced dechlorinations, a companion experiment was conducted using silica (sea sand) to fill the reactor column. With the optimized reaction conditions (450 ^{o}C / 25.3 MPa, 1.5 mL/min $scCO_2$), a feedstock of 10% (w/v) PCP in 1,2-dimethoxyethane, delivered at 0.1 mL/min was used to collect six traps of effluent. The resulting chromatograms indicated the loss of approximately 50% of the PCP peak area and the formation of 4-chlorophenol, 3,5-dichlorophenol, 2,3,5-trichlorophenol, 2,3,4-trichlorophenol and 2,3,4,5-tetrachlorophenol that comprised the remainder of the products. Only a trace of phenol was detected in any of the eluate solutions. Thus, thermally induced dechlorination is inefficient relative to the action of the Ag^o/Fe^o particles.

Dechlorination of a methanolic feedstock was appreciably more efficient. A variety of products were detected by GC-MS (Table 3) that included methylated benzenes (*m*- and *p*-xylene, 1,2,4-trimethylbenzene, pentamethyl- and hexamethylbenzene), phenol and methylated phenols (*o*- and *m*-cresol, 2,4-; 2,6-; 3,4- and 3,5-dimethylphenol; 2,3,6- and 2,4,6-trimethylphenol) and intermittently traces of 4-chlorophenol. Table 3 also provides a measure of the levels of repeatability that was achieved in these studies and indicates that 74.6 ± 2.7% of the substrate (mean mass balance) was accounted for among the products. Presumably other products were not detected because they co-eluted from the GC with the solvent. Alternatively, products were not trapped from the effluent or were not were not sufficiently volatile or thermally stable to survive GC *intact*.

It was also of interest to gain insight into the source of the methyl substituents in the dechlorinated products. Several patents have described the ortho /para methylation of phenolic substrates by methanol over metal oxide catalysts at elevated temperature (6-8). An alternate source of the methyl groups in the products might be the $scCO_2$ mobile phase. Carbon dioxide can be reduced at iron surfaces in the presence of water to form short chain alkanes. Approximately 90% of the products consisted of methane (9). A possible probe might be to use a different solvent to dissolve the substrate PCP. Table 3 records

Table 3. Mean product yields (± 1RSD) and mass balances observed in five or six sequential 30-min traps for a 20% (*w/v*) methanolic PCP feedstock or a 10% (*w/v*) PCP feedstock in propan-2-ol or a 5% (w/v) PCP feedstock in water methanol (1 + 4, v/v) delivered, at 0.1 mL/min, to a single Ag^o/Fe^o or a Pd^o/Mg^o column maintained at 450 °C/25.2 MPa or at 400 °C/22.5 Pa.

Products	MeOH [a]	i-PrOH [a]	20% $H_2O/$ MeOH[b]	20% $H_2O/$ MeOH[c]
Chlorinated	0.5 ±0.5	0.2 ±0.3	1.8 ± 2.4	8.1 ± 9.0
Cyclohexanone	N.D.	N.D.	1.2 ± 0.8	0.4 ± 0.4
O-methylated phenol	N.D.	N.D.	3.8 ± 1.8	10.1 ± 1.0
ring-methylated phenols	47.5±1.5	13.3±1.7	3.6 ± 1.5	4.6 ± 2.1
Phenol	46.9±1.0	50.4±1.0	79.0 ± 3.6	61.7 ± 8.6
Methylated benzenes	5.1±0.5	36.5±1.0	N.D.	N.D.
Mass balance	75± 3%	88± 2%	89 ± 2%	83 ± 3%

the product distribution when substrate PCP, in propan-2-ol, was delivered to the reactor (450 °C/25.3 MPa) at 0.1 mL/min. Although there were no differences in the product identities and the mean mass balance over the six traps accounted for 87.7 ± 2.1% of the PCP substrate, the distribution of products were appreciably feedstock delivered to the reactor column [a] filled with (2% w/w) Ag^o/Fe^o, [b] filled with (2% w/w) Pd^o/Mg^o or [c] filled with (1% w/w) Pd^o/Mg^o different from those observed for the methanolic carrier (Table 3). Whereas phenol accounted for 50.4 ± 1.0% of the products in propan-2-ol, (not appreciably different from the 46.9±1.0% observed in methanol), methylated phenols accounted for only 13.3 ± 1.7% vs. 47.5 ± 1.5% in the methanol carrier. The remainder of the products in the propan-2-ol carrier was 1,2,4-trimethylbenzene (36.5 ± 1.0%). Propan-2-ol solvent has been used a source of thermally induced hydrogen radicals and was anticipated to lower the contribution of methyl groups from the solvent. The ability to influence the selectivity/direction of the reaction would be helpful. The loss of the hydroxyl group from the substrate is considered undesirable as a detoxification route since the lipophilicity of the product is increased and a functional group, that can facilitate metabolic transformations, is lost. Thus, methanol is favoured as a solvent because of the formation of methylated phenols at the expense of methylated benzenes.

The behaviour of the methanolic feedstock was also investigated during extended operation. A total of 28 successive traps were collected during 14h (each trap corresponding to 30-min of operation). At 2.5-3h intervals, the flow of feedstock was interrupted and the column was washed with 30 mL methanol-water (1/1) to remove accumulated chloride from the metal surfaces then dried with $scCO_2$. As recorded in Table 4, chlorinated products were detected only just prior to the column regeneration sequence (traps 5, 11, 17 and 23). Post

Table 4. Percent distributions of recovered products in 30 min cumulative fractions of reactor eluate either prior to (traps 5, 11, 17, 22, or 28) or post (traps 6, 12, 18, or 23) an on-column wash with 30 mL methanol-water.

Product	Trap 5	Trap 6	Trap 11	Trap 12	Trap 17	Trap 18	Trap 22	Trap 23	Trap 28
methylated benzenes	4.8	2.6	3.4	3.7	3.8	3.8	3.2	3.0	1.8
Phenol	48.4	47.8	50.2	52.5	58.4	57.0	53.4	40.2	58.8
methylated phenols	45.9	49.6	46.3	43.8	36.9	39.3	42.2	56.5	39.3
chlorinated phenols	0.9	N.D.[a]	N.D.	N.D.	1.8	N.D.	1.2	N.D.	N.D.
mass balance	0.737	0.703	0.761	0.722	0.737	0.738	0.687	0.711	0.702

[a] N.D. = not detected

SOURCE: Adapted with permission from reference 5. Copyright 2001, Royal Chemical Society

treatment (traps 6-10, 12-16, 18-21, 23-28) chlorinated products were absent from the eluate.

Dechlorinations in Subcritical Water

The one solvent that would be less expensive than $scCO_2$ would be water. We have evaluated subcritical water (scH_2O) as a medium to dechlorinate PCP. In batch trials at 200 °C, complete dechlorination of 0.5 mg PCP was achieved after 20h of reaction with Ag^o/Fe^o (Table 5) but recoveries remained incomplete. One possible explanation for the observed selectivity might be that the interaction of the substrate PCP with the surface of the metal accelerator occurred predominantly via the π-electron system of the aromatic ring so that all the Cl substituents were accessible to the catalyst. The ortho – para directing character of the hydroxyl substituent did not seem to explain the relatively high level of selectivity that was associated with the observed sequential dechlorination process. As an alternative explanation, the removal of bulky chlorine substituents from the ring would provide relief of steric strain. The initial loss of Cl from the C-2 or C-6 position (*ortho* to the hydroxyl and a chlorine substituent) decreased steric strain and repulsion more than the loss of a

Table 5. Variations, with time, in the mean molar percent distribution [a] for the dechlorination of pentachlorophenol (PCP, 1.87 µmol) in scH_2O at 200 °C, in the presence of 200 mg of 2% (*w/w*) Ag^o/Fe^o bimetallic mixture.

Time (h)	triCls	diCls	4-Cl	phenol
0.5	39 ± 11	54 ± 7	2 ± 3	N.D. [b]
1.0	27 ± 8	74 ± 24	N.D.	N.D.
1.5	8 ± 11	81 ± 13	12 ± 13	N.D.
2.0	N.D.	50 ± 7	50 ± 11	N.D.
2.5	N.D.	9 ± 8	86 ± 16	3 ± 2
3.0	N.D.	8 ± 14	90 ± 13	2 ± 4
3.5	N.D.	12 ± 17	82 ± 9	6 ± 9
5.0	N.D.	10 ± 9	82 ± 6	8 ± 3
10.0	N.D.	N.D.	95 ± 3	5 ± 3
20.0	N.D.	N.D.	92 ± 5	8 ± 5

[a] ± one relative standard deviation based on two or three replicate runs
[b] N.D. = not detected

m-Cl substituent. The loss of a second chlorine substituent occurs at a position with two nearest Cl neighbours and the loss of the third chlorine again occurs from the position that is ortho to the hydroxyl group.

A magnesium-based bimetallic accelerator proved to be more reactive with efficient dechlorination within 2h using 200mg Pd°/Mg°. (10) Cyclohexanone and cyclohexanol (resulting from aryl ring reduction) were minor products with these conditions. Mg°-based accelerators followed a reaction course somewhat different from the Fe°-mediated dechlorinations for which only stepwise dechlorinations had been observed. For bimetallic Pd°/Mg°, phenol was the major product yet with limited loadings of this accelerator, the other major products were tetrachloro species. Based on the accumulation of limited quantities of partially dechlorinated compounds in the product mixture, Figure 2 is proposed as a probable dechlorination scheme. All three tetrachloro species were present in a ~2:1:1 ratio with 2,3,4,6 predominating. Two congeners dominated the trichloro fraction 2,3,5- and 2,4,5- (~5:4 ratio), two dichloro species (2,3- and 3,4- in approximately equal concentrations) dominated the dichloro fraction and two monochlorophenols *ortho* and *para* were present among the products in a ~ 2:1 ratio. In addition to the dechlorination products isolated from Fe°-mediated dechlorinations, the remaining chlorinated products (labeled with an * in Figure 2) are similar to the distribution of products reported by Shin and Keane (11) that included 2,3,5-TCP; 2,4,5-TCP and 2,5-DCP over Ni/zeolite and 2,3-DCP and 2-CP (but not 3,5-DCP) over Ni/SiO$_2$. Small quantities of cyclohexanone also accumulated in the presence of pre- formed bimetallic mixture whereas a larger excess of K$_2$PdCl$_6$ together with Mg° resulted only in cyclohexanol. Mg° alone or Ag°/Mg° were less efficient accelerators and caused only limited dechlorination.

Continuous PCP dechlorinations over Pd°/Mg°

Success with scH$_2$O encouraged us to evaluate/optimize the reactor column filled with Pd°/Mg° for continued PCP dechlorinations. (12) Dechlorination efficiency (with a single 25 cm reactor column) was influenced appreciably by the composition of the feedstock solvent. Although dechlorinations were more extensive on Pd°/Fe° surfaces than on Pd°/Mg° for PCP dissolved in methanol, dechlorinations proved to be appreciably more extensive in water methanol mixtures than in methanol which in turn was more efficient than in propan-2-ol. Presumably, interactions of the water with the Mg° surface generated hydrogen species [and Mg(OH)$_2$] that were more active at promoting the hydrogenolysis of aryl chlorine substituents. It also seems possible that the Mg° might be converted directly to MgO at 400 °C and that the liberated hydrogen species could also participate in the hydrodechlorinations.

A variety of products were detected by GC-MS from extended trials using a 5% (w/v) PCP feedstock dissolved in aqueous methanol (1 + 4 v/v). Conditions were 0.1 mL/min feedstock, delivered at 400 °C/22.5 MPa, to one reactor column containing 2% or 1% (w/w) Pd°/Mg°. In these extended trials, phenol was the dominant product and neither substrate PCP, polychlorinated

182

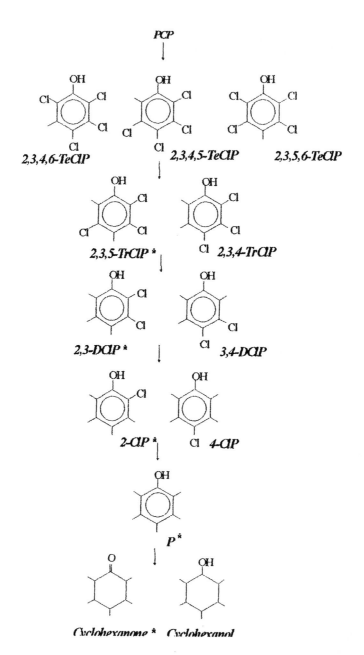

Figure 2. A probable reaction scheme for the Mg⁰-mediated dechlorination of PCP. Reproduced with permission from reference 10. Copyright 2002, Royal Chemical Society

phenols nor methylated benzenes were detected. Monochlorinated phenols, absent from the early traps, increased slowly over the course of the trials. Anisole (the result of oxygen methylation) and ring-methylated products accounted for the remaining products. Cyclohexanone was only a minor component of the product mixture. With time, the reactor lost activity gradually. None the less, over the 6h trial, the dechlorinating efficiency was 0.995 for the 2% loading of Pd^o on Mg^o and 0.984 over 5h for the 1% loading respectively.

Soil Decontamination

Efforts were then directed to mobilizing PCBs from a historically contaminated (33 ± 7 $\mu g/g$) soil. (13) From among seventeen commercial surfactants, four formulations (Brij 97, Triton CF54, Triton DF16 and Tween 85) were selected for further study based of their ability to mobilize Aroclor 1242 into $scCO_2$, a lack of appreciable foaming during the extraction process and efficient recovery of surfactant in the aqueous retentate. For extractions of aqueous 1% (v/v) Aroclor 1242 into $scCO_2$ at 14.2 MPa and 30-80 °C during 30 or 60 min, PCB mobilization efficiency increased with increased temperature and with increased duration of extraction but the quantity of surfactant remaining with the aqueous retentate also was decreased proportionately. Figure 3 presents results that were typical of all four surfactant formulations in the initial ranging study.

For mobilizations from the soil, two approaches were considered to mimic soil washing and continuous soil column flushing. Sonication-washing experiments consisted of sonicating 10g of air-dried soil suspended in 1, 3 or 5% (w/v) aqueous surfactant suspension for 10 min followed by filtration under gentle suction. For the flushing experiments, soil 30g, was packed into a 30 x 1 cm column (that was terminated at both ends with a coarse filter) and flushed continuously with 1, 3 or 5% surfactant suspension delivered from an HPLC pump. All four commercial non-ionic formulations were approximately equally efficient when used with procedures designed to mimic either *ex situ* washing or *in situ* soil flushing. The removal efficiency from the soil was predicted accurately with a simple exponential expression that included the number of successive treatments and a "time-constant" (λ) that was characteristic of either the washing or flushing experiments. Equilibrations with Triton DF16 (Figure 4) were typical; 5.74 and 2.94 washes were predicted to be required for the $t_{1/2}$ with 1% and 5% surfactant emulsions respectively. The mathematical model accurately predicted the course of the flushing process with the soil column. In this case, the PCB concentration was predicted to be reduced two-fold after 3.95h (237 mL) or 2.92h (~175 mL) of flushing. Comparable values for sonication–washing and for flushing were observed with similar concentrations of the other surfactants. Despite differences in the quantity of soil treated (10 g for sonication-washing *vs.* 30 g for soil flushing) comparable mobilization

184

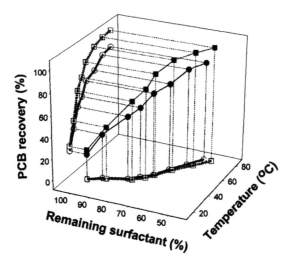

Figure 3. PCB mobilization efficiency (%) as a function of operating
temperature and surfactant remaining with the aqueous phase (%) for 30 min
(●, ○) or 60 min (■, □) of scCO₂ extraction, at 1.42 MPa, from suspensions
containing 3% (v/v) Triton DF16. (Adapted with permission from reference 13.
Copyright 2001, Royal Chemical Society.)

efficiencies were achieved with the sonication-washing and column flushing.
There were no apparent differences in the avidity with which separate fractions of
the total PCB burden interacted with the soil.

PCBs were mobilized from the soil cleaning emulsion by back-extraction
with scCO₂ during 30 min at 50 °C. The results indicated that >99% of PCBs
were mobilized from the soil extracts and 23-78% of surfactant remained in the
aqueous suspension post treatment. The cleaned surfactant suspension could then
have been reused to mobilize more PCB from the contaminated soil. In
subsequent trials (14), PCB formulation (Aroclor 1242 or 1248)-surfactant (Brij
97 or Triton CF54) emulsion (0.1 – 1.0% v/v) were extracted with scCO₂.
(Figure 5). The declinations curve that resulted for each loading of PCB substrate
was modeled accurately as a single exponential decay and the half life [$t_{1/2}$,
10.76 min ± 2% (Brij 97) or 10.0 min ±14% (Triton CF54)] was independent of
the level of loading. When the extraction was combined with *on line*
dechlorination, the $t_{1/2}$ was increased to 17.62 min ± 6% or 16.46 min ±5% (filled
vs. open symbols of Figure 5). The dechlorination reactor contributed
appreciably to the back pressure of the system so that under identical scCO₂ head
pressures, the flow rate at the exit of the system was reduced from 1500 mL/min
(as decompressed gas) to 900 mL/min. When corrected for the difference in flow

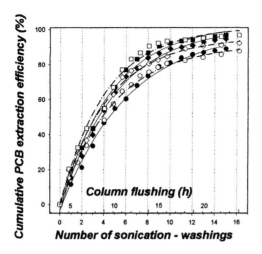

Figure 4. Cumulative PCB extraction efficiencies for mobilizations from 10g soil, into $scCO_2$,, by sonication (10 min)-washing with 1% (●), 3% (♦) or 5% (■) or by continuous flushing of 30g soil with 1% (○), 3% (◇) or 5% (□) aqueous suspension of Triton DF16. (Adapted with permission from reference 13. Copyright 2001, Royal Chemical Society.)

rates, the declination curves in the absence and presence of the reactor became virtually identical to each another. For surfactant suspension from a soil that was field-burdened   with only 6 ppm PCBs, extended operation (extraction-dechlorination) for 15h during which the substrate suspension was replaced each 30 min caused no loss of the dechlorination efficiency but, for 1% (v/v) Aroclor 1242   or 1248 in surfactant suspension, the reactor gradually lost reactivity over 5h of continued operation. However, reactivity could be restored virtually completely by (i.) purging the reactor column with $scCO_2$ for 3h or (ii.) washing the column at ambient temperature with 30 mL of methanol-water (1+1 v/v).

In summary, surfactant-mediated mobilizations from soil were efficient when extended to longer periods. The course of the extraction was predicted accurately with a single exponential decay model. However in addition to minor losses to the soil, the principal shortcoming of the approach was that an appreciable portion of the surfactant was co-mobilized into the $scCO_2$-MIBK solvent mixture during the back-extraction stage. The challenge remains to (i.) decrease the energy requirements for the post-processing dechlorination and (ii.) increase the retention of surfactant in the aqueous retentate   during the back extraction.

186

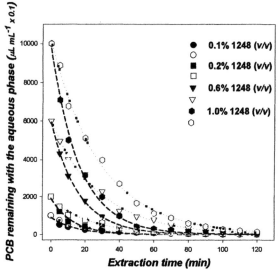

Figure 5. Aroclor-1248 content (initially 0.1-1.0 % v/v) remaining with the (1+9) MIBK-aqueous Triton CF-54 (3% v/v) suspension as a function of minutes of back-extraction with scCO₂ at 22.1 MPa.(Adapted with permission from reference 14. Copyright 2001, Royal Chemical Society.)

References:

1. URL http://www.nrtee-trnee.ca/Publications/SOD_Brownfield_E.pdf
2. URL http://clu-in.com/products/roadmap/home.htm
3. Wu, Q.; Majid, A.; Marshall, W. D. *Green Chem.* **2000**, *2*, 127-132.
4. Wu, Q.; Marshall, W. D. *J. Environ. Anal. Chem.* **2001**, *80*, 27-38..
5. Kabir, A.; Marshall, W. D. *Green Chem.* **2001**, *3*, 47-51.
6. Valiullina, G. A.; Kishkan, L.N.; Khlebnikov, B. M.; Krysin, A. P.; Egorova, T. G.; Titova, T. F. Russian Patent 2027694, 1995.
7. Imanari, M.; Inaba, M.; Inui, Y. Japanese Patent 60,218,345, 1985.
8. Adey, K. A.; Yates, F. S.; Young, J. H. U.S. Patent 4,471,149, 1984.
9. Gillham, R. W.; Hardy, L. I. *Environ. Sci. Technol.* **1995**, *30*, 57-65.
10. Marshall, W. D.; Kubátová, A.; Lagadec, A .J. M.; Miller, D .J.; Hawthorne, S.B. *Green Chem.* **2002**, *4*, 17-23.
11. Shin, E-J,; Keane, M. A. *Catal. Lett.* **1999**, *58*, 141-145.
12. Yuan, T.; Marshall, W. D. *J. Environ. Monit.* **2002**, *4*, 451-457.
13. Wu, Q.; Marshall, W. D. *J. Environ. Monit.* **2001**, *3*, 281-287.
14. Wu, Q.; Marshall, W. D. *J. Environ. Monit.* **2001**, *3*, 499-504.

Analytical and Modeling
Methods

Chapter 13

Time-Resolved Laser-Induced Fluorescence Characterization of Uranium Complexes and Processes in ScF CO_2

R. Shane Addleman[1] and Chien Wai[2]

[1]Environmental Technology Directorate, Pacific Northwest National Laboratory, Richland, WA 99352
[2]Department of Chemistry, University of Idaho, Moscow, ID 83844–2343

A time resolved laser induced fluorescence (TRLIF) system has been developed for the on-line measurement of uranyl chelates in supercritical carbon dioxide. This system has been applied to the study of dynamic supercritical uranium extraction processes. Fundamental physical parameters such as complex solubility and distribution coefficients can also be determined with TRLIF.

Supercritical fluid (ScF) extraction is generally thought of as applicable to only lipophilic species. However, recent studies have demonstrated the feasibility of metal extraction from aqueous and solid matrices with ScFs containing organic chelators. (*1-8*) Recent reports indicate that uranyl ions in aqueous solutions can be extracted by ScF CO_2 containing complexing agents such as tributyl phosphate (TBP) or β-diketones with efficiencies comparable to the conventional solvent extraction processes. (*1,3,9,10*) Furthermore, because of the high diffusivity and low viscosity of ScFs, uranium in solid materials such as soil, sediments, and mine tailings can be extracted by ScF CO_2 containing suitable ligands. (*9,11*) Uranium dioxide can even be dissolved directly in ScF CO_2 by TBP-solvated nitric acid and then subsequently extracted, resulting in a completely dry process for the extraction of nuclear materials. *(12)* Hence, supercritical fluid extraction (SFE) technology provides a new method of removing uranium and other metals from liquid or solid matrices for analytical, environmental remediation, and material processing applications. *(1-11)* ScFs have been shown to be an environmentally friendly alternative to the conventional solvent extraction processes that often utilize toxic chemicals. SFE technology has also been successfully scaled up to commercial industrial processes. *(13-15)* Consequently, SFE may allow industrial-scale uranium extraction without the waste typically associated with nuclear materials processing.

The need to understand metal extraction with ScFs, particularly uranium SFE, motivated the development of an on-line measurement technique for uranyl complexes in ScF CO_2. In SFE processes, the extracted material is usually collected by depressurizing the ScF, followed by chemical or spectroscopic analysis. On-line monitoring of SFE can be preferable because it is fast, free of collection loss, and allows monitoring of process dynamics. However, depending on the specific analyte and the measurement matrix, the desired levels of sensitivity and selectivity can be difficult to achieve with on-line methods, particularly with detectors that can be coupled with pressurized ScF systems.

Time resolved laser induced fluorescence (TRLIF) has been shown to be effective at measuring uranyl complexes in aqueous environments. *(16-23)* With the ability to select excitation wavelength, emission wavelength, and temporal region, TRLIF provides triple selectivity. Direct speciation of uranyl complexes is possible with TRLIF because of the characteristic changes in emission spectra and fluorescence decay time. *(22-23)* The long fluorescence decay time of uranyl compounds, in some environments over 200 µs, allows easy separation from the emissions of organic compounds such as polyaromatic hydrocarbons, which have fluorescence decay times typically less than 100 ns. TRLIF provides a large dynamic range and is a sensitive technique with detection limits reported at 1ng/L for uranium. *(19,21)* TRLIF's ability to perform measurements via optical fibers is particularly advantageous for nuclear processes, where radiation fields can be significant and personnel radiation exposure must be minimized.

This chapter presents a summary of the application of TRLIF to study uranium complexation and extraction processes in ScF CO_2. The experimental apparatus, methods, synthesis, and relevant safety issues used in this work have been described elsewhere. *(24-28)* In addition to TBP, the uranyl complexes investigated included the ligands TTA (thenoyltrifluoroacetone), TOPO (trioctyl phosphine oxide), TBPO (tributyl phosphine oxide), HFA (hexafluoroacetylacetone), and HTFA (trifluoroacetylacetone).

Spectroscopy

Absorption and Emission

Unlike some reported species of polyaromatic hydrocarbons, the absorption bands of the uranyl chelate complexes do not shift in position or strength with changes in ScF CO_2 temperature or pressure for the conditions investigated. (*28,29*) Further, the absorption spectra of the uranyl chelates are reasonably strong in the near UV. Consequently, the complexes provide stable effective coupling into the 355 nm excitation source. However, the uranyl β-diketone complexes have broad overlapping absorption spectra that lack any specific signature. Free ligand material, always present in excess for any extraction process, absorbs in the UV where the uranyl chelate absorbance is strong enough to provide the method with some sensitivity. Consequently, in-situ measurements of uranyl chelates in ScF CO_2 by UV-Vis absorption spectroscopy has limited sensitivity and selectivity and was not pursued as a process monitoring method.

Figure 1 shows normalized TRLIF spectra of uranyl chelates with identical adducts but different primary ligands in ScF CO_2, at 50°C and 200 atm. Table 1 organizes the various TRLIF parameters measured for uranyl complexes with different primary ligands. The primary ligand clearly has a large effect on the emission spectral structure and decay time of the uranyl complexes. Compared to the uranyl β-diketone complexes with a coordination number of 7 (CN 7), $UO_2(NO_3)_2 \cdot 2TBP$ (CN 8) has a fundamentally different spectra and a significantly longer fluorescence decay time (τ). Similar to the absorption band structure, it appears that the symmetry of the complex strongly affects the emission characteristics of the uranyl chelates.

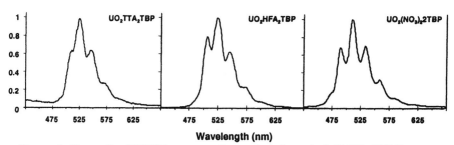

Figure 1. Normalized TRLIF spectra of uranyl chelates in ScF CO_2 (50°C, 200 atm, 10μM).

It can be seen in Table I that all complexes show a decrease in τ at higher ScF pressures. This is believed to be due to the increase in collisional interactions with the CO_2 at higher fluid densities. *(25,26,28)* Table I shows that the relative quenching due to increased pressure is higher for the CN 7 β-diketone uranyl chelates than the CN 8 $UO_2(NO_3)_2 \cdot 2TBP$. This difference might be a result of the CN 8 complex isolating the central uranyl ion to a greater degree than the CN 7 complexes, thereby reducing solvent interactions and making it less susceptible to ScF pressure quenching. Not all uranyl complexes are significantly luminescent. Some complexes, such as those with acetylacetone derivates, have substantially lower quantum yields, resulting in large reductions in analytical sensitivity. (28)

Table I. Primary Ligand Effects On Uranyl Chelates τ in ScF CO_2 at 50°C

Compound	Peaks (nm)	τ (ns) 100 atm	τ (ns) 200 atm	Coordination Number
$UO_2(NO_3)_2 \cdot 2TBP$	490, 511, 534, 558, 584	2480 ± 50	2200 ± 50	8
$UO_2TTA_2 \cdot TBP$	507, 522, 544, 569	50.7 ± 1.0	42.2 ± 0.4	7
$UO_2HFA_2 \cdot TBP$	506, 524, 545, 572	1075 ± 20	830 ± 18	7

SOURCE: Reproduced from reference 28. Copyright 2001 American Chemical Society.

The equatorially coordinated adduct ligand has little impact on the potential around the central uranium atom and therefor has a negligible effect on the structure of the emission spectra. *(30-33)* However, as shown in Table II, changing the adduct of a complex did result in small changes in the measured fluorescence decay time.

Table II. Effects of Uranyl Chelate Adducts on τ in ScF CO_2 at 50°C

Chelate	Adduct Formula	τ (ns) 100 atm	τ (ns) 200 atm
$UO_2TTA_2 \cdot TBP$	$-OP(OC_4H_9)_3$	50.7 ± 1.0	42.2 ± 0.4
$UO_2TTA_2 \cdot TBPO$	$-OP(C_4H_9)_3$	52.9 ± 1.0	44.0 ± 0.6
$UO_2TTA_2 \cdot TOPO$	$-OP(C_8H_{17})_3$	55.3 ± 1.0	45.9 ± 0.2

SOURCE: Reproduced from reference 28. Copyright 2001 American Chemical Society.

The data in Table II suggest the effect of the adduct on τ may be both chemical and steric in nature. Comparing TBPO with TOPO, which have nearly chemically identical bonding with the uranyl ion, shows the larger adduct (TOPO) has a slightly longer lifetime. This is consistent with an increase in steric shielding and subsequent reduction in solvent quenching. Temperature, pressure and solvent composition all significantly impact the fluorescence intensity and decay times of the ScF solvated uranyl chelates. The mechanisms for uranyl chelate quenching in ScFs, and various methods to correct for variable quenching, have been disussed in the literature. (25,26,28,34,35)

On-line Measurements

On-line monitoring of SFEs is preferable because it provides immediate in-situ data, avoids the difficulties of sampling from a pressurized system, and provides superior temporal resolution. On-line assay of uranyl chelates in ScF CO_2 is possible with TRLIF if the system calibration accounts for the various quenching factors and the emission characteristics of the analytes present in the process stream. (26) Measurement time is an important factor for on-line analysis in a dynamic system, since it limits the ability to detect changes in the sample flow stream. For any on-line system there is typically a compromise required between sensitivity and response time. Difficulty arises in using TRLIF for on-line work when uranyl chelate concentrations are very low (sub μM) and the features under observation change on time scales near or shorter than the time required for analysis. For instance, the TRLIF system used for the these studies has a lower limit of detection below 10^{-9} M, but if ScF assay is desired every 15 seconds, and two species are being measured, the limit of detection (10σ) for uranyl chelates increases to ~10^{-6} M.

The temporal capability of TRLIF provides an orthogonal parameter to the optical spectrum increasing the dimensionality of the data. In some, cases analytes can be directly isolated on the spectral-temporal map, eliminating the need for deconvolution. With judicious selection of timing parameters TRLIF allows on-line ScF speciation and quantitative analysis of mixtures of uranyl chelates with different primary ligands. For instance, the large difference in luminescent lifetime between $UO_2TTA_2 \cdot TBP$ and $UO_2(NO_3)_2 \cdot 2TBP$ allows these species to be directly resolved and analyzed on-line. Since these chelates can be temporally separated, the entire spectral peak area can be integrated, resulting in optimal signal to noise and limits of detection. On-line quantitative TRLIF speciation of a more complicated mix of uranyl chelates is possible if there is sufficient uniqueness to the species emission spectra and decay times.

Speciation with TRLIF of uranyl chelates with the same primary ligand but different adducts is not possible even with high-precision data. Uranyl chelates that differ only by adduct have identical emission spectra and, as shown in Table II, very similar fluorescence decay times. There is no region on the spectral-temporal map that is unique to the adduct differentiated chelates and, consequently, these uranyl species cannot be separated and analyzed with TRLIF.

Extraction Process Monitoring and Modeling

As noted, TRLIF can provide an effective method for on-line measurement of uranyl chelates. Figure 2 shows concentration profiles, measured on-line with TRLIF, from a mixed ligand extraction of 7.0 ml of 0.01 M UO_2^{2+} in 1 N nitric acid with ScF CO_2, at 150 atm and 50°C modified with 0.15M TBP and 0.06 M TTA. These are typical profiles for the 10 mL tubular extraction cell utilized in these studies. Initially, the concentration of uranyl complexes in the ScF phase is observed to increase rapidly as the uranium is transported into the headspace. After the initial equilibration period, the extraction profiles for both uranyl complexes exhibit first order extraction kinetics as the concentration of uranium in the aqueous phase is reduced. TTA and TBP were selected for multiligand SFE studies because they both allow selective uranium complexation and both of the resulting complexes (UO_2TTA_2·TBP and $UO_2(NO_3)_2$·2TBP) have high ScF solubilities. *(2,3,9-11,36,37)* Also, when used together, TTA and TBP are reported to have a synergistic effect on metal extraction efficiency. *(3,38,39)* Consequently, multiligand uranium extractions with TTA and TBP may have analytical and industrial applications.

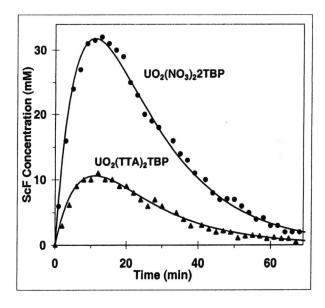

Figure 2. On-line TRLIF monitoring of the extractions of 0.1 M uranium in 6 N nitric acid with ScF CO_2(1ml/min, 150 atm, 50°C) modified with 0.15 M TBP and 0.06 M TTA. The solid lines are the modeled extraction response overlaying the data points from on-line concentration measurements.
(Reproduced from reference 28. Copyright 2001 American Chemical Society.)

If we assume this extraction process can be described by two mixed vessels in equilibrium contact then

$$V_L \frac{dC_{aq}}{dt} = -QC_I$$ (1)

and

$$V_{ScF} \frac{dC_{ScF}}{dt} = Q(C_I - C_{ScF}).$$ (2)

Where V_{ScF} = volume of ScF phase in extractor (ml), V_L = volume of liquid phase in extractor (mL), C_I = concentration of flow into ScF phase = DC_{aq}, D = distribution coefficient = C_{ScF}/C_{aq}, C_{aq} = concentration of metal in aqueous phase (mol/mL), C_{ScF} = concentration of metal complex in ScF phase (mol/mL), Q = flow rate through extractor (mL/min), and t = time (min).

Equation (1) equates the change in the amount of uranium in the aqueous phase with the amount of uranium complex moving into the ScF phase. Equation (2) equates the change in the amount of complex in the ScF phase with the difference in the amount of complex flowing into and out of the ScF phase. Substituting Equation (1) into (2) gives a linear nonhomogenous differential equation with a solution in the form of

$$C_{ScF}(t) = \frac{DC_{aq}^o}{1 - \frac{V_{ScF}}{V_L}} \left(e^{-\frac{DQ}{V_L}t} - e^{-\frac{DQ}{V_{ScF}}t} \right)$$ (3)

where C_{aq}^o is the initial concentration of uranium in the aqueous phase.

Equation (3) effectively describes single ligand SFE profiles and can be extended to multiligand SFE with an expansion of Equation (3) since

$$C_{ScF}^U(t) = \sum_n C_{ScF}^n(t, D_n).$$ (4)

For the extraction system studied here Equation (4) reduces to

$$C_{ScF}^U(t) = C_{ScF}^{NT}(t, D_{NT}) + C_{ScF}^{TT}(t, D_{TT})$$ (5)

where D_{NT} is the distribution coefficient (C_{ScF}/C_{aq}) for $UO_2(NO_3)_2 \cdot 2TBP$, and D_{TT} is the distribution coefficient for $UO_2TTA_2 \cdot TBP$.

From Equations (3) and (5) the extraction profiles of the individual species are found to be

$$C_{ScF}^{NT}(t) = \frac{D_{NT}C_{aq}^{o}}{1 - \dfrac{V_{ScF}}{V_L}}\left(e^{-\frac{DQ}{V_L}t} - e^{-\frac{DQ}{V_{ScF}}t} \right) \tag{6a}$$

$$C_{ScF}^{TT}(t) = \frac{D_{TT}C_{aq}^{o}}{1 - \dfrac{V_{ScF}}{V_L}}\left(e^{-\frac{DQ}{V_L}t} - e^{-\frac{DQ}{V_{ScF}}t} \right) \tag{6b}$$

where $D = D_{NT} + D_{TT}$ and both ligands are in excess.

Note that the predicted extraction profiles are strongly dependent on D, which is a function of the ligand, the ligand concentration, pH, and ScF conditions. The solid lines in Figure 2 represent the model-predicted response (6a, 6b) for uranium SFE with TTA and TBP under the given experimental conditions. Clearly the fit quality of the model to the experimental data indicates that the assumptions made for the model are valid for the physical system and within the tubular extractor there is equilibrium between the two phases and complete mixing within each phase. Deviation from the predicted behavior was observed if ligands were not in excess, flow rate was too high, or pressure was low enough for complex ScF solubility to become a limiting factor.

For a fixed TBP concentration, as the TTA concentration in the ScF CO_2 feed stream increased, the relative concentration of $UO_2TTA_2 \cdot TBP$ in the extracted flow subsequently rose and the total rate of uranium extraction increased. ScF concentrations of $UO_2(NO_3)_2 \cdot 2TBP$ fell slightly at higher TTA concentrations due to competition between the two complexes for the TBP ligand. Without TBP in the feed stream there was very little uranium extraction. Recent work has shown the ScF CO_2 solubility of the $UO_2TTA_2 \cdot X$ complexes to be largely dependent on the adduct. (40) The $UO_2TTA_2 \cdot H_2O$ complex is two orders of magnitude less soluble than the analogous TBP complex, which can be attributed to the increased shielding of the central metal ion by the bulky TBP adduct. When this solubility information is examined in context with the on-line TRLIF observations, it suggests that the synergistic effect observed in the extraction of uranium with these ligands may be largely due to the solubility of the complex.

At lower ScF pressures, solubility effects on the extraction process are observed. Figure 3 shows the on-line measurement of the extraction of 7.0 ml of 0.1 M uranium in 6 N nitric acid with ScF CO_2 at 50°C, at lower pressures, and

0.14 M TBP. The extraction profile in Figure 3 is clearly different from that observed in Figure 2 and obviously does not follow the behavior predicted by Equation (3). Figure 3 shows that increasing the pressure increases the amount of metal complex in the ScF phase. This behavior is interpreted as a solubility-limited SFE, that is, an extraction limited by the solubility of the $UO_2(NO_3)_2 \cdot 2TBP$ in the ScF CO_2. While the TBP ligand is a highly soluble and miscible liquid with ScF and subcritical CO_2 (24-25), the solubility of the extracted $UO_2(NO_3)_2 \cdot 2TBP$ depends strongly on, and increases with, ScF pressure (density). (11,37) At sufficiently high TBP and UO_2^{2+} concentrations, and at lower pressures, the limiting factor for the SFE is not the chemical equilibrium but rather the solubility of the metal complex in the CO_2 phase. Consequently, the extraction flow has a constant concentration, determined by the solubility of the complex under the ScF conditions. The exponential decrease in concentration predicted by Equation (3) is only observed towards the end of the extraction, when the concentration of UO_2^{2+} in the aqueous phase has been reduced to the point where chemical equilibrium results in an ScF phase that is not saturated with $UO_2(NO_3)_2 \cdot 2TBP$. For Figure 3, the behavior expected from an equilibrium controlled SFE, and predicted by Equation (3), can be observed to begin at approximately 130 minutes.

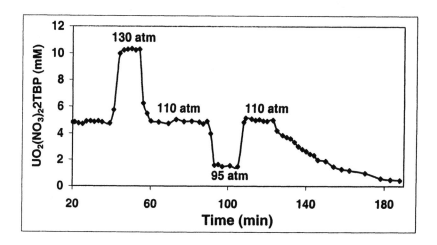

Figure 3. On-line TRLIF monitoring of a solubility-limited SFE of 0.1 M uranium in 6 N nitric acid with ScF CO_2 at 50°C, with 0.14 M TBP, at 3 lower pressures.
(Modified from reference 10. Copyright 1997 University of Idaho.)

Distribution Coefficients

For an extraction, which is not solubility limited, and for which the liquid phase is significantly larger ($V_L/V_{ScF} \geq 2$) than the ScF phase, the later portion of the extraction profile can be described by a simplified form of Equation (3):

$$C_{ScF}(t) = \frac{DC_{aq}^{o}}{1 - \frac{V_{ScF}}{V_L}} \left(e^{-\left[\frac{DQ}{V_L}\right]t} \right) \tag{7}$$

Equation (7) indicates that the majority of the extraction profile can be approximated by a single exponential function that could be interpreted as first order behavior. For instance, after 30 minutes, the data shown in Figure 2 fit an e^{-kt} function with correlation coefficients better than 0.995. Equation (7) allows D and an empirical extraction rate constant k to be easily determined. From Equation (7), the distribution coefficient can be found to be

$$D = -\frac{mV_L}{Q} \tag{8}$$

where m is the slope from a semilog plot of C_{ScF} versus t. The empirical extraction rate constant is simply $-m$ from a semilog plot of C_{ScF} versus t.

The complexation reaction is given by

$$UO_2^{2+} + 2NO_3^{2-} + 2TBP_{ScF} \rightarrow UO_2(NO_3)_2 \cdot 2TBP_{ScF} \tag{9}$$

Consequently, it is expected that D and the extraction rate will be proportional to $[TBP]^2$. Figure 4 shows D as a function of the ScF TBP concentration. Below 0.25 M TBP, it can be observed that even small changes in ScF ligand concentration make very large changes in D and consequently the rate of extraction. For equivalent conditions and TBP concentrations, the D values determined with on-line TRLIF of dynamic extraction match, within experimental error, previously reported values obtained with static extractions. (10,11) This supports the early assumption that this dynamic extractor is in chemical equilibrium. The curve in Figure 4 is $[TBP]^2$, the theoretically expected response for the complexation reaction. It can be observed in Figure 4 that increasing the TBP concentration above approximately 0.25 M does not increase the D value for this system. This would suggest that at higher TBP concentrations there is some process limiting the extraction of uranium into the ScF phase. A detailed discussion is given in the literature as to the possible reasons for this behavior. (27)

198

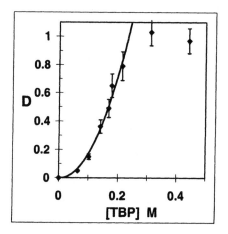

Figure 4. Distribution coefficient (D) as a function of TBP concentration for the extraction of 0.1 M uranium in 6 N nitric acid with ScF CO_2 at 150 atm and 50°C. The curve is $[TBP]^2$, the expected equilibrium response. (Modified from reference 10. Copyright 1997 University of Idaho.)

At pressures sufficient to avoid complex solubility issues, and within the experimental accuracy (± 15 %), the distribution coefficient was found to decrease with pressure. Figure 5 shows the measured D values at various pressures for ScF CO_2 at 50°C that are consistent with previously published values when adjusted for temperature and TBP concentration. (10,11)

The D values (and consequently the extraction rates) reported here, are similar to those reported for liquid solvent extraction (37,43), even though diffusion within

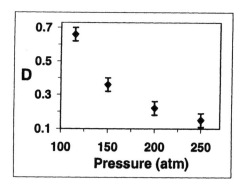

Figure 5. Distribution coefficient (D) as a function of pressure at 50°C for the extraction of 0.1 M uranium in 6 N nitric acid with 0.14 M TBP in ScF CO_2. (Modified from reference 10. Copyright 1997 University of Idaho.)

the ScF phase is known to be much faster. The insensitivity to ScF pressure (density) and similarity to the liquid phase indicates that the governing factor for this extraction is not a diffusional process within the ScF phase.

The reduction in D with increasing pressure cannot be attributed to complex solubility since the solubility of $UO_2(NO_3)_2 \cdot 2TBP$ increases with ScF density. D values, and extraction rates, have been observed to be independent of uranium concentration in this study and in previous work. (2,11) Consequently, pressure dependence of D in these dynamic extractions must be due to the behavior of TBP. This conclusion is supported by data from this study and others that show D increasing at higher TBP concentrations, particularly for lower ($\leq 50^{\circ}C$) temperatures. (10,11) TBP, and other organophosphorous ligands, are known to have increased ScF solubility with pressure. (24) As ligand solubility in the ScF phase increases, the amount of ligand distributing into the aqueous phase can be expected to decrease. The reduction of the aqueous TBP concentration most likely accounts for the observed decrease in D, and subsequent extraction rate, with pressure.

As shown in Figure 6, D increased with temperature for the extraction of aqueous uranium with TBP modified ScF CO_2. Why D increases rapidly with temperature is an interesting question. These were dynamic extractions, in a turbulent vessel shown to be at equilibrium, and consequently diffusional processes should not be a limiting factor.

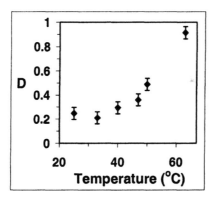

Figure 6. Distribution coefficient (D) as a function of temperature at 150 atm for the extraction of 0.1 M uranium in 6 N nitric acid with 0.17 M TBP in liquid and ScF CO_2.
(Modified from reference 10. Copyright 1997 University of Idaho.)

Compound solubility is known to change dramatically with temperature in ScF CO_2. It can be observed that D values always decrease with increasing ScF density, when either temperature or pressure is changed. It is reasonable to postulate that as temperature goes up (and ScF density falls) more TBP enters into the aqueous phase allowing more uranyl complex to be formed and subsequently extracted into the ScF phase. *(27)*

The increase in D between 33 and 25°C suggests that subcritical CO_2 is more effective at uranium extraction than the ScF phase. A similar increase in extraction efficiency for liquid CO_2 over ScF CO_2 has also been observed in the extraction of Ni^{2+} and Zn^{2+} with β-diketone modified CO_2. *(44)* It is difficult to explain the increase in extraction efficiency in terms of increased solubility, since the density difference between liquid and ScF CO_2 is small (0.86 – 0.83 g/mL at 150 atm). Further study is needed to ascertain if, and why, liquid CO_2 could have higher D values and extraction rates. If real, the increase of D, and subsequently the extraction efficiency, is very significant to metal extraction with compressed CO_2. Liquid CO_2 is preferable to ScF CO_2, since lower pressures are safer, energetically more favorable, and do not require expensive high pressure equipment.

Solubility

Solubility data for the metal complexes in ScF CO_2 are essential in order to develop models for the extraction processes and assess the feasibility of using ScF CO_2 to replace organic solvents in conventional extraction processes. ScF CO_2 solubility data for uranium complexes with TBP, β-diketones, or phosphine oxides are of particular interest for the development of a ScF based process for reprocessing spent nuclear fuels.

A variety of techniques have been used to measure the solubility of metal chelates in ScFs, including gravimetric, *(45,46)* spectroscopic, *(36,37,47,46,48)* and chromatographic methods. *(36,37,45-51)*. No method is ideal, and the relative merits of these techniques are discussed in the literature. As previously shown, laser induced fluorescence (LIF) is a fast, sensitive, and selective method for in-situ determination of uranyl chelates in ScF CO_2. In this section we discuss the methods and merits of using LIF to determine solubilities of uranyl chelates in ScF CO_2.

To quantitatively explain and predict solubility behavior of solutes in ScFs several models have been proposed. These models usually employ the Hildebrand solubility parameter *(52)* or Peng-Robinson equation of state *(53)* and require a knowledge of the thermodynamic properties of the solute, which are not accurately known for many compounds. Alternatively, a simpler approach was proposed by Chrastil *(54)* to relate the solubility to the density of the fluid:

$$\ln(S) = k\ln(D) + C \qquad (10)$$

where S and D are the solubility of the solute (g/L) and density of the fluid (g/L), respectively. The Chrastil equation (10) is derived from the association laws and the entropies of components, avoiding the complexities involved with using the equations of state. Equation (10) predicts a linear relationship between ln(D) and ln(S) with the slope (k) and intercept (C) values being a function of the solvent/solute system. The constant k is related to the solute-solvent interactions and decreases with temperature. C is also a temperature-dependent constant related to the volatility of the solute and independent of fluid density. Therefore, knowledge of k and C for the temperatures studied enables solubility data to be calculated over a range of densities. The Chrastil model has been shown to be valid for a number of metal chelates in ScF CO_2. (54,55)

The solubility of the chelates was determined by measuring the fluorescence intensity with stepwise decreases in pressure. (40) As the pressure is reduced the solvent strength of ScF CO_2 decreases until the solution becomes saturated. At this point, the complex begins to precipitate, resulting in a decrease in the measured concentration (or fluorescence intensity). Figure 7 shows the effect of pressure/density upon concentration, for a 10 µg/L mM solution of $UO_2(NO_3)_2$ 2TBP, as the pressure decreases from 160 to 80 atm at 50°C. Between 160 and 130 atm, the concentration (density normalized for clarity) remains constant because the solvent strength of the fluid is still sufficient to keep the solute in solution. At 130 atm the concentration begins to fall rapidly, indicating that the solution has now become saturated for the given solute concentration. The concentration continues to decrease as the pressure is reduced from 130 to 80 atm, and the measured ScF concentrations are solubility values for the metal complex. Similar curves can be obtained for any desired ScF temperature and measurable solute.

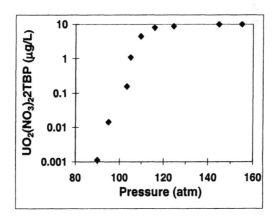

Figure 7. Precipitation of 10 µg/L $UO_2(NO_3)_2 \cdot 2TBP$ from 50°C ScF CO_2.
(Modified from reference 26. Copyright 2000 American Chemical Society.)

Comparative work has shown solubility data obtained by LIF at lower pressures can be used with the Chrastil equation for extrapolation of solubility values to higher pressures. However, this generalization should be used with care, especially with volatile liquid materials, since it has been shown that the solubility of some compounds deviates from the simple Chrastil model. (54) For instance, at high pressures the solubility of $UO_2(NO_3)_2 \cdot 2TBP$ increases dramatically and deviates from the Chrastil model as the complex (a viscous fluid) becomes miscible with very dense ScF CO_2. (37)

Figure 8 shows the solubility data for some uranyl chelates in ScF CO_2 at 50°C. The solubility of these complexes obviously follows the Chrastil relationship. As expected, solubility always increases significantly with density, although it is interesting to note that the relative solubility of the chelates can change. For instance $UO_2(NO_3)_2 \cdot 2TBP$ is less soluble than $UO_2(TTA)_2 \cdot TBP$ at low densities, but more soluble at high densities. Similarly, the solubility profiles for $UO_2(TTA)_2 \cdot H_2O$ and $UO_2(TTA)_2 \cdot TOPO$ can also be seen to cross. This crossing of solubility profiles can also be attributed to the competing effects of solute volatility and solvent strength of the fluid. It is most likely to be observed

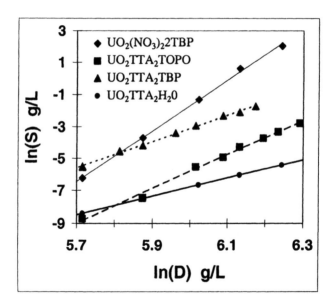

Figure 8. A comparison of lnS vs. lnD plots for selected uranyl complexes in ScF CO_2 at 50°C.
(Modified from reference 26. Copyright 2000 American Chemical Society.)

in the low density regions of an ScF where volatility makes a larger contribution to a compound's solubility.

The Chrastil constants k and C for the uranyl chelates at 50°C are shown in Table III. The more soluble the complexe, the more negative the C value and larger the k value. The k parameter is related to the average number of solvent molecules associated with the metal complex. The data in Figure 8 and values in Table III clearly show the adduct effect on the ScF solubility of the complexes. By increasing the average number of solvent molecules clustered around the metal complex, the adduct significantly increases the complex solubility. Since C is a function of the enthalpy of solvation, larger negative values probably indicate stronger ligand solvent interactions. (55) Thus, the adduct not only increases the number of solvent molecules around the metal complex but also the net strength of the interactions with the solvent sphere.

Table III. Chrastil Parameters for Selected Uranyl Complexes in ScF CO_2

Compound	LIF[a]	
	k	C
$UO_2(NO_3)_2 \cdot 2TBP$	15.9	-97.1
$UO_2TTA_2 \cdot H_2O$	5.6	-40.5
$UO_2TTA_2 \cdot TBP$	8.3	-52.8
$UO_2TTA_2 \cdot TOPO$	10.6	-69.5

[a] 85-110 atm, 50 °C

SOURCE: Modified from reference 26. Copyright 2000 American Chemical Society.

The adduct is critical to the completion of the chelation sphere, which allows solvation of charged species in ScF CO_2. Without adducts, uranyl chelates are practically insoluble in ScF CO_2. Attempts to measure the adduct-free complex, UO_2TTA_2, with LIF were unsuccessful since the amount of material in solution was below LIF detection limits. The addition of even a small polar adduct, such as H_2O, to the UO_2TTA_2 chelate allows the complex to become measurably soluble. Uranyl chelate complexes with large adducts such as TTA and TOPO are very soluble, behaving similar to ferrocene and polyaromatics in both trend and magnitude.(54) This behavior is attributed to the fact that the metal is fully coordinated and no sites are available for interaction with the ScF CO_2. It is not a coincidence that the most soluble uranyl chelate, $UO_2(NO_3)_2 \cdot 2TBP$, has a CN of 8, providing better shielding of the charged species than other complexes measured, which have shown CN of 7. The importance of the adduct to the complex solubility has obvious impacts on ScF processes utilizing metal complexes.

Conclusions

TRLIF is valuable and versatile tool for studying ScF systems with luminescent species. The method provides rapid in-situ data and has a large dynamic range. The uranyl chelates have widely variable fluorescent spectra and decay times in ScF CO_2. A variety of factors affect the emission from uranyl complexes in ScF CO_2, including ligand, adduct, pressure, and temperature. With judicious selection of instrument parameters and proper calibration, TRLIF allows on-line speciation and quantitative analysis of uranyl complexes within ScF CO_2. On-line analysis of mixed ligand SFE shows that uranium is being extracted as several different species. Further, it was found that the extraction profiles are not simple functions and could be significantly altered by many parameters. An effective model for dynamic SFE has been developed, assuming equilibrium between the two phases and complete mixing within each phase. TRLIF has also been shown to be capable of determining fundamental parameters of compounds in ScFs, such as the solubilities and distribution coefficients, with good sensitivity and precision.

TRLIF has a range of potential applications to other ScF investigations. The methods discussed in the chapter are not limited to uranyl chelates in ScF CO_2 and are applicable to other ScFs and other compounds such as polyaromatic hydrocarbons, lanthanide complexes, and luminescent chemical probes. Future investigations with TRLIF in ScFs could involve work on analytical and industrial extraction processes, determination of physical parameters such as solubility, or fundamental studies of solvent-solute interactions in ScF systems.

References

1. Smart, N.G.; Carleson, T.; Kast, T.; Clifford, A.A.; Burford, M.D.; Wai, C.M. *Talanta* **1997**, *44*, 137-150.
2. Lin, Y.; Smart, N.G.; Wai, C.M. *Environ. Sci. Technol.* **1995**, *29*, 2706-2708.
3. Laintz, K.E.; Tachikawa, E. *Anal. Chem.* **1994**, *66*, 2190-2193.
4. Wang, J.; Marshall, W.D. *Anal. Chem.* **1994**, *66*, 3900-3907.
5. Lin, Y.; Brauer, R.D.; Lantz, K.E.; Wai, C.M. *Anal. Chem.* **1993**, *65*, 2549-2551.
6. Wai, C.M.; Lin, Y.; Brauer, R.D.;Wang, S. *Talanta* **1993**, *40*, 1325-1330.
7. Laintz, K.E.; Wai, C.M.; Yonker, C.R.; Smith, R.D. *Anal. Chem.* **1992**, *64*, 2875-2878.
8. Laintz, K.E.;Yu, J.J.; Wai, C.M. *Anal. Chem.* **1992**, *64*, 311-315.
9. Toews, K.L.; Smart, N.G.; Wai, C.M. *Radiochimica Acta* **1996**, *75*, 179-184.

10. Phleps, C. L.; Doctoral dissertation, *"Extraction of Uranium from Uranium Oxides Using Beta-Diketones and Alkyl Phosphates Dissolved in Supercritical Carbon Dioxide"*, University of Idaho, **1997**.

11. Meguro, Y. Iso, S.; Takeishi, H.; Yoshida, Z. *Radiochimica Acta* **1996**, *75*, 185.

12. Samsonov, M.D.; Wai., C.M.;Lee, S.C.;Kulyako, Y.; Smart, N.G. *Chem. Commun.* **2001** 1868-1889.

13. Phelps, C.L.; Smart, N.G.; Wai, C.M. *J. Chem. Ed* **1996**, *73*, 1163-1168.

14. Lira, C.T.; *Supercritical Fluid Extraction and Chromatography: Techniques and Applications*; Charpentier, B.A., and Sevenants, M.R. Eds.; ACS Symposium Series 366, American Chemical Society, Washington DC, **1988**, Chap. 1.

15. Chester, T.L.; Pinkston, J.D.; Raynie D.E. *Anal. Chem.* **1996**, *68*, 487R

16. Moulin, C.; Rougeault, S.; Hamon, D.; Mauchien, P. *Appl. Spectros.* **1993**, *47*, 2007-2012.

17. Fujimori, H.; Matsui, T.; Suzuki, K. *J. Nuc. Science and Tech.* **1988**, *25*, 798-804.

18. Hong, K.B.; Jung, K.W.; Jung, K.H. *Talanta* **1989**, *36*, 1095-1099.

19. Moulin, C.; Beaucaire, C.; Decambox, P.; Mauchien, P. *Anal. Chim. Acta* **1990**, *238*, 291-296.

20. Moulin, C.; Decambox, P.; Moulin, V.; Decaillon, J.G. *Anal. Chem.* **1995**, *67*, 348-353.

21. Brina, R.; Miller, A.G. *Anal. Chem.* **1992**, *64*, 1413-1418.

22. Meinrath, G.; Kato, Y.; Yoshida, Z. *J. of Radioanal Nucl. Chem.* **1993**, *174*, 299-314.

23. Geipel, G.; Brachmann, A.; Brendler, V.; Bernhard, G.; Nitsche, H. *Radiochimica Acta* **1996**, *75*, 199-204.

24. Addleman, R.S.; Hills, J.W.; Wai, C.M. *Rev. Scien. Instruments* **1998**, *69*, 3127-3131.

25. Addleman, R.S.; Wai, C.M. *Physical Chemisty Chemical Physics* **1999**, *1*, 783-790.

26. Addleman, R.S.; Wai, C.M. *Anal. Chem.* **2000** ,*72*, 2109-2116.

27. Addleman, R.S.; Wai, C.M. *Radiochimica Acta,* **2001**, *89*, 27-33.

28. Addleman, R.S.; Carrott, M.J.; Wai C.M.; Carleson, T.E.; Wenclawiak, B.W. *Anal. Chem.* **2001**, *73*, 1112-1119.

29. Rice, J.K.; Niemeyer, E.D.; Bright, F.V. *Anal. Chem.* **1995**, *67*, 4354-4357.

30. Meinrath, G. *J. Radiodnal. Nucl. Chem.,* **1997**, *224*, 119-126.

31. Denning, R.G. *Electronic Structure and Bonding in Actinyl Ions. Structure Bond* **1992**, *79*, 215.

32. Goerrler-Walrand, CH.; Jaegere, S.D. *Spectrochim. Acta. Part A* **1972**, *A28*, 257-268.

33. Goerrler-Walrand, CH.;Vanquickenborne, L.G. *J. Chem. Phys.* **1971**, 54, 4178-4286.
34. Moulin, C.; Decambox, P.; Mauchien, P.; Pouyat, D. Couston, L. *Anal. Chem.* **1996**, 68, 3204-3209.
35. Lakowicz, J.R. *Principles of Fluorescence Spectroscopy*, 2nd Ed., Plenum Press, New York, **1999**.
36. Carrott, M.J.; Wai, C.M. *Anal. Chem* **1998**, 70, 2421-2425.
37. Orth, D.A.; Wallace, R.M.; Karraker, D.G. in *Science and Technology of Tributyl Phosphate*, Ed. Schulz, W.W.; Navratil, J.D.; Kertes, A.S., CRC Press, **1984**, Vol. 1.
38. Irving, H.M.; Edgington, D.N. *J. Inorg. Nucl. Chem.* **1960**, 15, 158-170.
39. Sagar, V.; Chetty, K.V. *Radiochim. Acta* **1994**, 68, 69-73.
40. Addleman, R.S.; Carrott, M.J.; Wai, C.M. *Anal. Chem.* **2000**, 72, 4015-4021.
41. Meguro, Y.; Iso, S.; Sasaki, T.; Yoshida, Z. *Anal. Chem.* **1998**, 70, 774-779.
42. Page, S.H.; Sumpter, S.R.; Goates, S.R.; Lee, M.L.; Dixon, D.J.; Johnston, K.P. *J of Supercritical Fluids* **1993**, 6, 95.
43. Farbu, L.; McKay, H.A.C.; Wain, A.G.: Transfer of Metal Nitrates Between Aqueous Nitrate Media and Neutral Organophosphorus Extractants. *Proc. Int. Solvent Extr. Conf. Soc. Chem. Ind. London*, **1974**, 2427.
44. Laintz, K.E.; Hale, C.D.; Stark, P.; Rouqnette, C.L.; Wilkinson, J. *Anal. Chem.* **1998**, 70, 400.
45. Wai, C.M.; Wang, S.; Yu, J.J. *Anal. Chem.* **1996**, 68, 3516.
46. Guigard, S.; Hayward, G.L.; Zytner, R.G.; Stiver, W.H. *Anal. Chem.* **1998**, 999.
47. Laglante, A.F.; Hansen, B.N.; Bruno, T.J.; Sievers, R.E. *Inorg. Chem.* **1995**, 34, 5781-5785.
48. Hansen, B.N.; Laglante, A.F.; Sievers, R.E.; Bruno, T.J. *Rev. Sci. Instrumen.* **1994**, 65, 2112.
49. Cowey, C.M.; Bartle, K.D.; Burford, M.D.; Clifford, A.A.; Zhu, S.; Smart, N.G.; Tinker, N.D. *J. Chem. Eng. Data* **1995**, 40, 1217.
50. Ashraf-Khorassani, M; Taylor, L.T.; Combs, M.T *Talanta*, **1997**, 44, 755-763.
51. Ashraf-Khorassani, M; Combs, M.T.; Taylor, L.T. *Supercritical Fluid Extraction of Inorganics in Environmental Analysis;* Encylopedia of Analytical Chemistry, John Wiley & Sons, 2000; Vol. 4, 3410-3423.
52. Laglante, A.F.; Hansen, B.N.; Bruno, T.J.; Sievers, R.E., *Inorg. Chem.* **1995**, 34, 5781-5785.
53. Cross, W.; Akerman, A.; Erkey, C. *Ind. Eng. Chem. Res.* **1996**, 35, 1765.
54. Chrastil, J. *J. Phys. Chem.,* **1982**, 86, 3016-3021.
55. Smart, N.G.; Carleson, T.; Kast, T.; Clifford, A.A.; Burford, M.D.; Wai, C.M. *Talanta* **1997**, 44, 137-150.

Chapter 14

Solubilization Study by QCM in Liquid and Supercritical CO_2 under Ultrasonar

Kwangheon Park, Moonsung Koh, Chunghyun Yoon, Hakwon Kim, and Hongdoo Kim

Green Nuclear Research Laboratory, Kyung Hee University, Kyung Hee 449–701, South Korea

QCM technique was applied to the solubilization study in CO_2. Basics of QCM were reviewed, and applications of QCM to experiments in static and dynamic systems were conducted. Roughness effect should be considered when measuring absolute mass change. Solubilization of $Cu(acac)_2$ in CO_2 was measured, and the diffusivity was estimated. Ultrasonic waves enhance solubilization mainly due to acoustic streaming. Co ion solubility in a cyanex-CO_2 solution was obtained by the QCM technique in a dynamic system

1. Introduction

Carbon dioxide(CO_2) has emerged these days as the most promising alternative for a green solvent. It provides several benefits: It is non-flammable, non-toxic, environmentally benign, and inexpensive. Also, CO_2 is a tunable fluid, i.e., its density and solubility can change by controlling pressure and temperature. Supercritical fluid technology has been applied in many and diverse fields ranging from extractions of important ingredients from natural products to material production processes such as polymerization and nano-particle generation [1,2,3]. Liquid and supercritical (L/SC) CO_2 have been used as an effective solvent for cleaning of mechanical parts, semiconductors, and fabrics. Cleaning mechanisms include mechanical and chemical removal. Most of the cleaning power of a solvent comes from chemical removal, i.e., the ability to dissolve the target contaminants into the solvent (solubilization).

The process of solubilization depends on the kinetics of contaminant dissolution into the solvent. So far, the cleaning ability of a solvent is found by comparing the differences in cleanliness before and after the cleaning action. In-situ observation is quite limited in the case of L/SC CO_2 due to the difficulty of obtaining quantitative measurements inside high-pressure systems. A quartz crystal microbalance (QCM) is suggested as a tool for in-situ measurements of a solubilization reaction in L/SC CO_2 in this paper. QCM comprises a thin quartz crystal sandwiched between two metal electrodes that establishes an alternating electric field across the crystal, causing vibration of the crystal at its resonant frequency. The resonance frequency is sensitive to mass changes of the crystal and the electrode [2]. The sensitivity of QCM corresponding to mass change is approximately $\sim ng/cm^2$. There are quite a few of experiments that have applied QCM techniques to the measurement of the reactions in L/SC CO_2 [4,5,6]. Expansion of QCM technique to other in-situ measurements in high-pressure systems seems promising, based on its simplicity of use and sensitivity.

Mechanical agitation generally enhances cleaning efficiency. Ultrasonic wave has been used in the aqueous system, and its effectiveness at cleaning is well known. However, the enhancement effect of ultrasonic waves on the solubilization reaction in L/SC CO_2 is not yet known in detail.

Hence, the objectives of this paper are to develop and explore the use of QCM technique to measure the solubilization rates in L/SC CO_2, and to measure the effects of ultrasonic impulses on reactions in L/SC CO_2 using the QCM technique.

2. Quartz crystal microbalance

QCM can be described as a thickness-shear mode resonator, since weight change is measured on the base of the resonance frequency change. The acoustic wave propagates in a direction perpendicular to the crystal surface. The quartz crystal plate has to be cut to a specific orientation with respect to the crystal axis to attain this acoustic propagation properties. AT-cut crystals are typically used for piezoelectric crystal resonators[7]. The use of quartz crystal microbalances as chemical sensors has its origins in the work of Sauerbrey[8] and King [9] who carried out micro-gravimetric measurements in the gas phase. It was assumed in their work that a thin film applied to a thickness-shear-mode device could be treated in sensor measurements, and a shift in the resonance frequency of an oscillating AT-cut crystal could be correlated quantitatively with a change in mass added to or removed from the surface of the device. This chapter deals with the basics of QCM and its application to the use in highly viscous fluid such as L/SC CO_2.

2.1 Basics of QCM.

When the resonant condition of a thickness-shear oscillation is satisfied for AT-cut piezoelectric crystals, a shear wave propagates through the bulk of the material, perpendicular to the faces of the crystals. The fundamental resonance frequency, f_0 is given by;

$$f_0 = \frac{V_{tr}}{2t_Q} = \frac{\sqrt{\mu_Q/\rho_Q}}{2t_Q} \qquad (1)$$

where, V_{tr} is the speed of sound, t_Q is thickness of the resonator; μ_Q and ρ_Q are the shear modulus (2.947×10^{11} g cm^{-1}sec^{-2}) and the density (2.648 g cm^{-3}) of the quartz resonator, respectively[2].

The resonance frequency change (Δf) according to the thin film formation on the surface of the quartz resonator is,

$$\frac{\Delta f}{f_0} = -\frac{\Delta t}{t_Q} = -\frac{2f_0}{V_{tr}}\Delta t \qquad (2)$$

where Δt is the thickness of the film formed on the surface of the resonator.

If the film thickness is negligibly small compared to the resonator thickness, the film has the same acoustic property as the quartz resonator. Then the change of resonance frequency due to surface film formation can be written as the following equation, which is called 'the Sauerbrey relationship'[8],

$$\Delta f = -\frac{2f_0^2}{\sqrt{\mu_Q\rho_Q}} \cdot \frac{\Delta m}{A} \qquad (3)$$

where Δm is the total mass added to both faces of the resonator and A is the area of only one face.

For a 5 MHz crystal, a decrease of 1 Hz corresponds to the film deposition of 17.7 ng/cm^2 according to eq.(3). Sauerbrey relation has been verified for the application of film deposition (or removal) up to a mass load of 2% of the mass per unit area of the unloaded quartz resonator[10].

The AT-cut quartz resonator can be modeled mechanically as a body containing mass, compliance, and resistance. Figure1-a) shows the mechanical vibration motion depicting the vibration of the quartz resonator. An electrical network called an 'equivalent electrical circuit' consisting of inductive, capacitive and resistive components can represent this mechanical model. Figure

210

1-b) is an equivalent electrical circuit describing a quartz crystal resonator. The inductance, L_1 is the inertial component related to the mass, capacitance C_1 is the compliance of the resonator representing the energy stored during oscillation, and R_1 is the energy dissipation due to internal friction, mechanical losses in the mounting system and acoustical losses to the surrounding environment [2]. C_0 is the static capacitance of the quartz resonator with the electrodes on both surfaces.

The equivalent electrical circuit has the advantage of expressing the mechanical properties of a quartz resonator during oscillation easily by the use of simple electrical network analysis. Impedance analysis can elucidate the properties of the quartz resonator as well as the interaction of the crystal with the contacting medium. Impedance analysis involves the measurement of current over a specified range of frequencies at a known voltage. The admittance (Y), the reciprocal of impedance, in the equivalent circuit shown in Fig.1 can be expressed as shown in eqn.(4) [11],

$$Y = G + jB = \frac{R_1}{R_1^2 + (2\pi fL_1 - \frac{1}{2\pi fC_1})^2} + j\left(2\pi fC_0 - \frac{(2\pi fL_1 - \frac{1}{2\pi fC_1})}{R_1^2 + (2\pi fL_1 - \frac{1}{2\pi fC_1})^2}\right) \quad (4)$$

where, G and B are conductance and susceptance respectively, f is frequency applied to the equivalent circuit, and j is $\sqrt{-1}$.

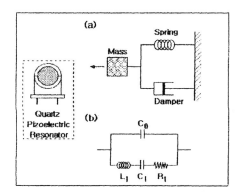

Figure.1. Equivalent circuit for piezoelectric crystal resonator.

Usually, the properties of a quartz crystal resonator with respect to frequency can be discerned from admittance plots, where the abscissa represents the real part of the admittance (conductance, G) and the ordinate the imaginary component (susceptance, B). Resonance occurs at two frequencies where the admittance locus crosses the real axis, f_s and f_p, which are the series and parallel resonance frequencies, respectively. If R_1 is negligible, the series and parallel frequencies at which resonance occurs are given by eqn.(5) [2],

$$f_s = \frac{1}{2\pi\sqrt{L_1C_1}}$$

$$f_p = \frac{1}{2\pi}\sqrt{(L_1C_1)^{-1} + (L_1C_0)^{-1}} \qquad (5)$$

Typically C_0 is much less than C_1, the maximum magnitude of acceptance exists at near f_s. This series resonance frequency is the one generally used in the measurement.

2.2 QCM behaviors in high-density fluid.

The QCM technique was originally developed for the measurement of film deposition on a surface in a vacuum or in air [8]. These days, the area of its application extends not only to the gas phase, but also in the liquid phase (or high density fluid in general). Under the liquid phase surrounding the resonator, the oscillation of the quartz crystal resonator becomes suppressed due to the viscous damping effects from the liquid. From the equivalent circuit, we can explain the resonance frequency change by the effects of both loading mass and viscous damping [2, 12]. First, considering the effect of added mass only upon the surface of QCM, the increase of the inertial mass component can be explained by introducing the inductance, L_f (Figure 2-(a)). If the mass change is negligibly small compared to the mass of the resonator, the series resonance frequency change (Δf) from the mass change on the surface can be given by,

$$\frac{\Delta f}{f_{s,0}} = -\frac{L_f}{2L_1} \qquad (6)$$

where, $f_{s,0}$ is the original series resonance frequency of the bare QCM (eq.(5)).

When a quartz resonator is in contact with a viscous liquid, viscous coupling is operative (Figure 2-(b)). The added effect of viscous liquid modifies the equivalent circuit to include both the mass loading density component (L_L), and a resistive viscosity component (R_L) of the liquid. The increase of the total value of the resistor ($R_1 + R_L$), results in reducing the radii in admittance circles, which affects the stability of oscillation. The series resonance frequency shifts its value when a QCM is immersed in the liquid. The new resonance frequency due to the increased inductance from the viscous liquid is,

$$f_{s,L} = 2\pi\sqrt{(L_1 + L_L)C_1} \cdot \qquad (7)$$

When a mass is added on the surface and the mass is negligibly small, the series resonance frequency change (Δf) due to the mass added on the surface is given by,

$$\frac{\Delta f}{f_{s,L}} = -\frac{L_f}{2(L_1 + L_L)} \qquad (8)$$

A relationship was derived by Kanazawa and Gordon [13], which expressed the change in oscillation frequency of a quartz crystal in contact with a fluid in terms of the material parameters of the fluid and the quartz. They coupled the crystal motion to a shear wave in the fluid, and the relationship of the shift in the series resonance frequency and the material properties of the liquid becomes,

$$\Delta f_L = f_{s,L} - f_{s,0} = -f_{s,0}^{3/2} \cdot \sqrt{\frac{\rho_L \eta_L}{\pi \rho_Q \mu_Q}} \qquad (9)$$

where, ρ_L and η_L are the density and the viscosity of a liquid, respectively.

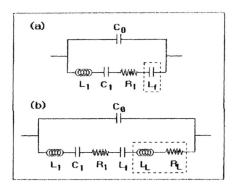

Figure 2. Equivalent circuits for a) added mass (L_f),
b) contact with viscous fluid (L_L, R_L).

Even though there is a big shift in the resonance frequency, the mass sensitivity to thin film is observed to be the same in a fluid as in a vacuum. In addition to the viscous motion, many other factors still affect the resonance frequency of QCM in a high-density fluid. Roughness of the surface is one important factor. Based on the measurement, resonance frequencies turned out to be dependent on the surface roughness [14,15]. The roughness seems to induce a shift of the resonant frequency by both the inertial contribution from the liquid mass rigidly coupled to the surface, and the viscous contribution from the viscous energy dissipation caused by the nonlaminar motion in the liquid [16].

3. Experimental

The quartz crystals used in this study were commercially available 5MHz AT-cut type resonators with gold electrodes on both sides, purchased from International Crystal Manufacturing Co., Oklahoma City, USA. Two different types of surface finish were prepared for experiments (etched-electrode and polished-electrode). Crystals of the etched electrode surface were used mainly for solubilization experiments, while polished-electrode surface crystals were additionally used to ascertain the effect of surface roughness on the resonance frequency change. The circuitry for oscillation of the resonator was made as described by Hwang and Lim [17]. The output frequency was measured by a frequency counter (FLUKE, PM6680B) connected to a PC for recording.

A quartz crystal resonator is located in a high-pressure cell, the volume of which is about 31 ml. To see the effect of ultrasonar on solubilization, we also made a high-pressure cell containing a horn that is connected to a sonar vibrator outside. The frequency of the horn was 20 kHz. The amount of energy dissipation by the horn inside the cell is not clearly known, however, we guess approximately 10 – 20W, which is the energy efficiency of 5-10% of 200W of the total energy consumption of sonar. The volume of the latter cell is about 75ml (Figure 3).

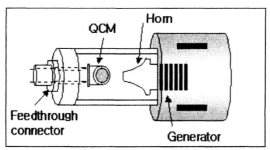

Figure 3. A QCM cell with a horn in it for ultrasonic
wave generation.

Two types of experimental systems – static and dynamic - were devised for solubilization experiments (Fig.4 and Fig.5). An ISCO 260D syringe pump was used to supply CO_2 at fixed pressure ranging from 100 to 250 bar. A computer connected to the RS-232 port of the pump constantly monitored the pressure and flow rate of CO_2 leaving the syringe pump.

In the static system (Figure 4), fixed pressure and temperature were applied during the measurement through to the end of the experiment. The solubilization rate of a target material deposited on the electrode surface was measured by the change of resonance frequency of the quartz crystal. The target material used for the experiment in the static system was $Cu(II)(acac)_2$ (99.99% purity, Aldrich Chemical Co.). This chelate was dissolved in methylene chloride. Drops of this solution were placed on both electrode surfaces of a clean quartz crystal resonator by a micrometer syringe, then fully dried in the oven at 60°C. The resonance frequencies before and after the placement of target materials were monitored in dry air. Typically 30 – 40 $\mu g/cm^2$ of deposition was observed. After the loaded crystal resonator was placed in the pressure cell, CO_2 was supplied to the cell at a given pressure. The pressure cell was immersed in the water bath at a fixed temperature. In the case of solubilization experiments under ultrasonar, a cell equipped with a horn was used (Figure 4). The temperature of the cell was controlled by a water jacket that surrounded the cell. The frequency of the quartz crystal resonator was monitored constantly and recorded to the PC throughout the entire experiment.

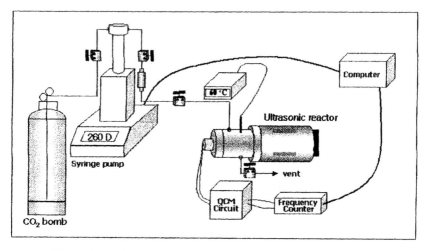

Figure 4. A schematic diagram of the static experimental system.

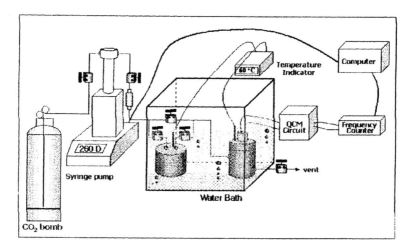

Figure 5. A schematic diagram of the dynamic experimental system.

In the dynamic system shown in Figure 5, CO_2 flows through the system at a fixed flow rate. The system set-up is similar to the static system except for the addition of a mixing reactor cell that mixes the containing additives and CO_2 to a solution. The mixing cell is located between a syringe pump and a solubilization reactor cell. In the dynamic solubilization experiment, Co spikes were used for the target material. $CoCl_2$ (99.99%-purity, Aldrich Chemical Co.) was dissolved in pure water, then placed on both electrode surfaces of the quartz resonator by the same method as was used in the static experiment. After placement of the loaded crystal resonator inside the solubilization cell, both the mixing and solubilization cells were charged with CO_2 by the liquid syringe pump up to a fixed pressure. After stabilization, CO_2 flowed through the mixing reactor cell first, then passed through the solubilization cell. The additives used in the experiment were Cyanex-272 and Cyanex-302. The solubilization of Co spikes in the Cyanex-CO_2 solution was measured by monitoring the frequency change of the quartz resonator. The flow rate of CO_2 is about 5ml(L-CO_2)/min.

4. Results and Discussion

4.1. Bare QCM behaviors in L/SC CO_2

The frequency change of a bare QCM in L/SC CO_2 was measured as a function of pressure. The temperature was fixed at 40°C, and pressures were raised to 19 MPa. The results are shown in Figure 6. The resonance frequency decreases with respect to increasing pressure. Near the critical point, the frequency abruptly changes due mainly to the sudden change of material properties of CO_2 in this region.

These results are very similar to those of Otake, et al.[18]. Otake considered three factors that affect the resonance-frequency change with applied pressure; i) the change of crystal elastic modulus, ii) damping effect of viscous fluid, and iii) adsorbed mass on the electrode surface. Giguard, et al., adapted this explanation for the analysis of solubility as measured by the QCM technique. However, one thing noticeable is the dependence of the resonance frequency on the surface roughness of the resonator. The resonator having a rough surface shows about three times more sensitivity to the pressure change than the one with a polished surface. This roughness dependency cannot be explained by a simple damping model [13] nor adsorption theory [18]. The roughness effect has to be included[19] in addition to the other three factors suggested by Otake for its use in L/SC CO_2. It seems difficult to measure the absolute mass change during the process in a high-density fluid by QCM only. Hence, it is necessary to measure the reference resonance-frequency using a bare QCM during the process to prevent misinterpretation of experimental results.

216

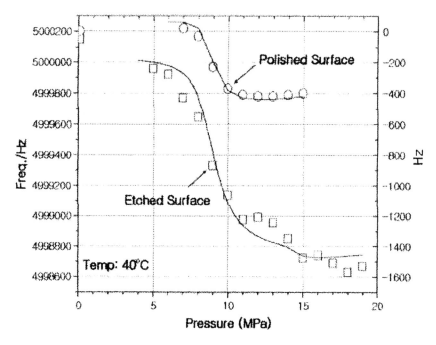

Figure 6. Resonance frequency change of a bare QCM in CO_2
with respect to pressure. Symbols are measured values
and lines are derived from calculation[19].

4.2. Solubilization rate measurement in a static system

Figure 7 shows the results of solubilization experiments in a static system. Each line indicates the resonance frequency change over time under different conditions. There are two types of lines – solid lines and lines with symbols. Solid lines are the results attained in the liquid state of CO_2 (10 MPa, 20°C), while lines with symbols are in the supercritical state (10 MPa, 45°C). Since the resonance frequency of a quartz crystal is directly affected by the properties of the ambient fluid, reference measurements using a bare quartz crystal resonator were performed before the experiment. Noisy signals were observed when CO_2 was purged into the system. The results of the reference measurements indicate that about 70-80 seconds are required until the stabilization of the system. The data recorded during this stabilization period in ignored.

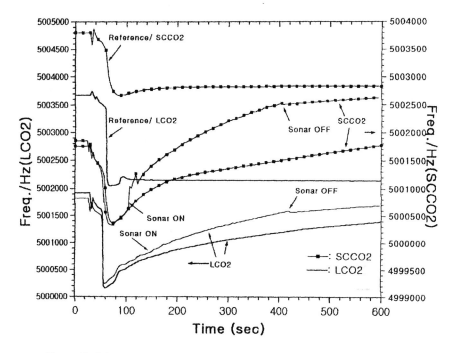

Figure 7. QCM measurements in a static system during the experiment
Solid line: Liquid state, Line with symbols: Supercritical state.

Decrease of the resonance frequency due to dissolution of $Cu(acac)_2$ is noticeable under both the liquid and supercritical CO_2 states. The solubilization rate in the supercritical state looks slightly larger than that of the liquid state. When the sonar is on, i.e., the horn in the cell starts to vibrate, the solubilization rate suddenly increases. Under the ultrasonic pulses, the rapid dissolution of $Cu(acac)_2$ into ambient CO_2 fluid is clearly visible. Once the sonar is off, the dissolution rate returns to normal value.

QCM measurements directly give in-situ kinetic information of reactions inside the cell. Figure 8 shows the tight correlation of dissolution with the square root of time. The resonance frequency change was converted to that of surface mass using eq.(3). This linear dependency indicates a strong possibility of diffusion-controlled dissolution. The rate-limiting step, in this case, seems to be diffusion of $Cu(acac)_2$ to the ambient fluid.

Assuming the 'one-dimensional diffusion' theory is applicable, the released mass per unit area in one direction up to time 't', M(t) can be written in the following form, if the solute surface concentration is constant [20].

$$M(t) = \sqrt{\frac{2D}{\pi}} \Phi_0 \sqrt{t} \qquad (9)$$

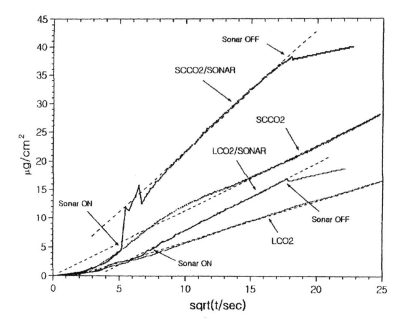

Figure 8. Dissolution of Cu(acac)₂ as a function of the square root of time at each condition.

where, D is the diffusivity of the solute in the solution, Φ_0 is the concentration of the solute at the surface.

We can estimate the diffusivity of Cu(acac)₂ in CO_2 from the slope in the released-mass line with respect to $\sqrt{t}$, if Φ_0 is known. In this experiment, Φ_0 is the equilibrium solubility of Cu(acac)₂ in CO_2 at the given condition. Table 1 summarizes the obtained diffusivity and data used for estimation. The solubility of Cu(acac)₂ in liquid CO_2 was calculated by the equation of the state suggested by Cross, et al.[21]. The diffusivities obtained in the liquid and the supercritical states are in the range of the value of the self-diffusion coefficient of CO_2 [1,22,23].

Table 1. Diffusivity of Cu(acac)₂ in CO_2 and data used for the estimation.

Experiment condition	CO_2 Density (g/cm³)	Φ_0 (mol/mol-CO_2)	Slope ($\mu g/cm^2 \cdot \sqrt{sec}$)	Diffusivity (cm²/sec)
SC CO_2 (10MPa, 45°C)	0.50	3.8×10^{-6} [6]	1.1	3.7×10^{-3}
L CO_2 (10MPa, 20°C)	0.86	6.0×10^{-6} [21]	0.71	2.1×10^{-4}

Ultrasonic wave clearly enhances the solubilization process (Figure 7, 8). Figure 8 also indicates that diffusion-controlled dissolution maintained even under ultrasonic pulses. The diffusivity under ultrasonic pulses was measured to increase about 3-4 times at the given geometry and power used in this experiment. There are two known effects of ultrasonic waves on solubilization; i) acoustic streaming, and ii) cavitation/implosion[24,25].

Cavitation/implosion is the most effective mechanism known in ultrasonic cleaning. Cavity formation needs a negative pressure that is generated in the incompressible fluid. The isothermal compressibility of CO_2 in the conditions of this experiment is about 2-4 orders higher than that of water at room temperature (Table 2). In other words, CO_2 at the current experimental conditions seemed to be no longer an incompressible fluid nor did it induce cavitation/implosion.

Table 2. Compressibility and surface tension in water and L/SC CO_2

Properties	Water at 20°C	L CO_2	SC CO_2
Isothermal compressibility(MPa^{-1})	4.58 x 10^{-4}	1.4 x 10^{-2} (20°C, 10MPa)	0.3 (45°C, 10MPa)
Surface Tension (dyne/cm)	72	0.05 (30°C, 7.2MPa)	-

Based on the observation through a view cell, we found a strong streaming from the vibrating horn, however, we could not find any cavity nor any bubble in the fluid. The surface tension of CO_2 is very small compared to water, and this good 'wetting' property also prevents nuclide formation for cavities at the surface of structures inside the cell. Based on visual observation and the material properties of CO_2, we conclude that the enhancing effect of ultrasonic waves on solubilization of Cu(acac)$_2$ in L/SC CO_2 was mainly from the acoustic streaming, not from cavitation/implosion, and that cavitation may rarely occur in L/SC CO_2 compared to that in water.

4.2. Solubilization rate measurement in a dynamic system

Solubilization of Co ion into the solution of Cyanex-CO_2 was studied by QCM technique. Co ion spikes were deposited on the both surfaces of a clean crystal resonator, then this resonator with Co ion spikes was used for the measurement. The results are shown in Figure 9. To get the reference in the measurement, we put a bare crystal resonator in the cell and let L/SC CO_2 flow to the system after compression. The reference frequency did not change during the mock-up process. And, the same procedure was applied for a resonator having Co ion spikes on the surface. Co spikes did not fall out of the surface under the constant flow of CO_2.

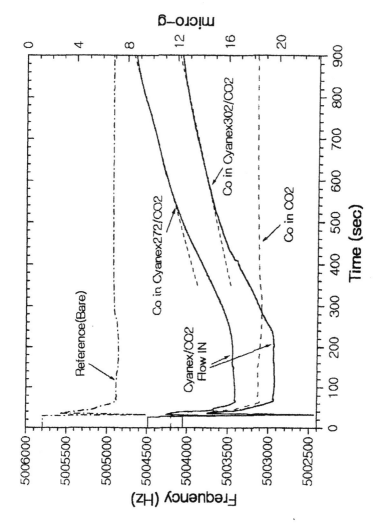

Figure 9. Frequency change of a quartz crystal resonator in a dynamic process.

When a cyanex-CO_2 solution flowed around the resonator, Co ions started to dissolve into the solution. The dissolution rate reaches a constant value after a stabilization period. We took these slopes to be equilibriums. The solubility of Co ion in Cyanex in the solution can be obtained from the slope of the solubilization line. The solubility of Co ion in Cyanex-272 and that in Cyanex-302 are estimated and shown in Table 2 with the data used for calculation. The solubility of Co is expressed as the extracted mass of Co per the amount of Cyanex used in the fluid. Real solubility can be higher, considering the possibility of incomplete .mixing of Cyanex in CO_2 in a dynamic system. As shown in Table 2, very fine and minute solubility measurements are possible with the QCM technique, due to the high sensitivity of QCM on mass change.

Table 2. Solubility of Co ion in cyanex, and data for calculation.

Type	Flow rate (CO_2)	Solubility of cyanex	Slope (ng-Co/sec)	Solubility $\mu g(Co)/g(cyan)$
Cyanex 272	5ml(LCO_2)/min	0.022 g/g(CO_2)	9.3	4.4
Cyanex 302	or, 98mg(CO_2)/sec	[26]	7.4	3.5

5. Conclusion

QCM technique was applied to the solubilization study in CO_2. Based on the experiments, the following conclusions were obtained.

1. The roughness effect was observed in addition to viscous damping in high-density fluid. Absolute mass change under the process in L/SC CO_2 may not be obtainable. So, reference measurement using a bare QCM is recommended.

2. QCM technique can supply in-situ kinetics of the reaction inside the cell. Solubilization of Cu(acac)$_2$ in L/SC CO_2 was diffusion-controlled. Diffusivities of Cu(acac)$_2$ in L/SC CO_2 were also obtained. Ultrasonic waves enhance dissolution of Cu(acac)$_2$ in CO_2 mainly due to acoustic streaming. Cavitation/implosion in L/SC CO_2 does not occur as often as in water.

3. QCM technique enables low solubility measurements to be made. Co ion solubility in the cyanex-CO_2 solution was measured in a dynamic system. Solubility as small as 3-4 $\mu g(Co)/g(cyan)$ was measured.

Reference

1. M. A. McHugh and V. J. Krukonis, 'Supercritical Fluid Extraction – Principles and Practice, Butterworth-Heinemann, **1994**
2. D.A.Buttry and M.D.Ward, *Chem. Rev.* **1992**, 92 1355

222

3. Ya-Ping Sun and H.W. Rollins, *Chemical Physics Letters,* **1998**, 288, 585
4. J.H.Aubert, *J. Supercritical Fluid,* **1998**, 11, 163
5. K.Miura, K.Otake, S.Kurosawa, T.Sako, T.Sugeta, T.Nakane, M.Sato, T.Tsuji, T.Hiaki, M.Hongo, *Fluid Phase Equil.,* **1998**, 144, 181
6. S.E.Guigard, G.L.Hayward, R.G.Zytner, W.H.Stiver, *Fluid Phase Equil.,* **2001**, 187-188, 233
7. C.K.O'Sullivan and G.G.Guilbault, *Biosensors & Bioelectronics,* **1999**, 14, 663
8. Sauerbrey, *G.Z.Phys.,* **1959**, 155, 206
9. W.H.King, Jr., *Anal. Chem.* **1964**, 36, 1735
10. C.S.Lu and A.W.Czanderna, 'Applications of Piezoelectric Quartz Crystal Microbalnces, Elsevier, Amsterdam, **1984**
11. V.E.Bottom, 'Introduction to Quartz Crystal Unit Design, Van Nostrand Reinhold, New York, **1982**
12. A.L.Kipling and M.Thompson, *Anal. Chem.* **1990**, 62, 1514
13. K.K.Kanazawa and J.G.Gordon II, *Analytica Chimica Acta,* **1985**, 175 99
14. S.J.Martin, G.C.Frye, and A.J.Ricco, *Anal. Chem.* **1993**, 65, 2910
15. S.Bruckenstein, A.Fensore and Z.Li, *J. Electroanalytical Chem.* **1994**, 370, 189
16. M.Urbakh and L Daikhin, *Colloids and Surfaces A: Physicochemical and Engineering Aspects,* **1998**, 134, 75
17. E.Hwang and Y.Lim, *Bull. Korean Chem. Soc.* **1996**, 17, 39
18. K.Otake, S.Kurosawa, T.Sako, T.Sugeta, M.Hongo, and M.Sato, *J. Supercritical Fluid,* **1994**, 7, 289
19. K.Park, H.Kim, H.Kim, M.Koh, and C.Yoon, 'The behavior of QCM in high pressure CO_2', to be published.
20. H.S.Carslaw and J.C.Jaeger, 'Conduction of Heat in Solids', Oxford, **1959**
21. W.Cross, Jr., A.Akgerman, and Can Erkey, *Ind. Eng. Chem. Res.,* **1996**, 35, 1765
22. H.Fu, L.A.F.Coelho, M.A.Matthews, *J. Supercritical Fluid,* **2000**, 18, 141
23. 'Superciritical Fluid Cleaning – Fundamentals, Technology and Applications', edited by J.McHardy and S.P.Sawan, Noyes Publication, **1998**
24. O.V. Abramov, 'Ultrasound in liquid and solid metals', CRC Press, **1994**
25. S.Awad, 'Ultrasonic cavitations and precision cleaning', Precision Cleaning Magizine, November, **1996**
26. R.D.McMurtrey, 'SCF Extraction of heavy metals from solid matrices using a dynamically added ligand', M.S. Thesis, University of Idaho, **1997**

Chapter 15

Uranyl Extraction by TBP from a Nitric Aqueous Solution to SC-CO$_2$: Molecular Dynamics Simulations of Phase Demixing and Interfacial Systems

Rachel Schurhammer and Georges Wipff[*]

Institut de Chimie, 4 rue B. Pascal, 67 000 Strasbourg, France
*Corresponding author: email: wipff@chimie.u-strasbg.fr

In relation with the liquid-liquid extraction of uranyl nitrate from an acidic aqueous phase to supercritical CO$_2$, we present a series of molecular dynamics (MD) simulations on the "interfacial" systems involving UO$_2$(NO$_3$)$_2$ species and high concentrations of TBP and nitric acid. We compare the distribution of solvent and solutes at the interface which forms upon the demixing of "chaotic mixtures" of water / CO$_2$ solutions. The simulations highlight the importance of interfacial phenomena in uranyl extraction to CO$_2$. In most cases, demixing leads to separation of aqueous and CO$_2$ phases which form an interface. At low concentrations, TBP and the neutral form HNO$_3$ of the acid adsorb at the interface, while the uranyl salt and ionic species sit in water. Spontaneous complexation of uranyl salts by TBP is observed, leading to UO$_2$(NO$_3$)$_2$(TBP)(H$_2$O) and UO$_2$(NO$_3$)$_2$(TBP)$_2$ species of 1:1 and 1:2 stoichiometry, respectively, which adsorb at the interface. As the TBP concentration is increased, the proportion of 1:2 species, more hydrophobic than the 1:1 species, increases, following the Le Chatelier principle. Nitric acid competes with the uranyl complexation by TBP which forms hydrogen bonds with H$_3$O$^+$ or HNO$_3$ species. Thus, at high acid:TBP ratio, the concentration of 1:1 and 1:2 complexes decreases. Also noteworthy is the evolution of the interface from a well-defined border at low acid and TBP concentrations, to a mixed microscopic "third phase" containing some 1:2 complexes which can be considered as "extracted". We believe that such heterogeneous microphase is important for the stabilization and extraction of uranyl complexes by TBP and, more generally, in the extraction of highly hydrophilic cations (e.g. lanthanides or actinides) to organic media.

INTRODUCTION

Supercritical CO_2 "SC-CO_2" can be used as a promising ecological alternative in liquid-liquid extraction systems.(*1-6*) This is of particular interest in the context of nuclear waste partitioning, which is generally initiated from aqueous solutions of the metal ions, obtained by dissolution of the irradiated material in concentrated nitric acid solutions. Examples of metal extraction to SC-CO_2 from solid or liquid matrices involve extraction of metallic, lanthanide and actinide cations by β-diketonate ligands,(*7-9*) of strontium by crown ethers (*8*) or of UO_2^{2+}, Th^{4+}, (*10-15*) lanthanides (*16, 17*) or heavy metals (*18*) by organophosphorus ligands. Reviews can be found in ref. (*2, 19*). Recently, a new method for dissolving solid uranyl dioxide in SC-CO_2 with the CO_2-philic TBP·HNO_3 complexant without requiring dissolution by acid in water has been reported.(*11*)

In this paper we focus on the uranyl extraction by TBP (tri-*n*-butylphosphate) in acidic conditions, with the aim to investigate the interfacial behavior of the partners involved in the extraction process to SC-CO_2. As metallic ions are quasi-insoluble in SC-CO_2 while uncomplexed ligands used in liquid-liquid extraction are not soluble in water, it is stressed that the border region ("interface") between the two liquids should play a key role in the transfer of the complexed ions to the CO_2 phase. TBP is used in the industrial PUREX process for nuclear waste processing to extract uranyl as $UO_2(NO_3)_2(TBP)_2$ complexes (*20-24*) and also extracts the uranyl cation from acidic aqueous solutions to SC-CO_2. Owing its amphiphilic character, TBP is also surface active and should thus concentrate at the surface of water. Similar features are anticipated for the neutral form HNO_3 of nitric acid, which, according to MD simulations (*25, 26*) and surface spectroscopy studies,(*27*) is also surface active and accumulates at the water/"oil" interfaces, which display marked analogies with the water/air interface.(*28*) On the other hand, hydrophilic metallic ions are "repelled" by the interface, preventing their capture by water insoluble extractants.(*29*) It is thus important to understand at the molecular level what happens at the liquid boundaries (*30*) and we report new microscopic insights into this question obtained from molecular dynamics (MD) simulations with explicit representation of the solvents and solutes.

In our first investigations of water "/oil" extraction systems, the "oil" phase was modeled by chloroform, focusing on the distribution of ligands (e.g. calixarenes, crown ethers, cryptands, podants, CMPO, TBP) and their complexes, as well as all kinds of ionic species.(*31-33*) Recently, we reported the first MD investigations on water / SC-CO_2 interfaces, either neat, or in the presence of salts or extractants molecules.(*26, 28, 34*) These studies considered pH neutral systems. Here, we focus on uranyl extraction in acidic conditions. The simulated systems, noted **A** to **J**, are described in Table 1 and Figure 1. The water / CO_2 interfacial system **A** (no solute) and the acidic systems **B** and **C** (corresponding to about 1 and

2 mol.L^{-1}) have been previously reported (*28, 34*) and are briefly described as references for more complex and concentrated systems. The systems **F-J** contain 5 or 6 $UO_2(NO_3)_2$ uranyl salts (about 0.05 mol l^{-1}), and increasing amounts of TBP (from 30 to 120 molecules) and of nitric acid in the simulation box. The 60 TBP solutions **E**, **H** and **I** correspond roughly to the concentration used experimentally in the PUREX process. One important issue in modeling studies concerns the representation of acidity, as force field methods cannot properly account for proton transfer processes, and require an *a priori* choice of the protonated / deprotonated species. As this is not known from experiment for heterogeneous systems, we decided to model nitric acid as $(HNO_3, H_3O^+ NO_3^-)_m$, i.e. an equimolar mixture of neutral and ionic forms, which represent the dominant contributions in organic and aqueous phases, respectively. Thus, $m = 18$ in systems **F** and **H** (corresponding to about 1M aqueous solutions) and $m = 36$ in systems **G** and **J**.

The MD simulations **D** - **J** focus on the demixing of completely mixed water/ CO_2/TBP/acid systems, which are "chaotic arrangements", also prepared by MD simulations (see methods). These systems, "perfectly mixed" at the microscopic level (probably more than they are in reality) are highly unstable. We want to investigate how they spontaneously evolve and relax, and in particular to which extent the aqueous and CO_2 phases will separate, and how the acid and uranyl nitrate species will distribute once the equilibrium is reached. One critical issue, in relation with assisted extraction, is whether and under which conditions uranyl will be complexed by TBP and finally extracted to CO_2. The pH-neutral **D** and **E** systems, described in more details in ref. (*28*) are presented here for a purpose of comparison with the acidic ones **F** - **J**.

METHODS

The simulations were performed with the modified AMBER5.0 software (*35*) where the potential energy is described by a sum of bond, angle and dihedral deformation energies, and pairwise additive 1-6-12 (electrostatic + van der Waals) interactions between non-bonded atoms.

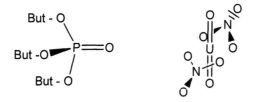

Figure 1. The simulated species: TBP and the $UO_2(NO_3)_2$ salt.

$$U = \sum_{bonds} K_r \left(r - r_{eq} \right)^2 + \sum_{angles} K_\theta \left(\theta - \theta_{eq} \right)^2 + \sum_{dihedrals} V_n \left(1 + \cos n\phi \right)$$

$$+ \sum_{i<j} \left(\frac{q_{ij}}{R_{ij}} - 2\varepsilon_{ij} \left(\frac{R^*_{ij}}{R_{ij}} \right)^6 + \varepsilon_{ij} \left(\frac{R^*_{ij}}{R_{ij}} \right)^{12} \right)$$

The $UO_2{}^{2+}$ parameters are from Guilbaud and Wipff. (36) Water was represented with the TIP3P model. (37) For SC-CO_2, we used the parameters of Murthy et al. (38) : charges $q_C = 0.596$, $q_O = -0.298$ e and van der Waals parameters $R^*_O = 1.692$, $R^*_C = 1.563$ Å and $\varepsilon_O = 0.165$, $\varepsilon_C = 0.058$ kcal/mol. All O-H, O-H bonds and the C=O bonds of CO_2 were constrained with SHAKE, using a time step of 2 fs. As in ref. (28), the $UO_2(NO_3)_2$ salts were constrained to remain bound and neutral in order to allow for their possible extraction to an hydrophobic medium.

The temperature was monitored by separately coupling the water, CO_2 and solutes subsystems to a thermal bath at the reference temperature (350 K) with a relaxation time of 0.2 ps for the solvents and 0.5 ps for the solutes. Non-bonded interactions were calculated with a residue-based twin cutoff of 12/15 Å for all systems, excepted for **D** and **E** for which we used a 13 Å cutoff with a reaction field correction for the electrostatic interactions.

Table 1

Systems [a]	Box size $(Å^3)$ [b]	$N_{CO2} + N_{water}$	Time (ns)	[TBP] / [acid] $(mol.L^{-1})$ [c]
A Free interface	28×28×(28+28)	241+726	1.0	-
B 18 (HNO_3, H_3O^+, NO_3^-)	43×36×(23+38)	409+1802	1.0	-/1.0
C 36 (HNO_3, H_3O^+, NO_3^-)	44×56×(31+31)	813+2450	1.0	-/1.6
D 5 $UO_2(NO_3)_2$ + 30 TBP	40×40×(43+35)	595+1832	3.0	0.72/-
E 5 $UO_2(NO_3)_2$ + 60 TBP	42×39×(43+45)	619+1992	5.6	1.41/-
F 6 $UO_2(NO_3)_2$ + 30 TBP + 18 acid	42×42×(47+47)	750+2441	4.0	0.60/0.7
G 6 $UO_2(NO_3)_2$ + 30 TBP + 36 acid	42×41×(57+52)	905+2621	4.0	0.51/0.7
H 6 $UO_2(NO_3)_2$ + 60 TBP + 18 acid	43×41×(63+56)	957+2980	4.0	0.89/1.2
I 6 $UO_2(NO_3)_2$ + 60 TBP + 36 acid	43×41×(63+56)	957+2850	5.0	0.89/1.2
J 6 $UO_2(NO_3)_2$ + 120 TBP + 36 acid	44×43×(69+56)	950+2800	7.3	1.52/1.1

[a.] $acid = HNO_3 + NO_3^- + H_3O^+$
[b.] $x \times y \times (z_{CO2} + z_{water})$
[c.] Concentration of TBP in the organic phase and of nitric acid in the aqueous phase.

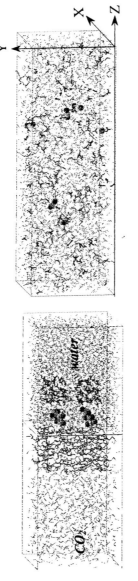

Figure 2. The simulation box, with the starting arrangement of system **I** (6 $UO_2(NO_3)_2$ + 60 TBP + 36 (HNO_3, NO_3^- H_3O^+) species (left). Snapshot of the mixed system (right).

The CO_2 / water interface has been built as indicated in ref. (*39*) starting with adjacent boxes of CO_2 and pure water (Figure 2). The corresponding densities are 0.80 and 1.0 respectively. All systems were represented with 3D periodic boundary conditions, thus starting with alternating slabs of water and CO_2 separated by one CO_2 / water interface.

The solutes were initially placed at the interface, as shown in Figure 2. After energy minimization, MD was run at 350K at constant pressure for 50 ps. This was followed by a mixing step of 1 to 1.5 ns, during which the system was heated at 700 K and the electrostatic interactions were scaled down by a factor of 100 in order to enhance the mixing of hydrophobic and hydrophilic species. This lead to "chaotic mixtures" of water, CO_2 and of solutes, as shown in Figure 2. The demixing simulation was then initiated be resetting the temperature to 350 K and the dielectric constant of the medium to 1.0.

The results have been analyzed as described in ref. (*40*). The interface position is instantaneously defined by the intersection of the solvent density curves. The percentage of species "near the interface" was calculated from the average number of species, which sit within a distance of 7 Å from the interface, which corresponds to about half of the interfacial width.

RESULTS

In most cases, the "demixing" process was rapid, but its rate and completeness diminished with the complexity and concentration of the solutes. For instance, the acidic system **C** (no TBP) evolved to two well separated phases in less than 0.5 ns while, for systems like **I** or **J**, no full phase separation was achieved after 5 ns. Anyway these events remain extremely fast at the extraction experiment time scale, and we thus mainly focus on the systems at the end of the simulation. More detailed analysis can be found in ref. (*41*). The main features are illustrated by final snapshots, while density curves of the different species provide a statistical description during the last 0.2 ns.

Comparison of the pure neutral water / CO_2 system to the acidic one to which uranyl salts and increasing amounts of TBP and acids have been added reveals important evolutions which are of particular significance in the context of assisted extraction of uranyl.

1- The pure water / CO_2 system in pH neutral *versus* nitric acidic conditions (A - C)

The systems **A** - **C** have been simulated with "standard" MD simulations, starting from a prebuilt interface (i.e. without mixing / demixing procedures). As shown in Figure 3, the pH-neutral neat water / CO_2 system **A** consists of two well defined phases, separated by an interface of about 12 Å thick. A few CO_2 molecules diffused to water, while the CO_2 phase remains dry, in agreement with the corresponding low miscibility of these liquids.

When nitric acid is added as $(HNO_3, H_3O^+ NO_3^-)_m$, to form systems **B** ($m = 18$) or **C** ($m = 36$), the two phases remain well separated and no acid is found in the CO_2 phase. The acid mainly dilutes in water, as expected, but neutral HNO_3 molecules concentrate at the interface, where their concentration increases with the total nitric acid concentration (from 26 % in **B** to 40 % in **C**). This contrasts with the ionic NO_3^- and H_3O^+ components, which are not surface active (only 6 % of them sit within 7 Å from the interface).

2- Demixing simulations of pH-neutral "chaotic mixtures" of the 5 $UO_2(NO_3)_2 / n$ TBP / SC-CO_2 / water (systems D and E)

During the MD simulations of "chaotic mixtures" of water, CO_2, TBP, $UO_2(NO_3)_2$ and 30 TBP (**D**) or 60 TBP (**E**), demixing occurred, leading to separated aqueous and CO_2 phases and (in relation to the imposed 3D periodicity) to two interfaces (Figure 4). In both systems, the bulk aqueous phase finally contains some CO_2 molecules, but neither TBP nor uranyl species.

Three remarkable features appear. First, *TBP's mostly concentrate at the interfaces*, while some fraction dilutes in CO_2. At the lowest concentration (system **D**), only about 10% of the TBPs are in CO_2, while at higher concentration (system **E**), about 50% are at the interface and the other 50% in CO_2. Most of them are hydrogen bonded via the phosphoryl oxygen to one or two H_2O molecules. One can also notice that the interface becomes more "rough" and perturbed at higher concentration of TBP and that, in no case, it is fully covered by the TBPs.

The second remarkable feature concerns the *spontaneous complexation of uranyl ions by TBP at the demixing stage, forming 1:1 or 1:2 complexes*, namely $UO_2(NO_3)_2(TBP)(H_2O)$ and $UO_2(NO_3)_2(TBP)_2$ species in which UO_2^{2+} is hexacoordinated in its equatorial plane by oxygen atoms (Figure 5). The 1:1 complexes involve one H_2O molecule and are therefore more hydrophilic than the 1:2 complexes which correspond to the stoichiometry of the extracted system. At the end of the simulation, all complexes are 1:2 in **E**, while in **D**, two are 1:1 are two are 1:2 (one remains uncomplexed), thus following trends expected from Le Chatelier principle.

Third, as a result of complexation by TBP, *all uranyl complexes sit at the interface*, presumably due to two opposite features: the lipophilic character of TBP and the attraction by interfacial water molecules. Thus, the 1:1 complexes sit somewhat deeper in water than the 1:2 ones that generally sit more on the CO_2 side. In the absence of TBP, the uranyl salts completely diluted in bulk water, forming $UO_2(NO_3)_2(H_2O)_2$ species which are too hydrophilic to approach the interfacial region, thus preventing their extraction.(*29*)

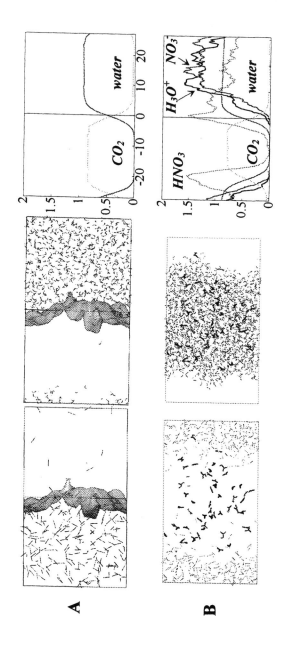

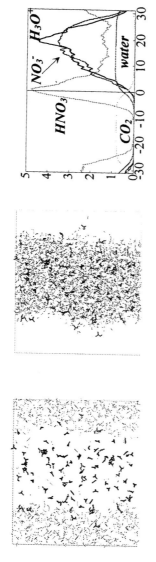

Figure 3. The neat water / CO_2 interface without nitric acid (A) and with 18 (HNO$_3$, NO$_3^-$, H$_3$O$^+$) (B) or 36 (HNO$_3$, NO$_3^-$, H$_3$O$^+$) species (C) after 1 ns of MD simulation. Final snapshots (with water and CO$_2$ solvents shown side by side, instead of superposed, for clarity) and average density curves (right; averages during the last 0.2 ns). The surface of water at the interface is shown for the A system.

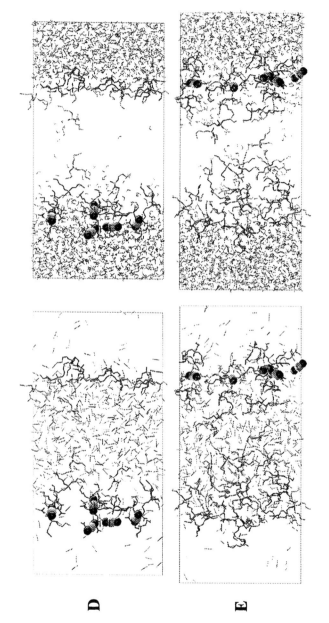

*Figure 4. Demixing of the pH-neutral water / CO₂ mixtures containing 5 UO₂(NO₃)₂ and 30 TBP (system **D**) or 60 TBP (system **E**) at the end of the demixing simulations. Water (right) and CO₂ (left) solvents shown side by side, instead of superposed, for clarity.*

*Figure 5. $UO_2(NO_3)_2 \cdot TBP \cdot H_2O$ (left) and $UO_2(NO_3)_2 (TBP)_2$ (right) complexes formed at the water / CO_2 interface during the demixing simulations of systems **D** or **E**.*

3- Demixing simulations of acidic "chaotic mixtures" of 6 $UO_2(NO_3)_2$ / n TBP / m (HNO_3, NO_3^-, H_3O^+) / SC-CO_2 / water containing systems (F-J).

In this section, we consider systems to which nitric acid has been added, at different TBP concentrations. They all contain 6 $UO_2(NO_3)_2$ salt molecules. Final snapshots are given in Figure 6 and density curves in Figure 7.

System **F** is similar to **D**, i.e. contains 30 TBPs, and additional 18 (HNO_3, NO_3^-, H_3O^+) species. It also demixes and, at 2.5 ns, the aqueous and CO_2 phases are well separated. As in the above systems, most TBPs and HNO_3 neutral species adsorb and spread onto these two interfaces, while five of the six $UO_2(NO_3)_2$ salts are complexed by one or two TBPs and adsorb at the interface. Most ionic NO_3^- and H_3O^+ species are immersed in water. From 2.5 ns until the end of the simulation (4 ns), the system **F** fluctuates without major reorganization. Finally, the two interfaces are somewhat different, with different distributions of the adsorbed species. The "left interface" (Figure 6) is more flat and less perturbed than the "right one", in relation with the lower concentration of adsorbed species. The most perturbed interface contains four of the five uranyl complexes (three are 1:1 and one is 1:2), and a large concentration of TBP and HNO_3 species (see density curves). Many of them form hydrogen bonded TBP-HNO_3 supermolecules (Figure 8), three of which moved to the CO_2 side of the interface. Thus, increasing the acidity with 30 TBPs does not promote the extraction of uranyl to the CO_2 phase, but induces some water/TBP/HNO_3/CO_2 mixing at the border between the aqueous and CO_2 phases. Such mixing should thus decrease the interfacial pressure, thus facilitating the extraction of the complexes to the CO_2 phase.

Trends observed in system **F** are confirmed when the acid concentration is doubled (system **G**). At the end of the demixing simulation, one finds a well defined interface ("left" in Figure 6) without complexes and little TBP or acid, while the right "interface" is broader, due to significant mixing of water/CO_2/TBP/acid, which may be viewed as a microscopic *"third phase"*. As far as uranyl extraction is concerned, it appears that the latter is less favored in **G** than in systems **C** - **F**. Indeed, two uncomplexed uranyl salts finally remain in bulk water, as do two 1:1 complexes with TBP, while no 1:2 complexes formed. The main reason can be found in the competition of acid *vs* uranyl interactions with TBP. Thus, about 1/3 of the TBPs are hydrogen bonded with H_3O^+ (Figure 8), thus preventing their complexation to uranyl. As a result, the number of neutral TBP-HNO_3 dimers also drops (to about 2 on the average), compared to system **F**. Another difference, compared to neutral system **D** is the increased water content of the CO_2 phase: the latter is no longer dry, but contains some hydrogen bonded water molecules.

System **H** contains 60 TBP and 18 (HNO_3, NO_3^-, H_3O^+) species, and thus a larger TBP / acid ratio (about two), compared to **F** or **G**. After 4 ns, the resulting demixed state is similar, as far as the aqueous phase is concerned: it contains mainly the ionic components of the acid, plus one uncomplexed uranyl salt, but no TBP. Important differences appear, however, as far as the uranyl extraction is concerned (Figure 6). Indeed, the CO_2 phase contains more TBP (about 20 %) than in systems **F** or **G**, and most importantly, two 1:2 $UO_2(NO_3)_2(TBP)_2$ complexes, which sit beyond the cutoff distance from the interface, and thus *may be considered as "extracted"*. Another clear evolution, concerns the *disruption of the water / CO_2 interfaces*, which are less well defined than at lower TBP concentrations. Interfacial TBPs thus adopt "random" instead of "amphiphilic" orientations. This may be related to the formation of TBP-HNO_3 dimers near the interface, thus reducing the affinity of the polar head of TBP for water. About 1/3 of the HNO_3 species are thus co-extracted to the CO_2 phase by TBP. Increasing the acid concentration with 60 TBP's (i.e. from system **H** to **I**) leads to similar features, as far as the extent of phase separation is concerned, but one finally finds two water slabs and two CO_2 slabs in the simulation box (Figure 6), thus leading to four interfaces instead of two in the above systems. This arrangement remained when the MD was pushed further for one ns (i.e. 5 ns for **I**), presumably because the cohesive solvent - solvent forces are not strong enough, and the cutoff distance is smaller than the width of these slabs. As the interfacial area is nearly doubled, compared to systems **E** - **H**, most TBPs concentrate at the interfaces. As far as uranyl complexation is concerned, one finds four 1:2 $UO_2(NO_3)_2(TBP)_2$ complexes at the interface, one 1:1 complex on the waterside of the interface, and one uncomplexed uranyl salt. Thus, like at lower concentration of TBP or acid, no complex is extracted, which hints at the *importance of saturation of the interface by extractant and acid molecules to promote the extraction process*. Again, about eight TBP·HNO_3 dimers formed, half of which sitting near the interface, and the other half in CO_2. Some TBP complexes with H_3O^+ of up to 3:1 stoichiometry $(TBP)_3·H_3O^+$ also sit in the interfacial region, thus attracting some NO_3^- anions from the aqueous phase to the interface.

In order to investigate the effect of box shape on the outcome of the demixing process, this system **I** was simulated further, by stepwise modifying the box dimensions, in order to make it less elongated at constant volume (Figure 9). After 7.5 ns of dynamics, the final box was nearly "cubic", and the solvents rearranged to form two slabs only and two interfaces. This arrangement remained for two more ns. As the final interfacial area was close to the initial one, not surprisingly, the distribution of the solutes was similar, i.e. most of them (HNO_3 and four 1:2 uranyl complexes) adsorbed at the interface, while two 1:1 complexes were on the water side of the interface. Thus, no complex was extracted to CO_2.

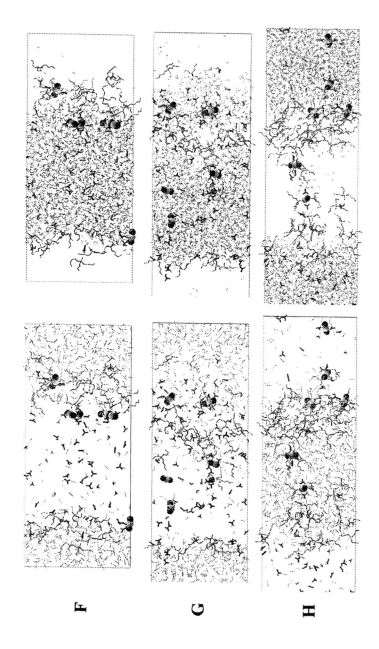

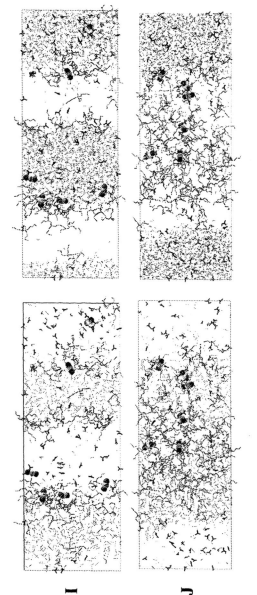

Figure 6. Demixing of the acidic water / CO_2 mixtures containing 6 $UO_2(NO_3)_2$, TBP and nitric acid (system F to J). Snapshots at the end of the demixing simulations. The water (right) and CO_2 (left) solvents are shown side by side, instead of superposed, for clarity.

238

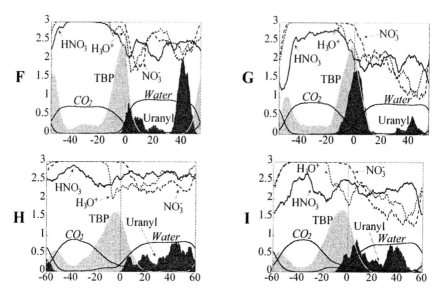

*Figure 7. Demixing of the acidic water / CO₂ mixtures (systems **F** to **I**). Averages density curves during the last 0.4 ns.*

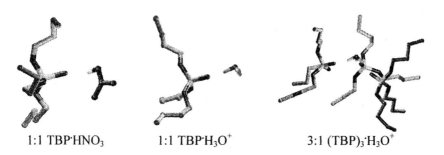

1:1 TBP·HNO₃   1:1 TBP·H₃O⁺   3:1 (TBP)₃·H₃O⁺

Figure 8. Typical TBP-acid complexes found in the CO_2 phase at the end of demixing simulations.

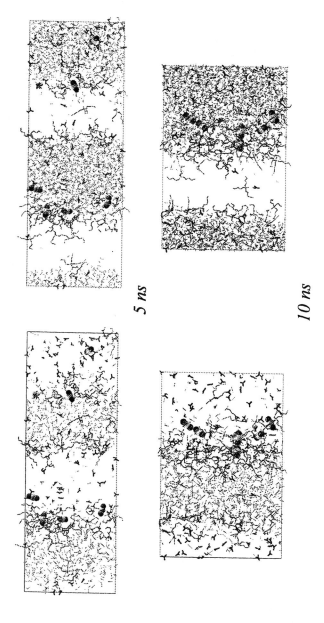

5 ns

10 ns

Figure 9. Evolution of system I as the box z-dimension is reduced (total volume is conserved).

The most concentrated system we simulated is **J**, which contains 120 TBPs and 1.1 M acid, i.e. 36 (HNO_3, H_3O^+ NO_3^-) species. Thus, the TBP concentration reaches about 30% of the total volume. During the demixing simulation, the separation of the aqueous phase was very slow (about 5 ns) and until the end of the simulation (7.3 ns), it contained only NO_3^- and H_3O^+ ions. Yet, the final picture is quite different from the above ones as far as the CO_2 component, the interfaces, and uranyl extraction are concerned. The organic phase now consists of a binary mixture of TBP and CO_2 and surprisingly, the *TBP concentration at the water boundaries is quite lower (< 15%) than in other simulated systems.* More than 85% of the TBPs are thus in the supercritical phase, where about 30% of them are hydrated. The organic phase now contains neutral and ionic forms of the acid, strongly interacting with TBPs. About 50% of H_3O^+ ions sit in the organic phase and form $(TBP)_3H_3O^+$ complexes (Figure 8), without dragging nitrate counterions. Also about 40% of HNO_3 acid species form 1:1 adducts with TBP. Thus, the organic phase is strongly mixed, with a high ion content. *Of particular significance is that it also contains five of the six uranyl complexes, forming 1:2 $UO_2(NO_3)_2(TBP)_2$ species which can thus be considered as extracted.* A sixth 1:2 complex sits near the interface, i.e. on the pathway for extraction. Thus, system **J** is the most "efficient one", as far as uranyl extraction is concerned. When compared to other systems in which TBP and acid are less concentrated, it allows one to better understand at the microscopic level the requirements for assisted extraction to SC-CO_2.

DISCUSSION AND CONCLUSION

We report MD simulations on the extraction of uranyl by TBP to supercritical CO_2 in acidic conditions. They point to a number of analogies with the water / "oil" interfacial systems simulated in standard conditions, where "oil" was modeled by chloroform, and with the acid-free systems: the aqueous and organic phases generally form distinct phases, separated by a liquid-liquid interface, onto which extractant molecules and their complexes adsorb. With CO_2 as organic phase, the interface is somewhat thicker, in relation with the higher mobility of the CO_2 molecules under supercritical conditions. Simulations of the neat interface under several thermodynamic conditions have recently been reported.(*42*) From the size of the simulated systems (less than 80 Å in the z-direction), it cannot be concluded whether one is dealing with macroscopic or microscopic interfaces only (e.g. at the periphery of microdroplets, microemulsions, or micelles (*24, 43-46*)). We believe however that this interface is an important feature for the mechanism of ion extraction. Species adsorbed at the interface (lipophilic anions, extractants and their complexes, solvent modifiers like fluorinated additives) reduce the interfacial tension, and thus lower the energy required for the complex to migrate to the supercritical phase. The local dielectric constant at the interface is also smaller than in the source (aqueous) phase, thus enhancing the stability of the cation complexes. As pointed out previously, amphiphilic topologies are not mandatory for interfacial activity, as even "quasi-spherical" ions like $AsPh_4^+$ or BPh_4^-,(47) bicylic cryptates (48, 49) or even tetracharged tetrahedral tetraammonium cages (50) also concentrate at the interface. Acids are also somewhat surface active, as shown here for nitric acid, and, a fortiori, for weaker acids used in solvent extraction (e.g. picric acid, carboxylic acids). The conjugated bases may also adsorb at the interface, therefore creating a negative potential which attracts the hard cations near the interface and facilitate their capture by interfacial ligands. Such a "synergistic interfacial anion effect" likely operates in the enhanced lanthanide cation extraction by CMPO in the presence of picrate anions.(51) In the systems reported here, nitrate anions behave differently as they mostly dilute in water instead of concentrating near the interface.

A different process is observed here in the absence of such lipophilic anions. We see that large amounts of TBP are necessary to promote the formation and extraction of the complexes. TBP indeed acts not only as a complexant and surface active molecule, but also as a co-solvent with CO_2. Interestingly, in our simulations, uranyl complexation takes place during the metastable demixing step, i.e. after the aqueous and organic phases were forced to mix. Otherwise, the salts would not approach the interface enough to be captured by the interfacial TBP molecules. The $UO_2(NO_3)_2(TBP)_2$ complexes form at the border of the two forming phases during the demixing process and, as phase separation goes on, finally adsorb at the interface, instead of spontaneously migrating to CO_2. In our simulation, one clearly sees that high concentrations of TBP (about 30% in volume) are needed to promote the migration to the CO_2 phase. This increases the concentration of extractable 1:2 complexes and, more importantly, *changes the nature of the interface, which becomes a "third phase" region where CO₂, TBP, some acid and complexes mix.* This heterogeneous microenvironment stabilizes the complexes, which can be considered as "extracted" from water.

As far as the effect of acidity is concerned, we generally find that the acid competes with the complexation of uranyl, due, to the formation of strong hydrogen bonds between TBP and HNO_3 or H_3O^+ species. Protonated $TBP·H^+$ species, not considered here, are also likely to form and to be surface active. Thus, *a large excess of TBP is required (e.g. in system J), to efficiently complex uranyl.* This may seem in contradiction with experimental data, according to which the uranyl distribution ratio D between the aqueous and CO_2 phases increases with the increased concentration of nitric acid.(52) These data follow the global extraction equation (I) which assumes that the uncomplexed uranyl salt is mostly dissociated and hydrated in the aqueous phase:

$$(UO_2^{2+})_{aq} + 2\ TBP_{CO2} + 2\ (NO_3^-)_{aq} \rightleftharpoons (UO_2(NO_3)_2(TBP)_2)_{CO2} \quad (I)$$

242

They also follow expected effects of added ions (e.g. Li^+ NO_3^-) via the reduction of the water activity ("salting-out effect"). The apparent discrepancy comes from our choice to model uranyl by its neutral $UO_2(NO_3)_2$ salt instead of UO_2^{2+}, in order allow for its possible extraction. In fact, it is very difficult to predict the status of ion pairs in pure solutions (see e.g. ref (*53*) for lanthanide salts in acetonitrile) and, a fortiori, at liquid-liquid interfaces. The formation of neutral salts may also be too slow to occur at the simulated time scales. Thus, this facet cannot be addressed by our simulations. In this context, the spontaneous formation of uranyl complexes with TBP is remarkable, and points to the *importance of (micro)-heterogeneities of the systems and metastable conditions to form the complexes and promote their extraction to organic or supercritical phases.*

Acknowledgements. The authors are grateful to IDRIS, CINES and Université Louis Pasteur for computer resources and to PRACTIS for support. RS thanks the French Ministry of Research for a grant. This work has been stimulated by the EEC FIKWCT2000-0088 project. Acknowledgement is made to the Donors of The Petroleum Research Fund, administered by the American Chemical Society, for partial travel support for WG to Orlando.

References

1- Laintz, K. E.; Wai, C. M.; Yonker, C. R.; Smith, R. D., *Anal. Chem.* **1992**, *64*, 2875.

2- Wai, C. M.; Wang, S., *J. Chromatogr. A* **1997**, *785*, 369-383.

3- Smart, N. G.; Carleson, T.; Kast, T.; Clifford, A. A.; Burford, M. D.; Wai, C. M., *Talanta* **1997**, *44*, 137-150.

4- Babain, V.; Murzin, A.; Shadrin, A.; Smart, N. *5th Meeting on Supercritical Fluids*, **1998**, 155-160.

5- Lin, Y.; Brauer, R. D.; Laintz, K. E.; Wai, C. M., *Anal. Chem.* **1993**, *65*, 2549-2551.

6- Ashraf-Khorassani, M.; Combs, M. T.; Taylor, L. T., *J. Chromatogr. A* **1997**, *774*, 37-49.

7- Lin, Y.; Wai, C. M.; Jean, F. M.; Brauer, R. D., *Environ. Sci. Technol.* **1994**, *28*, 1190.

8- Wai, C. M.; Kulyako, Y.; Yak, H.-K.; Chen, X.; Lee, S.-J., *Chem. Commun.* **1999**, 2533-2534.

9- Toews, K. L.; Smart, N. G.; Wai, C. M., *Radiochim. Acta* **1996**, *75*, 179-184.

10- Carrott, M. J.; Waller, B. E.; Smart, N. G.; Wai, C. M., *Chem. Commun.* **1998**, 373-374.

11- Samsonov, M. D.; Wai, C. M.; Lee, S. C.; Kulyako, Y.; Smart, N. G., *Chem. Comm.* **2001**, 1868-1869.

12- Lin, Y.; Smart, N. G.; Wai, C. M., *Environ. Sci. Technol.* **1995**, *29*, 2706-2708.

13- Meguro, Y.; Iso, S.; Takeishi, H.; Yoshida, Z., *Radiochim. Acta* **1996**, *75*, 185-191.

14- Addleman, R. S.; Wai, C. M., *Phys. Chem. Chem. Phys.* **1999**, *1*, 783-790.

15- Addleman, R. S.; Carrott, M. J.; Wai, C. M., *Anal. Chem.* **2000**, *72*, 4015-4021.

16- Dehgani, F.; Wells, T.; Cotton, N. J.; Foster, N. R., *J. Supercrit. Fluids* **1996**, *9*, 263.

17- Joung, S. N.; Yoon, S. J.; Kim, S. Y.; Yoo, K. P., *J. Supercritic. Fluids* **2000**, *18*, 157-166.

18- Smart, N. G.; Carleson, T. E.; Elshani, S.; Wang, S.; Wai, C. M., *Ind. Eng. Chem. Res.* **1997**, *36*, 1819-1826.

19- Erkey, C., *J. Supercrit. Fl.* **2000**, *17*, 259-287.

20- Choppin, G. R.; Nash, K. L., *Radiochimica Acta* **1995**, *70/71*, 225-236.

21- Horwitz, E. P.; Kalina, D. G.; Diamond, H.; Vandegrift, G. F.; Schultz, W. W., *Solv. Extract. Ion Exch.* **1985**, *3*, 75-109.

22- den Auwer, C.; Lecouteux, C.; Charbonnel, M. C.; Madic, C.; Guillaumont, R., *Polyhedron* **1998**, *16*, 2233-2238.

23- den Auwer, C.; Charbonnel, M. C.; Presson, M. T.; Madic, C.; Guillaumont, R., *Polyhedron* **1998**, *17*, 4507-4517.

24- Jensen, M.; Chiarizia, R.; Ferraro, J. R.; Borkowski, M.; Nash, K. L.; Thiyagajan, P.; Littrell, K. C. *ISEC 2002*, **2002**, 1137-1142.

25- Schurhammer, R.; Wipff, G., *New J. Chem.* **2002**, *26*, 229-233.

26- Vayssière, P.; Wipff, G., *Phys. Chem. Chem. Phys.* **2003**, *5*, 127-135.

27- Schnitzer, C.; Baldelli, S.; Campbell, D. J.; Shultz, M. J., *J. Phys. Chem.* **1999**, *103*, 6383-6386.

28- Baaden, M.; Schurhammer, R.; Wipff, G., *J. Phys. Chem. B* **2002**, *106*, 434-441.

29- Berny, F.; Schurhammer, R.; Wipff, G., *Inorg. Chim. Acta; Special Issue* **2000**, *300-302*, 384-394.

30- Watarai, H., *Trends in Analytical Chemistry* **1993**, *12*, 313-318.

31- Wipff, G.; Lauterbach, M., *Supramol. Chem.* **1995**, *6*, 187-207.

32- Berny, F.; Muzet, N.; Schurhammer, R.; Troxler, L.; Wipff, G. in *Current Challenges in Supramolecular Assemblies, NATO ARW Athens*; G. Tsoucaris Ed.; Kluwer Acad. Pub., Dordrecht, 1998; pp 221-248.

33- Berny, F.; Muzet, N.; Troxler, L.; Wipff, G. in *Supramolecular Science: where it is and where it is going*; R. Ungaro; E. Dalcanale Eds.; Kluwer Acad. Pub., Dordrecht, 1999; pp 95-125 and references cited therein.

34- Schurhammer, R.; Berny, F.; Wipff, G., *Phys. Chem. Chem. Phys.* **2001**, *3,* 647-656.
35- Case, D. A.; Pearlman, D. A.; Caldwell, J. C.; Cheatham III, T. E.; Ross, W. S.; Simmerling, C. L.; Darden, T. A.; Merz, K. M.; Stanton, R. V.; Cheng, A. L.; Vincent, J. J.; Crowley, M.; Ferguson, D. M.; Radmer, R. J.; Seibel, G. L.; Singh, U. C.; Weiner, P. K.; Kollman, P. A., *AMBER5, University of California, San Francisco* **1997**.
36- Guilbaud, P.; Wipff, G., *J. Mol. Struct. THEOCHEM* **1996**, *366,* 55-63.
37- Jorgensen, W. L.; Chandrasekhar, J.; Madura, J. D.; Impey, R. W.; Klein, M. L., *J. Chem. Phys.* **1983**, *79,* 926-936.
38- Murthy, C. S.; K. Singer; McDonald, I. R., *Mol. Phys.* **1981**, *44,* 135-143.
39- Muzet, N.; Engler, E.; Wipff, G., *J. Phys. Chem. B* **1998**, *102,* 10772-10788.
40- Lauterbach, M.; Engler, E.; Muzet, N.; Troxler, L.; Wipff, G., *J. Phys. Chem. B* **1998**, *102,* 225-256.
41- Schurhammer, R., Thesis, **2001**, Université Louis Pasteur de Strasbourg.
42- da Rocha, S. R. P.; Johnston, K. P.; Westacott, R. E.; Rossky, P. J., *J. Phys. Chem. B* **2001**, *105,* 12092-12104 ; da Rocha, S.R.P.; Johnston, K. P.; Rossky, P. J., *J. Phys. Chem. B* **2002**, *106,* 13250-13261.
43- Erlinger, C.; Gazeau, D.; Zemb, T.; Madic, C.; Lefrançois, L.; Hebrant, M.; Tondre, C., *Solv. Extract. Ion Exch.* **1998**, *16,* 707-738.
44- Lefrançois, L.; Delpuech, J.-J.; Hébrant, M.; Chrisment, J.; Tondre, C., *J. Phys. Chem. B* **2001**, *105,* 2551-2564.
45- Osseo-Asare, K., *Advances in Colloid Interface Sci.* **1991**, *37,* 123-173.
46- Stoyanov, E. S., *Phys. Chem. Chem. Phys.* **1999**, *1,* 2961-2966.
47- Schurhammer, R.; Wipff, G., *New J. Chem.* **1999**, *23,* 381-391.
48- Jost, P.; Galand, N.; Schurhammer, R.; Wipff, G., *Phys. Chem. Chem. Phys.* **2002**, *4,* 335-344.
49- Jost, P.; Chaumont, A.; Wipff, G., *Supramol. Chem., in press.*
50- Chaumont, A.; Wipff, G., *J. Comput. Chem.* **2002**, *23,* 1532-1543.
51- Naganawa, H.; Suzuki, H.; Tachimori, S.; Nasu, A.; Sekine, T., *Phys. Chem. Chem. Phys.* **2001**, 2509-2517.
52- Meguro, Y.; Iso, S.; Yoshida, Z., *Anal. Chem.* **1998**, *70,* 1262-1267.
53- Baaden, M.; Berny, F.; Madic, C.; Wipff, G., *J. Phys. Chem. A* **2000**, *104,* 7659-7671.

Chapter 16

Experimental Measurement and Modeling of the Vapor–Liquid Equilibrium of β-Diketones with CO_2

Chris M. Lubbers[1,2], Aaron M. Scurto[1], and Joan F. Brennecke[1,*]

[1]Department of Chemical Engineering, University of Notre Dame, South Bend, IN 46556
[2]Current address: Goodyear Chemical Research and Development, 1485 East Archwood Avenue, Akron, OH 44316

Isothermal bubble points and saturated liquid molar volumes for binary mixtures of the β-diketones: pentane-2,4-dione, 2,2,6,6-tetramethylheptane-3,5-dione, 2,2,7-trimethyloctane-3,5-dione, 1,1,1-trifluoropentane-2,4-dione, and 6,6,7,7,8,8,8-heptafluoro-2,2-dimethyloctane-3,5-dione + carbon dioxide have been measured in the temperature region 30°C-45°C and at pressures up to 80 bar. The fluorinated ligands showed a higher affinity for CO_2 than the alkylated ligands at lower pressures, with this trend diminishing at higher pressures. The Peng-Robinson equation of state with van der Waals-1 mixing rules and estimated critical properties predicted liquid phase compositions and liquid molar volumes quite well.

Introduction

Supercritical CO_2 has rapidly been gaining attention as a viable replacement for many traditional organic solvents used for the extraction of metal ions from contaminated soils, sludges, and wastewater streams due to increasing environmental legislation regulating the use of organic solvents. This is because CO_2 is non-toxic, non-flammable, readily available, inexpensive, and has moderate critical properties, $T_c=31^{\circ}C$ and $P_c=73.8$ bar. Additionally, supercritical fluids in general, unlike liquid solvents, have tunable solvation power, with small changes in pressure and/or temperature producing large changes in density, viscosity, and diffusivity. Therefore, a solute extracted into the CO_2 phase can be easily recovered and the CO_2 recycled by just reducing the system pressure which precipitates the solute. Additionally, liquid extraction processes can result in cross contamination of the solid matrix, aqueous solution or sludge stream with the organic solvent, thus requiring additional unit operations such as steam-stripping or distillation to recover this residual solvent. The use of supercritical CO_2 as an extraction solvent ameliorates this problem.

Research has already been conducted examining the solubility of pre-formed metal chelate complexes, as well as the efficacy of a variety of chelating agents in extracting metal ions from solid and liquid phases, using supercritical CO_2. Dithiocarbamates, crown ethers, organophosphorous compounds, functionalized picolylamines, dithiols, and dithiocarbamates, and β-diketones are some of the major classes of chelating compounds investigated. Smart *et al.* (*1*) presents a comprehensive review of the solubilities of chelating agents and metal chelates in supercritical CO_2 that have been presented in the literature. Erkey (*2*) reviews research on the supercritical fluid extraction of metals from aqueous solutions using CO_2. Additional reviews on the subject of supercritical fluid extraction of metal ions are presented by Ashraf-Khorassani *et al.* (*3*) and Wai and Wang (*4*).

It should be noted that very little phase behavior data is available for ligands with CO_2. Since β-diketones have been shown to be viable for the supercritical fluid extraction of a variety of metals, ranging from transition metals to lanthanides and actinides, we will focus on this set of compounds. In particular, the goal of this research is to determine the binary phase behavior of several β-diketones with CO_2 because the phase behavior of the ligand/CO_2 systems is necessary for the design of *in-situ* chelation processes. Table I lists the IUPAC and abbreviated names for the β-diketones investigated here. These particular β-diketone ligands were chosen because they could be obtained commercially and have been used in a number of supercritical fluid extraction studies (1-4).

To date little or no thermodynamic modeling of the phase behavior of the ligand/CO_2 or metal chelate/CO_2 systems has been conducted. However, in order for supercritical fluid extraction to be considered as a possible replacement for organic solvent extraction, accurate models must be developed to predict the phase behavior of these systems to allow for both equipment and process design. Equation of state (EOS) modeling was chosen here to model the vapor-liquid equilibrium of the β-diketone/CO_2 systems studied. Cubic EOSs are the most widely used in modeling high pressure and supercritical fluid systems. This is due to the fact that cubic EOSs are fairly simple, flexible, and have the ability to capture the correct temperature and pressure dependence of the density and all density-dependent properties, such as solubility (5,6).

Table I. IUPAC and Common Names of β-Diketones Studied

IUPAC Name	Abbreviation
Pentane-2,4-dione	ACAC
2,2,6,6-Tetramethylheptane-3,5-dione	THD
2,2,7-Trimethyloctane-3,5-dione	TOD
1,1,1-Trifluoropentane-2,4-dione	TFA
6,6,7,7,8,8,8- Heptafluoro-2,2-dimethyloctane-3,5-dione	FOD

Experimental

Liquid phase solubility and molar volume measurements were taken using a static equilibrium measurement apparatus, which is described in detail by Lubbers (7). Liquid β-diketone samples were loaded into a high pressure, graduated borosilicate glass cell with a teflon stir-bar coupled to an external magnet for agitation. The glass cell was immersed in a constant temperature water bath. Carbon dioxide was metered into the sample cell using a positive displacement pump. The liquid phase compositions were determined by stoichiometry knowing the pressure and temperatures (density) in the cell and tubing, and the volumes of the lines, headspace and liquid volume.

Materials

Pentane-2,4-dione (CAS 123-54-6), 1,1,1-trifluoropentane-2,4-dione (CAS 367-57-7), 2,2,6,6-tetramethylheptane-3,5-dione (CAS 1118-71-4), and 6,6,7,7,8,8,8-heptafluoro-2,2-dimethyloctane-3,5-dione (CAS 17587-22-3) containing 0.1% 2.6-di-*tert*-butylphenol were purchased from Aldrich Chemical Co. with purities of 99.8, 99+, 97.8, 98+, and 98%, respectively. They were used without further purification. 2,2,7-trimethyloctane-3,5-dione (CAS 69725-37-7) was purchased from Oakwood Products, Inc. with a purity of 96.5% and it was used without further purification, as well. Coleman instrument grade CO_2 was purchased from Mittler Supply, Inc. with a purity of 99.99% and it was used as received.

Modeling

Modeling of the ligand/CO_2 phase behavior is one of the key areas that must be addressed when considering the design of a full-scale extraction process. For process design calculations, cubic EOSs are widely used in modeling high-pressure fluid systems. In this study, the Peng-Robinson equation of state (EOS) (8) was employed with the simple van der Waals-1 (vdW-1) mixing rules to extend the EOS to mixtures (8). The Peng-Robinson EOS with vdW-1 mixing rules has only one adjustable interaction parameter, k_{ij}, for each binary mixture. This parameter should be small (-0.2< k_{ij} <0.2) and, in most cases, it can be regressed from experimental data to provide quantitative agreement with the data.

The equation of state uses corresponding states theory to determine the attractive and volume parameters of each species. Therefore, the pure component critical temperature, T_c, and critical pressure, P_c, are required. The EOS uses a third parameter, *viz.* Pitzer's acentric factor, ω, (9)

$$\omega = -\log P_r^{vap}\Big|_{T_{r=0.7}} - 1 \qquad (1)$$

where subscript r represents reduced properties, to enhance the prediction. However, no critical property data exists for any of the ligands studied. Therefore, the critical properties were estimated using either the Gani (10) or Joback (*11*) group contribution method. These methods use the chemical groups or substituents to predict many different physical properties. A detailed description of how we used available physical property data and these group contribution methods to estimate critical properties and acentric factors for the β-diketone compounds can be found elsewhere (*12*).

In this study, a modified Simplex method was used to regress the binary interaction parameter, k_{ij}, using a packaged algorithm, *DBCPOL* (*13*). The objective function minimized by the optimization routine was the percent absolute average relative deviation (%*AARD*)

$$\%AARD = \frac{100}{n}\sum_{i=1}^{n}\frac{\left|x_{1,i}^{exp} - x_{1,i}^{pred}\right|}{x_{1,i}^{exp}} \qquad (2)$$

where n is the number of experimental points, and x_1^{exp} and x_1^{pred}, respectively, are the experimental and calculated compositions of CO_2 in the β-diketone rich liquid phase for each of those points.

The vapor-liquid equilibrium was computed from the EOS model using the reliable and robust method of Hua et al. (14-16) based on interval analysis. Their method can find the correct thermodynamically stable solution to the vapor-liquid equilibrium problem with mathematical and computational certainty. Additionally, the tangent plane distance method (17,18) was used to test the predicted liquid and vapor phase compositions for global thermodynamic phase stability.

Results and Discussion

In this section the experimental liquid phase compositions and liquid molar volumes determined for the β-diketone/CO_2 systems studied will be presented coupled with the modeling results using the Peng-Robinson EOS with van der Waals-1 mixing rules. Table II lists the estimated critical constants, T_c and P_c, and the acentric factor, ω, needed for modeling these systems.

Table II. Estimated Critical Properties of β-diketones

Ligand	T_b	$\ln P^{vap}[Pa]=A-$		T_c	P_c	ω
		$B/T[K]$				
	$[°C]$	A	B	$[K]$	$[bar]$	
ACAC	140.4[a]	21.885[c]	4303.0[c]	619.71[g]	30.82[i]	0.406
THD	215[a]	22.074[d]	5149.5[d]	686.02[g]	21.21[i]	0.397
TFA	107[a]	23.308[a]	4478.9[a]	519.53[h]	23.90[i]	0.711
TOD	238	24.891[e]	6697.2[e]	670.47[h]	24.15[h]	0.770
FOD	167.5[b]	24.649[f]	5778.7[f]	580.89[h]	24.23[i]	0.852

a Ref. (19), b Ref. (20), c Regressed from ref. (21), d Regressed from ref. (22) and references therein, e Regressed from refs. (12,23,24), f Regressed from refs. (20,23) g Ref. (10), h Ref. (11), i Extrapolation of P^{vap} to T_c, j Equation 1.

β-Diketone/CO_2 Vapor-Liquid Equilibrium

Experimental liquid phase compositions and liquid molar volumes for all β-diketone/CO_2 systems studied are listed in Table III. Figure 1 shows the

Table III. Liquid Phase Compositions and Molar Volumes for β-diketone/CO₂ Systems

System with CO_2	T [°C]	P [bar]	Liquid Mole Fraction, x_i Ligand	CO_2	$\underline{V}^L$ [cm³/mol]
ACAC	30.0	28.1	0.571	0.429	79
		41.9	0.391	0.610	68
		54.1	0.233	0.768	61
		60.2	0.143	0.857	57
		62.8	0.104	0.896	56
	35.0	26.6	0.619	0.381	82
		40.7	0.444	0.556	70
		54.9	0.285	0.716	62
		62.6	0.196	0.805	58
		68.1	0.137	0.864	57
		71.8	0.103	0.898	57
	45.0	28.6	0.657	0.343	87
		43.7	0.505	0.495	77
		57.4	0.368	0.633	68
		70.5	0.244	0.756	62
		77.1	0.185	0.815	59
		81.7	0.136	0.865	58
THD	30.0	29.9	0.522	0.478	127
		44.1	0.356	0.644	103
		57.6	0.198	0.802	80
		64.5	0.101	0.899	66
		67.9	0.043	0.957	60
		69.1	0.030	0.970	60
	35.0	28.6	0.609	0.391	150
		44.4	0.422	0.578	119
		57.8	0.274	0.726	93
		70.0	0.123	0.877	70
		74.5	0.059	0.941	63
	45.0	28.8	0.634	0.366	154
		43.4	0.494	0.506	131
		58.4	0.356	0.645	108
		72.0	0.233	0.767	87
		78.2	0.173	0.827	75

Table III. *Continued*

System	T	P	Liquid Mole Fraction, x_i		$\underline{V}^L$
TOD	30.0	29.7	0.530	0.470	132
		43.4	0.362	0.638	103
		57.0	0.204	0.796	80
		64.6	0.088	0.912	64
		67.0	0.046	0.954	60
	35.0	29.5	0.566	0.435	141
		43.5	0.420	0.580	115
		56.9	0.282	0.718	93
		69.5	0.138	0.862	71
		72.8	0.088	0.912	64
	40.0	29.5	0.593	0.407	145
		43.3	0.450	0.550	120
		57.3	0.327	0.673	100
		71.5	0.191	0.809	78
		79.9	0.098	0.903	64
TFA	30.0	28.9	0.505	0.496	86.6
		42.4	0.331	0.669	72
		54.6	0.179	0.821	62
		62.0	0.092	0.908	57.8
		64.6	0.067	0.933	57.5
	35.0	28.6	0.544	0.456	89.6
		42.4	0.379	0.622	76
		55.3	0.235	0.765	66
		62.3	0.161	0.839	62
		67.5	0.111	0.889	60
	40.0	29.1	0.578	0.422	92
		43.1	0.411	0.589	80
		56.5	0.272	0.728	70
		73.0	0.112	0.888	61
FOD	30.0	43.6	0.267	0.734	98
		57.0	0.132	0.868	73
		64.8	0.058	0.943	63
		67.8	0.031	0.969	61
	35.0	21.2	0.597	0.403	164
		29.5	0.479	0.521	140
		43.0	0.321	0.680	110
		57.0	0.185	0.815	85
	40.0	29.5	0.505	0.495	141
		43.0	0.352	0.649	114
		56.7	0.228	0.772	92
		71.2	0.130	0.870	81
		76.9	0.073	0.927	66

isothermal bubble point data for the ACAC/CO$_2$ system at 30.0°C, 35.0°C, and 45.0°C, coupled with the modeling results using the Peng-Robinson EOS with van der Waals-1 mixing rules. Additionally, the bubble point data of Anitescu *et al.* (25) for this system at 25.0°C are included for comparison. This is the only other liquid phase solubility data for a β-diketone/CO$_2$ system that could be found in the open literature. Note that a rigorous analysis of the systematic error indicated that the liquid phase compositions could be measured to within ±0.002 mole fraction; consequently, the error bars are smaller than the size of the symbols in the figure. There is significantly more error, on the order of ±5-10 cm^3/mol, in the measurement of the liquid molar volumes. As expected, the solubility of CO$_2$ in ACAC decreased with increasing temperature at a fixed pressure. This same trend was observed for all of the β-diketone/CO$_2$ systems.

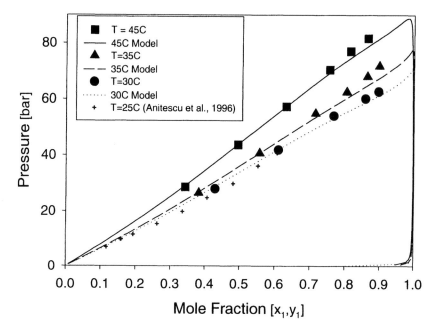

Figure 1. Vapor-liquid equilibrium of the ACAC/CO$_2$ system.

The Peng-Robinson EOS, with only one adjustable parameter, k_{ij}, and estimated critical properties, was able to reproduce the experimental liquid phase compositional data for this system quite well, with the relative deviation between the experimental values and the model ranging from 1.87% to 7.17%. Modeling results for all the β-diketone/CO₂ systems studied are presented in Table IV.

The %AARD values for all the other β-diketone systems ranged from only 0.86% to 4.59%, further showing the ability of the Peng-Robinson EOS to accurately model these systems using estimated critical properties. Figure 2 shows the experimental saturated liquid volumes for the THD/CO₂ system at 30.0°C, 35.0°C, and 45.0°C. Note that the Peng-Robinson EOS tends to overpredict the liquid molar volumes for this system with relative deviations ranging from 9.44% to 10.41%. This is not entirely unexpected as the adjustable parameter in the model was only fit to the liquid phase compositions (see equation 2).

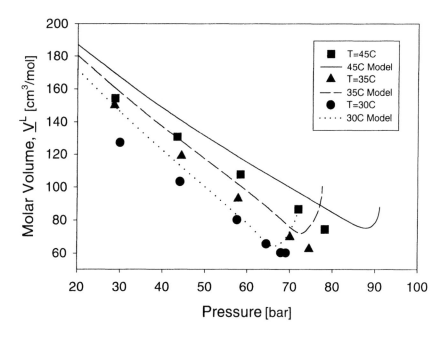

Figure 2. Liquid molar volume vs. pressure for the THD/CO₂ system

Figure 3 shows the isothermal bubble point data for the aliphatic β-diketone/CO₂ systems at 35.0°C. This figure clearly shows that the type of alkyl substituent on the β-diketone moiety has little effect on the liquid phase behavior. Similar bubble point compositions were seen for all these systems at 30.0°C as well.

Table IV. Peng-Robinson EOS Modeling Results

Ligand	T [°C]	k_{12}	%AARD x_{CO_2}	$\underline{V}^L$
ACAC	30	0.0517	7.17	23.93
	35	0.0460	2.97	24.36
	45	0.0570	1.87	26.59
THD	30	0.0423	2.18	10.41
	35	0.0602	1.11	9.44
	45	0.0565	0.86	11.34
TOD	30	0.0127	4.59	6.06
	35	0.0182	1.56	5.44
	40	0.0140	1.13	5.87
TFA	30	0.0323	1.46	18.38
	35	0.0454	2.98	21.50
	40	0.0378	1.46	22.55
FOD	30	-0.0191	0.12	8.35
	35	-0.0222	1.02	21.61
	40	-0.0290	0.91	15.97

Figure 4 shows a comparison of the experimental and predicted liquid phase compositions for the β-diketone/CO_2 systems studied at 35.0°C. This plot reveals all the salient features of the equilibrium phase behavior of these alkylated and/or fluorinated ligand/CO_2 systems. Namely, the bubble point compositions of the TFA and FOD/CO_2 systems are higher at lower pressures; i.e., at a given pressure more CO_2 dissolves in the two fluorinated ligands, TFA and FOD, than in the other β-diketones. However, at higher pressures the TFA and FOD equilibrium compositions are very similar to those of the ACAC, THD, and TOD/CO_2 systems.

With the liquid phase compositions and liquid molar volumes determined for these β-diketone/CO_2 systems, it is now possible to specify operating conditions and dimensions of a pre-extraction ligand saturation vessel such that a one-phase mixture of ligand and CO_2 is delivered to the extraction vessel. Again, it cannot be overstated that knowledge of the phase behavior of the ligand is essential when designing a complete extraction process. Furthermore, it has been shown that the Peng-Robinson EOS, with estimated critical properties and acentric factor for the ligand and incorporating standard van der Waals-1 mixing rules, is a viable model that could be used for the process design and control of this section of an *in-situ* extraction process using β-diketones as the chelating agents.

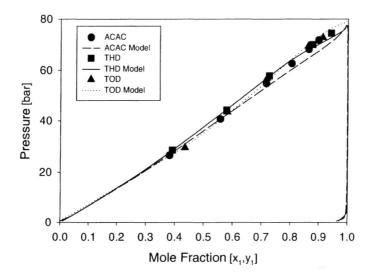

Figure 3. Vapor-liquid equilibrium of aliphatic β-diketone/CO₂ systems at 35.0°C

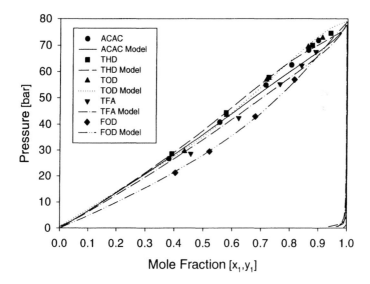

Figure 4. Vapor-liquid equilibrium of all β-diketone/CO₂ systems at 35.0°C

256

Summary

We have presented isothermal bubble points and saturated liquid molar volumes for three alkylated β-diketones (pentane-2,4-dione, 2,2,6,6-tetramethylheptane-3,5-dione, and 2,2,7-trimethyloctane-3,5-dione) and two fluorinated β-diketones (1,1,1-trifluoropentane-2,4-dione and 6,6,7,7,8,8,8-heptafluoro-2,2-dimethyloctane-3,5-dione) with CO_2 at temperatures between 30°C and 45°C and at pressures up to 80 bar. As anticipated, the fluorinated ligands showed a higher affinity for CO_2 than the alkylated ligands at lower pressures, but surprisingly this trend diminished completely at higher pressures so that the mixture critical points appear to be virtually identical. Thus, the pressure needed to form a single phase ligand/CO_2 mixture of any desired composition is no greater for the alkylated ligands. However, other factors, such as the solubility of the metal chelate complexes in CO_2 or the binding constants may make some of these ligands more attractive than others for supercritical fluid extraction of metals. We have also found that the Peng-Robinson equation of state with van der Waals-1 mixing rules and estimated critical properties modeled liquid phase compositions and liquid molar volumes quite well and, thus, can be used for process design of β-diketone/CO_2 systems.

Acknowledgements

Financial support from the Department of Energy (grants DE-FG07-96ER14691 and DE-FG02-98ER14924) and a Bayer Predoctoral Fellowship (A.M.S.) is gratefully acknowledged. We also wish to thank Professor. James P. Kohn for his assistance with the experimental apparatus and Dr. Gang Xu for help with the implementation of the completely reliable computational technique involving interval analysis.

References

1. Smart, N. G.; Carleson, T.; Kast, T.; Clifford, A. A.; Burford, M. D.; Wai, C. M. Solubility of Chelating Agents and Metal Containing Compounds in Supercritical Fluid Carbon Dioxide. *Talanta*, **1997**, *44*, 137-150.
2. Erkey, C. Supercritical Carbon Dioxide Extraction of Metals from Aqueous Solutions: A Review. *J. Supercrit. Fluids*, **2000**, *17*, 259-287.
3. Ashraf-Khorassani, M.; Combs, M. T.; Taylor, L. T. Supercritical Fluid Extraction of Metal Ions and Metal Chelates from Different Environments. *J. Chromatogr. A*, **1997**, *774*, 37-49.

4. Wai, C. M.; Wang, S. Supercritical Fluid Extraction: Metals as Complexes. *J. Chromatogr. A*, **1997**, *785*, 369-383.
5. McHugh, M. A.; Krukonis, V. J. *Supercritical Fluid Extraction: Principles and Practice;* Butterworth-Heinemann: Oxford, U.K., 1994.
6. Brennecke, J. F.; Eckert, C. A. Phase Equilibrium for Supercritical Fluid Process Design. *AIChE J.*, **1989**, *35*, 1409-1427.
7. Lubbers, C. M. M.S. thesis, University of Notre Dame, Notre Dame, IN, 2001.
8. Peng, D. Y.; Robinson, D. B. A New Two-Constant Equation of State. *Ind. Eng. Chem. Fund.*, **1976**, *15*, 59-64.
9. Pitzer, K. S.; Lippmann, D. Z.; Curl, R. F.; Huggins, C. M.; Petersen, D. E. The Volumetric and Thermodynamic Properties of Fluids. II. Compressibility Factor, Vapor Pressure and Entropy of Vaporization. *J. Am. Chem. Soc.*, **1955**, *77*, 3433-3440.
10. Constantinou, L.; Gani, R. New Group-Contribution Method for Estimating Properties of Pure Compounds. *AICHE J.*, **1994**, *40*, 1697-1710.
11. Reid, R. C.; Prausnitz, J. M.; Poling, B. E. *The Properties of Gases and Liquids, 4th Ed.;* McGraw Hill: New York, NY, 1987.
12. Scurto, A. M. Ph.D. dissertation, University of Notre Dame, Notre Dame, IN, 2002.
13. IMSL, Inc. *IMSL Libraries*, **1979**, *1*, p 872.
14. Hua, J. Z.; Brennecke, J. F.; Stadtherr, M. A. Reliable Prediction of Phase Stability Using an Interval-Newton Method. *Fluid Phase Equil.*, **1996**, *116*, 52-59.
15. Hua, J. Z.; Brennecke, J. F.; Stadtherr, M. A. Enhanced Interval Analysis for Phase Stability: Cubic Equation of State Models. *Ind. Eng. Chem. Res.*, **1998**, *37*, 1519-1527.
16. Hua, J. Z.; Brennecke, J. F.; Stadtherr, M. A. Reliable Computation of Phase Stability Using Interval Analysis: Cubic Equation of State Models. *Comput. Chem. Eng.*, **1998**, *22*, 1207-1214.
17. Baker, L. E.; Pierce, A. C.; Luks, K. D. Gibbs Energy Analysis of Phase Equilibria. *Soc. Pet. Eng. J.*, **1982**, *22*, 731-742.
18. Michelsen, M. The Isothermal Flash Problem. Part I. Stability. *Fluid Phase Equil.*, **1982**, *9*, 1-19.
19. Irving, R. J.; Ribeiro da Silva, M. A. V. Enthalpies of Vaporization of Some β-Diketones. *J. Chem. Soc. Dalton Trans.*, **1975**, 798-800.
20. Fluka Chemical, Inc. Catalogue, 2001-2002.
21. Askonas, C. F.; Daubert, T. E. Vapor Pressure Determination of Eight Oxygenated Compounds. *J. Chem. Eng. Data*, **1988**, *33*, 225-229.
22. Beilstein, F. K.; *Beilstein Handbook for Organic Chemistry;* Springer: Berlin, NY, 2001.
23. Oakwood Inc. Chemical Catalogue, 2001-2002.

24. Wenzel, T.J.; Williams, E.J.; Haltiwanger, R.C.; Sievers, R.E. Studies of Metal Chelates with the Novel Ligand 2,2,7-trimethyl-3,5-octanedione. *Polyhedron*, **1985**, *4*, 369-378.
25. Anitescu, G.; Gainar, I.; Vilcu, R. Solubility of Carbon Dioxide in Some Solvents Containing Ketone Group at High Pressures. *Rev. Roum. Chim.* **1996**, *41*, 713-716.

Chapter 17

Thermodynamics of the Baylis–Hillman Reaction in Supercritical Carbon Dioxide

Anthony A. Clifford, Paul M. Rose, Katherine Lee, and Christopher M. Rayner

School of Chemistry, University of Leeds, Leeds LS2 9JT, United Kingdom

The Baylis-Hillman reaction is a potentially important carbon-carbon bond-forming reaction. The neat reaction gives good yields, but this would give mixing and heat-control problems in production. In solution, however the yields are poor. Better yields were found in supercritical carbon dioxide and these were found to be pressure-dependent. Standard thermodynamic methods have been used to analyze this pressure dependence and to provide a physical explanation of the effect. A method for optimizing conditions for the Baylis-Hillman has been developed.

Introduction

The Baylis-Hillman (B-H) reaction is a synthetically useful carbon-carbon bond-forming reaction between an aldehyde and an electrophilic alkene, usually in the presence of a tertiary amine (1,2). One of its main attractions is the high degree of functionality present in the products and their resultant potential transformations. The reaction gives good yields when performed neat, but this would be difficult on a production scale because of mixing problems (the reagents are often solid) and because the reaction is exothermic. However the typical B-H reaction is notoriously slow in liquid solution unless particularly reactive substrates are chosen, and often fails to reach completion due to unfavourable equilibria (*3*). Kinetic studies have shown that the reaction is a complex equilibrium (*4*). The general B-H reaction scheme is given below.

$X = CO_2R, CN, COR$

The reaction is usually catalysed by a tertiary amine, most commonly 1,4-diazabicyclo[2.2.2]octane (DABCO), but also 3-hydroxyquinuclidine (HQD) (*5*) and derivatives thereof (*6*), diazabicyclo[5.4.0]undecene (DBU) (*7*) and phosphines (*8*). The reaction can be accelerated by Lewis acids and/or various protic additives or solvents (*9,10*), alongside physical methods such as microwave irradiation (*11*) and ultrasound (*12*). Recent reports have also shown that the use of ionic liquids as solvents can be beneficial (*13*). The B-H reaction is known to have a very high negative activation volume, rates are known to be increased by high pressures (typically 5-15 kbar) (*4,14*) and increasing pressure also brings about stereocontrol in appropriate cases (*15*).

Because of the problems the B-H reaction has both when performed neat and in liquid solution, and the success of some other reactions in supercritical carbon dioxide (scCO$_2$) (*16*), this reaction is an ideal candidate for investigation in scCO$_2$. For our initial studies we chose to investigate DABCO as catalyst, with methyl acrylate (MA) and a variety of aromatic aldehydes as substrates. The results of these studies have been reported (*17*) and a detailed study of one particular reaction using *p*-nitrobenzaldehyde (NBA) revealed an interesting pressure dependence on the observed yield. It was the purpose of the present study to try to model this effect thermodynamically, to see if the model gave physical insight into the pressure effects, and to develop a method of finding conditions to optimize the yield of the B-H reaction in scCO$_2$.

Experimental and Results

All the reactions reported on here used 1 mmol NBA, 2 mmol MA and 1 mmol of DABCO. When the reaction was performed neat a 94% yield of the Baylis-Hillman product (BHP) was obtained after 2 hours at 20°C. When the reaction was carried out in 20 mL toluene at 20°C for 24 hours, no product was obtained. The volume of the solvent was the same as that later used for scCO$_2$ studies. When the volume of toluene used was reduced to 2.5 mL a 10% yield was obtained under the same conditions. Thus this particular B-H reaction exhibits the general problems described earlier.

Reactions in scCO$_2$ were carried out in a 20 mL view cell. The same millimolar quantities of the reagents were used and CO$_2$ added to bring the pressure up to a desired value. In all cases the reaction mixture was observed to be homogeneous initially. However, at lower pressures a liquid layer formed during the reaction. When this happened, a dimer (an ether) of the B-H product was formed (17). The pressure dependence of the yields of the initial B-H product (filled circles) and the sum of the yields of the B-H product and its dimer (open circles) are shown in Figure 1. They are seen to rise as the pressure is lowered and the dimer is formed around 110 bar and below.

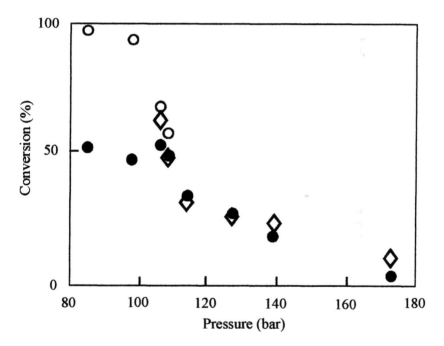

Figure 1. Percentage conversion of NBA into BHP in supercritical carbon dioxide at 50 °C, shown as a function of pressure. Filled circles give the conversion into BHP, open circles give the sum of the conversions into BHP and its dimer, and diamond give the predicted conversion into BHP.

Thermodynamic Modeling Methods

The aim was to use standard methods for the modeling and for the prediction of necessary parameters. Thus an exact agreement with experiment was not expected. The hope was that the modeling would show the same trends as the experimental results. The Peng-Robinson (P-R) equation (*18*) was chosen for modeling both the phase behaviour and the reaction equilibria. For this, critical temperatures, T_c, pressures, p_c, and volumes, V_c, acentric factors, ω, and binary interaction parameters, k_{ij}, are required. The values used are given in Table 1. Critical volumes are required for calculating binary interaction parameters.

Table 1. Critical parameters, acentric factors and binary interaction parameters used in the calculations.

	CO_2	MA	NBA	BHP	DABCO
p_c (bar)	73.8	43.0	59.0	31.1	46.4
T_c (K)	304.2	536	691	788	645
ω	0.239	0.350	0.951	1.717	0.334
$10^6 V_c$ (m^3 mol^{-1})	93.9	26.5	44.7	62.9	35.6
$k_{i,CO2}$		0.059	0.119	0.634	0.323

For CO_2 and MA, values for the critical parameters and acentric factors have been published (*19*). For the other compounds, no data are available and group contribution methods must be used. For the critical temperatures the Fedors method (*20*) was used. This is less accurate than the Lyderson method (*20*), but had to be used, as boiling point data for the compounds are also unavailable. For the critical pressures and volumes, the Lyderson method was used (*20*). For the acentric factors, the equation suggested by Reid, Prausnitz and Poling (*20*)

$$\omega = \frac{3(T_b/T_c)}{7[1-(T_b/T_c)]}\log p_c - 1$$

where T_b is the boiling point, was used with the ratio T_b/T_c calculated using the Lyderson method (*20*). Finally, the binary interaction parameters for the solutes with CO_2 were obtained using a published correlation (*21*). Binary interaction parameters between the components other than CO_2 had little effect on the calculations, because the frequency of their interaction is far smaller than those with CO_2, which is present with mole fraction 2-3 orders of magnitude greater than those of the other components. Binary interaction parameters between the components other than CO_2 were therefore set to zero.

Most of the values given in Table 1 are unremarkable, but the critical temperature, acentric factor and binary interaction parameter are high for the product BHP. This may be explained by the fact that BHP is a larger long molecule with an –OH group.

Modeling of Phase Behavior

The number of moles of the reagents, the product and the catalyst were obtained from amounts added and the yield measurements. The number of moles of CO_2 was obtained by increasing the value used until the pressure calculated from the P-R equation in a 20 mL cell equalled the experimental pressure. At pressures below 100.6 bar, the calculation showed the presence of two phases. Experimentally the existence of two phases was observed at 106 bar and 109 bar. Thus predicted and observed phase behaviour did not exactly agree, but there was approximate agreement.

Modeling of Chemical Equilibria

Calculations of chemical equilibrium were carried out for all pressures above 106 bar. Although two phases were observed experimentally at 106 and 109 bar, they were not predicted theoretically using the same parameters and equation of state, and so calculations at these pressures were thought worthwhile. The Baylis-Hillman reaction is believed to be a reversible equilibrium (*4*), in which case the conversion should be represented by the following equation containing the equilibrium constant, $K_f^{\ominus}$.

$$K_f^{\ominus} = \frac{(f_{BHP}/p^{\ominus})}{(f_{MA}/p^{\ominus})(f_{NBA}/p^{\ominus})}$$

where f_{MA}, f_{NBA} and f_{BHP} are the fugacities of MA, NBA and BHP, respectively, and $p^{\ominus}$ the standard pressure (here 1 bar). If conditions were ideal (patently not the case with supercritical carbon dioxide), the fugacities could be replaced by partial pressures, p_{MA}, p_{NBA} and p_{BHP} and the following equation would apply, containing the pressure equilibrium constant, $K_p^{\ominus}$.

$$K_p^{\ominus} = \frac{(p_{BHP}/p^{\ominus})}{(p_{MA}/p^{\ominus})(p_{NBA}/p^{\ominus})}$$

That this is not the case can be seen from Table 2, where $K_p^{\ominus}$ is calculated from the experimental results, by multiplying the mole fractions of the reagents and products at equilibrium with the total pressure. $K_p^{\ominus}$ is seen to fall by nearly a factor of 100 over the pressure range.

Table 2. Values of $K_p^{\ominus}$ and $K_f^{\ominus}$ calculated from the experimental results.

Pressure (bar)	$K_p^{\ominus}$	$K_f^{\ominus}$
106	3.61	6.29×10^3
109	2.21	7.43×10^3
115	7.53×10^{-1}	5.89×10^3
128	5.08×10^{-1}	5.34×10^3
140	2.82×10^{-1}	3.68×10^3
175	4.12×10^{-2}	1.58×10^3

To calculate $K_f^{\ominus}$, it is necessary to calculate the fugacity coefficients of the components, φ_i, from the P-R equation of state to convert partial pressures into their fugacities, $f_i = \varphi_i p_i$. This was carried out and the values of $K_f^{\ominus}$ obtained are also given in Table 2. The equilibrium constants are seen to still fall with increase in pressure, but now only by a factor of 4. Because of the uncertainties in the parameters used, this may be regarded as a satisfactory result, considering that it is the logarithm of the fugacity coefficient that is given by an equation of state. In fact small adjustment in some of the parameters, such as an increase to 0.7 for the binary interaction parameter between CO_2 and BHP, gives values of $K_f^{\ominus}$ which show scatter, but no rising or falling trend over the pressure range. There are also uncertainties in the experimental data and the reactions may not always reach equilibrium at the highest pressures.

Prediction of Conversions

If we assume the model, with its uncertainties is working approximately well, it can be used to predict conversions and yields. For this we assume that the equilibrium constant, $K_f^{\ominus}$ is truly constant and use the average of the values given in Table 2, i.e. $K_f^{\ominus} = 5.03 \times 10^3$. The number of moles of the reagents and product is adjusted to give this value of the equilibrium constant at all pressures, while at the same time adjusting the number of moles of CO_2 to give the experimental pressure. The outcome of this procedure are predictions of the yields from the reaction at the different pressures predicted by the P-R equation, the estimated parameters and the average value of the equilibrium constant. These are shown as the diamonds in Figure 1. The figure shows that the thermodynamic modeling, in spite of its uncertainties, predicts the general trends observed.

Physical Explanation of the Calculations

To interpret this calculation in physical terms, the fugacity coefficients of the reagents and products as a function of pressure are examined and these are shown in Table 3. The fugacity coefficients of the two reagents are seen to fall with increase in pressure. This indicates that the energy (chemical potential) of the reagents is falling with higher pressure, as the higher density solvates these molecules more effectively. For the product, however, the fugacity coefficient increases with increase in pressure, indicating that it is less effectively solvated at higher pressure. This is because it is a larger molecule with an –OH group that is of a different molecular character to CO_2 and has poor interactions with it.

Table 3. Fugacity coefficients for the reagents and product calculated from the Peng-Robinson equation of state.

Pressure (bar)	φ_{MA}	φ_{NBA}	φ_{BHP}
106	2.43×10^{-2}	2.31×10^{-3}	9.76×10^{-2}
109	2.13×10^{-2}	1.65×10^{-3}	1.18×10^{-1}
115	1.57×10^{-2}	1.46×10^{-3}	1.80×10^{-1}
128	1.49×10^{-2}	1.35×10^{-3}	2.11×10^{-1}
140	1.75×10^{-2}	1.06×10^{-3}	2.43×10^{-1}
175	1.39×10^{-2}	6.53×10^{-4}	3.48×10^{-1}

It is worth noting that, whereas for most substances solubility in CO_2 increases continuously with pressure, for some poorly soluble molecules solubility increases with pressure at lower pressure, but falls again at higher pressures (*22*). This is a direct indication that solvation can decrease at higher pressures and solutes are 'squeezed' out of solution.

Thus, increasing pressure decreases the energy of the reagents and increases the energy of the product by solvation. Increasing pressure will therefore increase the proportion of the reagents in the mixture at equilibrium and decrease the yield.

Predictions for other B-H reactions and at other temperatures

Benson (23,24) has developed a group contribution method for predicting the enthalpies and entropies of organic compounds. This can then be used to calculate the enthalpy change, ΔH° and the entropy change, ΔS°, for a reaction. From these the Gibbs energy change for the reaction, ΔG° and its equilibrium constant at any temperature, K_f°, can be estimated. In the case of the B-H reaction (and any reaction), it is not necessary to calculate the enthalpies and entropies of the reagents and products, but only to observe which groups are lost during the reaction and which are gained. During the B-H reaction the Benson groups CB(C), C(CB)(H)(O)(CD), O(C)(H) and CD(CO)(C) are gained and CD(CO)(H), CB(CO) and CO(CB)(H) are lost. (Those not familiar with the Benson groups can consult the references given, but briefly O(C)(H) indicates the molecule contains an oxygen atom which is connected to a carbon atom and a hydrogen atom and CD and CB are carbon atoms with a double bond or in a benzene ring, respectively.) According to the Benson approximation, this is the case whatever the particular substituents in the reagents are. Thus his method predicts that for all B-H reactions, the theoretical equilibrium constant, K_f° (although not the observed equilibrium involving concentrations) and the enthalpy change for the reaction are the same.

Values for the ΔH° contributions are available in compilations for all groups lost or gained except for C(CB)(H)(O)(CD). If the contribution for C(CB)(H)(O)(C) is used instead, a value for ΔH° of -52.8 kJ mol^{-1} is obtained. Unfortunately, no values for ΔS° contributions for any of these groups are available, so K_f° cannot be predicted. However, as K_f° values for all B-H reactions are predicted to be the same in this approximation, the average experimental value of K_f° = 5.03 × 10^3 could be assumed. This value corresponds to a value of ΔG° = -22.9 kJ mol^{-1} from which a value of ΔS° = -92.6 J mol^{-1} K^{-1}, a value which is consistent with a synthetic reaction.

Thus an approximate method is available using this average value of K_f°, the estimated value of ΔH°, and the Peng-Robinson parameters for particular reagents and products to predict yields from all B-H reactions at any temperature. This method can be used to predict conditions for maximum yield from any reaction, which could then be tested experimentally. Studies using this method are being currently made.

Conclusions

Modeling of the Baylis-Hillman reaction, using parameters from standard formulae and the Peng-Robinson equation of state reproduces the general experimental trends observed for the pressure dependence of a particular Baylis-Hillman reaction. The calculations show that the effect has a physical explanation, which is that the product stabilisation by solvation is reduced at higher pressures, reducing its equilibrium amount. The work has led to a general method, which can be used to predict conditions for optimum yield for any Baylis-Hillman reaction and by adaptation to any organic equilibrium reaction.

Acknowledgements

The authors gratefully acknowledge financial support for the Engineering and Physical Sciences Research Council, Aventis, GlaxoSmithKline, Pfizer, Solvay and the University of Leeds.

References

1. Ciganek, E. *Organic Reactions* **1997**, *51*, 201-350.
2. Basavaiah, D.; Rao, P.D.; Hyma, R.S. *Tetrahedron* **1996**, *52*, 8001-8062.
3. Fort, Y.; Berthe, M.C.; Caubere, P. *Tetrahedron* **1992**, *48*, 6371-6384.
4. Hill, J.S.; Isaacs, N.S. *Tetrahedron Lett.* **1986**, *27*, 5007-5010.
5. Drewes, S.E.; Freese, S.D.; Emslie, N.D.; Roos, G.H.P *Synth. Commun.* **1988**, *18*, 1565-1572.
6. Barrett, A.G.M; Cook, A.S.; Kamimura, A. *J. Chem. Soc. Chem. Commun.* **1998**, 2533-2534.
7. Aggarwal, V.K.; Mereu, A. *J. Chem. Soc. Chem. Commun.* **1999**, 2311-2312.
8. Rafel, S.; Leahy, J.W. *J. Org. Chem.* **1997**, *62*, 1521-1522.
9. Aggarwal, V.K.; Dean, D.K.; Mereu, A.; Williams, R. *J. Org. Chem.* **2002**, *67*, 510-514.
10. Yu,C.; Liu, B.; Hu, L. *J. Org. Chem.* **2001**, *66*, 5413-5418.
11. Kundu, M.K.; Mukherjee, S.B.; Balu, N.; Padmakumar, R.; Bhat, S.V. *Synlett* **1994**, 444.
12. Roos, G.H.P.; Rampersadh, P. *Synth. Commun.* **1993**, *23*, 1261-1266.
13. Rosa, J.N.; Afonso, C.A.M.; Santos, A.G. *Tetrahedron* **2001**, *57*, 4189-4193.
14. Hill, J.S.; Isaacs, N.S. *J. Chem. Res.*, **1988**, 330-331.

268

15. Oishi, T.; Oguri, H.; Hirama, M. *Tetrahedron Asymm.* **1995**, *6*, 1241-1244.
16. Oakes, R.S.; Clifford, A.A.; Rayner, C.M. *J. Chem.Soc. Perkin Trans. 1,* **2001**, 917-941.
17. Rose, P.M.; Clifford, A.A.; Rayner, C.M. *J. Chem. Soc. Chem. Commun.* **2002**, 968-969.
18. Peng, D.Y.; Robinson, D.B. *Ind. Eng. Chem. Fundam.* **1976**, *15*, 59-64.
19 Reid, R.C.; Prausnitz, J.M.; Poling, B.E. *The Properties of Liquids and Gases* McGraw Hill: New York, NY, 1987, pp 656-732.
20. Reid, R.C.; Prausnitz, J.M.; Poling, B.E. *The Properties of Liquids and Gases* McGraw Hill: New York, NY, 1987, pp 1-26.
21. Bartle, K.D.; Clifford, A.A.; Shilstone, G.F. *J. Supercrit. Fluids* **1992**, *5*, 220-225.
22. Bartle, K.D.; Clifford, A.A.; Shilstone, G.F. *J. Phys. Chem. Ref. Data* **1991**, *20*, 713-755.
23. Benson, S.W. *Thermochemical Kinetics* John Wiley & Sons: New York, NY, 1968; pp 178-215.
24. Reid, R.C.; Prausnitz, J.M.; Poling, B.E. *The Properties of Liquids and Gases* McGraw Hill: New York, NY, 1987; pp 173-190.

Novel Materials

Chapter 18

Dissolving Carbohydrates in CO_2: Renewable Materials as CO_2-philes

Poovathinthodiyil Raveendran and Scott L. Wallen[*]

Department of Chemistry, Kenan and Venable Laboratories and the NSF
Science and Technology Center for Environmentally Responsible Solvents
and Processes, The University of North Carolina, Chapel Hill, NC 27599–
3290
[*]Corresponding author: email: wallen@email.unc.edu

The design and synthesis of inexpensive, hydrocarbon based
CO_2-philic materials are of current interest as they are
applicable to the utilization of liquid and supercritical CO_2 as
an environmentally benign solvent.   In this section, we
describe the molecular interactions involved in enhancing the
solubility of carbohydrates in CO_2.   We propose the
peracetylation of these compounds as a simple method for the
preparation of inexpensive, environmentally benign CO_2-
philes. Utilization of these materials as CO_2-philes combines
the merits of two important green chemistry principles-
environmentally benign solvents and renewable materials.

1. Introduction

The chemical industry in the U.S. has been tremendously successful as demonstrated by the enormous trade surplus it commands ($20.4 billion in 1995), the fact that it is the largest manufacturing sector in terms of value added product, and the tremendous amount of money (~$17.6 billion) that is reinvested in research and development annually (*1*). Obviously many of the products that consumers rely on in daily life are a direct result of the success of this manufacturing sector. These include, but are not limited to, pharmaceuticals, medical products, communications technologies, transportation products and an enormous number of other consumer goods. In spite of these phenomenal successes there are major challenges facing the U.S. chemical manufacturing industry in the next century. Foremost among these challenges are the implementation of new processes which focus on environmental stewardship, sustainability and the recycling and/or reduction of waste. A large percentage of the current environmental problems that our country, and other industrialized nations, face is due to the reliance on standard organic solvents and organic-inorganic catalysts for the manufacture of feedstock chemicals, specialty chemicals and a vast array of consumer products. In fact the majority of products made today rely on traditional synthetic and separation technologies that were developed over 40 to 50 years ago (*1*). An additional consideration when examining chemical manufacturing processes is the waste generated after the useful lifetime of a product or the packaging that accompanies the product.

Green chemistry is defined as, "the utilization of a set of principles that reduces or eliminates the use and generation of hazardous substances in the design, manufacture and application of chemical products." (*2*) Over the past decade there has been an increasing emphasis on the introduction of green chemical methods in the analytical, synthetic and process chemistry segments of industry (*3-7*). One measure of the impact a particular process has on the environment is the environmental factor (E factor) introduced by Sheldon (*8*). The E factor is defined as the amount of waste generated per kilogram of product produced. The pharmaceutical industry has one of the highest E factors with values as high as 100 (*9*). The majority of waste generated is in the form of inorganic salt by-products and organic solvents for reaction, separation and purification. Two of the most effective approaches to lowering the E factor is by increasing reaction efficiency through the incorporation of catalysts or through the replacement and/or elimination of hazardous solvents (*9*). The latter approach can also have significant economic incentives when one considers the ever-increasing costs associated with waste stewardship and disposal.

One of the key areas within the realm of green chemistry is the identification of environmentally benign, alternative solvents that can replace the conventional, toxic, volatile organic solvents. Supercritical fluids (SCF) offer great opportunities in this arena due to the ease of solvent removal, recyclability and tunability of solvent parameters (4). Among SCFs, supercritical carbon dioxide ($scCO_2$) is advantageous to the green chemist due to its non-toxicity, non-flammability, low cost, abundance and easily achievable critical conditions. In fact $scCO_2$ is widely regarded as an environmentally responsible solvent and has become almost synonymous with green chemistry. Investigations to explore the utilization of this solvent system have been widely examined in the past decade (10-25).

The most important limitation of utilizing CO_2 as an industrial solvent has been the extremely poor solubility (26) of most non-fluorous, polar and amphiphilic materials in CO_2. Thus, the fundamental principles for the design of CO_2-philic molecules including amphiphiles have attracted great interest, and different molecular level approaches have been used to 'CO_2-philize' compounds that are otherwise insoluble in CO_2. The first, and so far the most widely applied method is the introduction of fluorocarbons (10-12). Though the interaction between CO_2 and fluorocarbons are rather weak, fluorocarbons exhibit high solubility in liquid and $scCO_2$. Fluorocarbon-based CO_2-philes are, however, expensive and have the potential to cause environmental problems on thermal degradation (27). Therefore, there is current interest in the development of inexpensive, hydrocarbon-based CO_2-philes. It has been identified in the past that the specific interaction of CO_2 molecules with Lewis base groups (28-30), especially carbonyl groups can be used in the design of CO_2-philic materials (15, 16).

Another focus in green chemistry currently is the use of renewable materials as a replacement for non-renewable resources (2, 6). Carbohydrates are an important class of inexpensive, abundant, and renewable biomaterials that has been the focus of research recently in a range of chemical, food and pharmaceutical applications. Thus, the utilization of carbohydrate based materials as CO_2-philes as well as the utilization of CO_2 for the synthesis, isolation, purification and the processing of carbohydrate based materials, and for the dispersion of a wide range of materials in carbohydrate based matrices such as cyclodextrins, are of great importance in green chemistry today. The CO_2-phobic nature of these polyhydroxy systems, however, makes them insoluble in liquid and $scCO_2$. Thus one needs to replace the hydroxyl groups with CO_2-philic functionalities to make the carbohydrates miscible with CO_2. Our ultimate vision is the industrial and academic implementation of chemistry based on environmentally benign solvent and renewable, functionalized solute systems. The following is a presentation of the fundamental ideas concerning this issue including the exploration of the nature of CO_2-philic interactions, the application of these ideas to dissolve significant quantities of renewable materials in CO_2 and a brief overview of future research targets in this area.

2. Interaction of CO_2 with Carbonyl Compounds

An understanding of the charge distribution, self-interaction (*31-33*) and the solvent-solute interactions of CO_2 are essential for the rational design of CO_2-philic materials. Our view of CO_2 as a solvent comes from quantum mechanical *ab initio* calculations of CO_2 (isolated as well as interacting with itself and other simple carbonyl containing systems). Historically, CO_2 was considered as a non-polar solvent owing primarily to the zero dipole moment and low dielectric constant and it was predicted that supercritical or liquid CO_2 should be comparable to hexane in its solvent properties. However, this notion was discarded in later years as the majority of materials that are miscible with hexane were reported to be insoluble in scCO$_2$. In fact, at this symposium an excellent historical perspective was provided by Beckman regarding the solubility of materials in CO_2 (*34*). CO_2 posseses a large quadrupole moment (*35*) and from an interaction point of view, CO_2 is considered a Lewis acid (LA) since the electron deficient carbon atom can interact with a Lewis base (LB) group on a solute molecule (e.g. a carbonyl group). Several experimental and theoretical studies have been carried out to estimate the nature and extent of these interactions. Kazarian et al. (*28*) used IR spectrsoscopy to study the specific interaction between CO_2 and carbonyl groups. The researchers observed the lifting of the degeneracy of the CO_2 bending modes, v_2, as a result of interactions with polymethylmethacrylate. These researchers even suggested that these interactions could be responsible for the swelling of polyacrylates in contact with CO_2. Following these experimental studies, Nelson and Borkman used *ab initio* calculations to quantify the splitting of the v_2 mode of CO_2 upon interaction with carbonyl groups in simple molecules (*29*). Based on these spectroscopic observations (*28*) and the work of McHugh and coworkers (*15*) concerning the solubility of vinyl acetates in CO_2, Beckman and coworkers synthesized poly-(ether-carbonate) co-polymers having 1-2 wt% solubility in liquid scCO$_2$ (*16*).

As mentioned previously, CO_2, though having zero dipole moment, is a charge-separated molecule with partial positive charge on the carbon atom and partial negative charges on the oxygen atoms. Thus, CO_2 can act as an electron donor as well as an electron acceptor. The situation is analogous to the case of H_2O, which can act both as a Lewis acid and a Lewis base. For a comparison, the Mulliken, as well as the CHELPG charges (calculated by fitting the electrostatic potentials) on the individual atoms of CO_2 and H_2O, calculated at the MP2/aug-cc-pVDZ level using Gaussian98 (*36*) are presented in Table I. It is observed that the charge separation is similar for CO_2 and H_2O except for the reversal of the signs of the charges on the end atoms. Although partial atomic charges are only an abstract representation of a more complex electron density picture, at this point it is clear that Table I suggests that, as a solvent, CO_2 has the capacity to act as a Lewis acid or a Lewis base and that the oxygen atoms can take part in significant interactions with other chemical moeities. We should make clear that although the charge separation in CO_2 and H_2O are similar the intermolecular associations in H_2O make it a unique solvent system due to the extended networks of cooperative hydrogen bonds.

Table I. Comparison of the partial charges on the individual atoms of H_2O and CO_2. Mulliken charges as well as the charges derived by fitting the electrostatic potentials (CHELPG charges) in electrons (e).

H_2O			CO_2		
Atom	Mulliken (e)	CHELPG (e)	Atom	Mulliken (e)	CHELPG (e)
H (1)	0.16	0.36	O (1)	-0.18	-0.36
O	-0.32	-0.72	C	0.36	0.72
H (2)	0.16	0.36	O (2)	-0.18	-0.36

3. Cooperative C-H···O Interactions

The CO_2 oxygen can act as a weak Lewis base unit for interaction with Lewis acid groups or acidic hydrogen atoms while the carbon atom, as mentioned previously, can act as a Lewis acid. This observation is of relevance since most of the carbonyl compounds encountered have hydrogen atoms attached to the carbonyl carbon or to an α-carbon atom. These hydrogen atoms are somewhat acidic (due to the electron withdrawal by the carbonyl group) and carry partial positive charges. Our calculations indicate, in all cases studied to date, that these C-H bonds may be involved in weak C-H···O hydrogen bonds (37, 38) with oxygen atoms of the CO_2 molecule while the carbonyl oxygen acts as a Lewis base unit through interacts with the electron deficient carbon atom of the CO_2 molecule (39, 40) as shown in Figure 1. These two interactions also act cooperatively, each reinforcing the other, and contribute to further stabilization of the solute-solvent complex. In order to understand which among the carbonyl containing functionalities will be more suitable for the CO_2-philization of molecular systems we investigated the interaction energies of CO_2 with simple carbonyl containing molecules such as formaldehyde (HCHO), acetaldehyde (AcH), acetic acid (AcOH) and methyl acetate (MeOAc) in different interaction geometries. The optimized geometries for CO_2 dimer and the complexes of CO_2 with HCHO, AcH, and MeOAc are presented in Figures 2 and 3.

For all these systems the optimizations are carried out at the MP2 level of theory using the 6-31+G* basis set and the depth of the potential energy well are calculated by single point calculations at the optimized geometries using Dun-

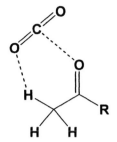

Figure 1. *Typical interaction geometry of CO_2 complexes involving a cooperative C-H···O interaction associated with a typical LA-LB interaction between CO_2 and a Lewis base system (carbonyl group) as in the interaction between CO_2 and an acetate group.*

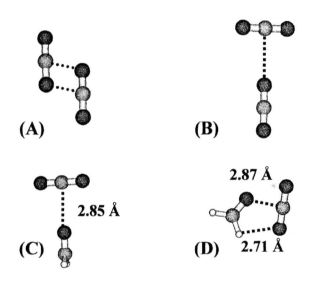

Figure 2. *Optimized structures (MP2/ 6-31+G*) for (A) the 'slipped' parallel (C2h symmetry) and (B) T-geometries (C2v symmetry) of the CO_2 dimer; (C) the T-structure (C2v symmetry) and (D) the proton side (Cs symmetry) configuration of the HCHO-CO_2 complexes (39).*

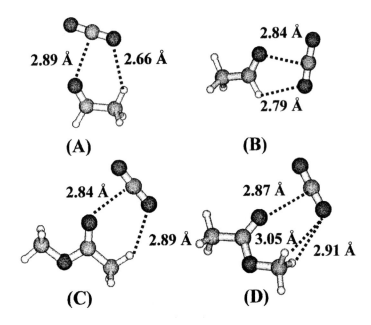

2.89 Å   2.66 Å

(A)

2.84 Å

2.79 Å

(B)

2.84 Å

2.89 Å

(C)

2.87 Å

3.05 Å
2.91 Å

(D)

Figure 3: Optimized structures (MP2/6-31+G*) for of AcH-CO₂ complex for the methyl side (A) and proton side (B) interaction geometries and the MeOAc-CO₂ complexes for the methyl side (C) and ester side (D) configurations (39).

ning's correlation consistent basis sets augmented by diffuse functions (aug-cc-pVDZ). Vibrational frequencies and other molecular parameters including bond distances, bond angles were also investigated to understand the nature of the interaction in more detail *(39)*. The interaction energies, corrected for basis set superposition error, are presented in Table II. For a comparison the interaction energies for the CO_2 dimer in the T-geometry and the slipped parallel configuration are also calculated at the same level of calculation.

In all the complexes studied, the CO_2 molecule is bent from its otherwise linear geometry as a result of this interaction. The two C=O bonds of the 'bound' CO_2 are non-identical. The C=O bond involved in the C-H···O interaction is consistently longer than the 'free' C=O, strongly supporting the presence of these interactions. Charge transfer and electron density changes in the systems upon complexation provides conclusive evidence for the existence of the C-H···O hydrogen bond *(39)*. A molecular orbital diagram of the methylacetate-CO_2 complex (Figure 4) also supports this view. The question remains as to the relative contribution of the C-H···O interaction to the overall stabilization, but such an estimate is elusive at present considering the intricate nature of the cooperativity of these interactions.

Solvation depends on the relative strengths of solvent-solvent, solute-solvent and solute-solute interactions apart from entropic considerations. While the interaction energy of the CO_2 dimer alone cannot represent exactly the solvent-solvent interaction cross-section in liquid and $scCO_2$ due to the pre-

Table II. BSSE corrected interaction energies (ΔE^c) for the CO_2 complexes.

Molecular species	ΔE^c (MP2) (kcal/mol)
HCHO-CO_2 (T)	-1.92
HCHO-CO_2 (P)	-2.43
AcH-CO_2 (M)	-2.52
AcH-CO_2 (P)	-2.69
MeOAc-CO_2 (M)	-2.82
MeOAc-CO_2 (E)	-2.64
AcOH-CO_2 (M)	-2.80
CO_2 dimer (‖)	-1.10
CO_2 dimer (T)	-0.94
H_2O---H_2O*	-4.11*

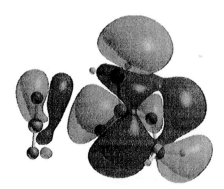

Figure 4. The highest occupied molecular orbital (HOMO) for the optimized geometry of the CO_2-methyl acetate complex (M) at the MP2/6-31+G* level (40).

sence of many-body interactions, they provide some measure of the solvent-solvent interactions. Our results for the complexes also are an approximation to the enthalpic contribution to solvation. The calculated interaction energies for all systems are presented in Table II (*39*). The results indicate that for all the simple carbonyl systems studied, the solute-solvent interaction is stronger than the solvent-solvent interaction, enthalpically favoring solvation of these materials in liquid and $scCO_2$.

Further, hydrocarbons functionalized with these carbonyl moieties may exhibit good CO_2-philicity. However, the solvation also depends on the entropic contribution and in the case of solids the lattice energies of materials functionalized with these carbonyl moieties. Table II indicates that among the various carbonyl compounds studied, methyl acetate has the strongest interaction with CO_2. This suggests that acetylation of polyhydroxy systems could be an amenable approach to increase the CO_2-philic nature of a compound. Carbohydrates represent a class of inexpensive, renewable, abundant and naturally occurring polyhydroxylated molecular systems and these calculations point to acetylation of carbohydrates as a method to soubilize this CO_2-phobic systems in CO_2. Acetylated carbohydrates are quite commonly available and they offer tremendous opportunity for the synthesis of inexpensive, renewable, environmentally benign CO_2-philes. This is the present approach that we are taking in the design of functional CO_2-philic systems.

4. Peracetylated Sugars

Based on these design principles, three peracetylated carbohydrate derivatives were selected as model systems (*40*) for our studies: the α- and β-forms of 1, 2, 3, 4, 6-pentaacetyl D-glucose (AGLU and BGLU, respectively) and 1, 2, 3, 4, 6-pentaacetyl β-D-galactose (BGAL). The chemical structure of these three stereoisomers are given in Figure 5. BGLU is a white solid that melts at 132°C under atmospheric pressure conditions (Figure 6A). However, as BGLU is exposed to CO_2 near room temperature (23.0°C) in a conventional high-pressure view cell, it absorbs CO_2 and becomes "wetted" with CO_2 at a pressure of 35-40 bar. The white solid appears as a salt does in a humid environment. At a gaseous CO_2 pressure of 55.9 bar a solid to liquid transition (deliquescence) occurs (Figure 6B) that is analogous to the deliquescence of hygroscopic materials absorbing atmospheric moisture. The carbohydrate melt continues to absorb CO_2 and swells to many times its original volume with addition of gaseous CO_2 pressure. At the liquid-vapor equilibrium pressure, the liquid CO_2 forms a separate layer on top of the viscous melt containing CO_2. However, the melt easily mixes with the upper layer of liquid CO_2 on stirring

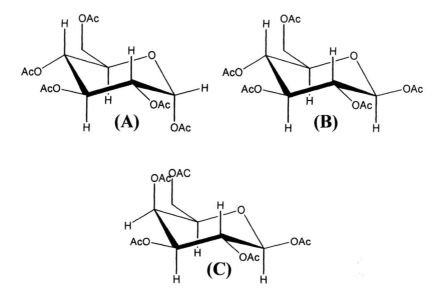

Figure 5. Structures of the stereoisomers (A) AGLU, (B) BGLU, and (C) BGAL.

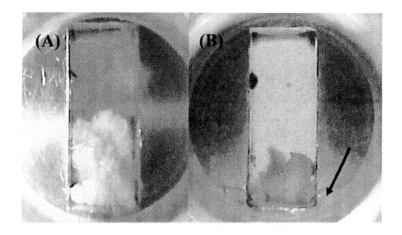

Figure 6. Photographs demonstrating the deliquescence of BGLU in CO_2 at 23.0 °C (A) solid material (B) in contact with gaseous CO_2 at the deliquescence pressure of 55.9 bar undergoing the solid-to-liquid transition as indicated by the black arrow.

and forms a single-phase liquid mixture in contact with the gaseous CO_2 phase. Further addition of CO_2 dilutes this liquid phase until beyond the liquid-vapor equilibrium a single phase is observed.

An approximate estimate of the BGLU concentration in the final melt reveals that the system contains more than 80 wt% of BGLU and can be diluted with liquid or $scCO_2$ in any proportion desired. This suggests the use of these systems or other derivatives (e.g. peracetylated trehalose) as viscosity enhancers in liquid and supercritical CO_2 solutions. The deliquescence point of AGLU is lower than that of BGLU by about 6-7 bar. BGAL does not exhibit deliquescence though it is readily soluble in liquid CO_2. These observations can be directly correlated to the differences in lattice energies as reflected in the melting points of the three compounds, AGLU, BGLU, and BGAL which are 109, 132 and 142°C respectively.

The cloud-point pressures of these systems dissolved in $scCO_2$ were examined at 40.0°C. As in the subcritical case, initially the solid melts and swells (for AGLU and BGLU) and all three peracetylated sugars readily go into a single phase, $scCO_2$ system. A plot of the cloud-point pressure versus the wt% for AGLU, BGLU and BGAL dissolved in supercritical CO_2 at 40°C is given in Figure 7 (40).

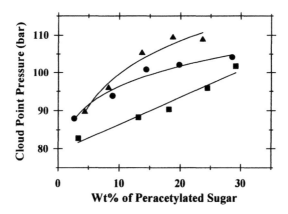

Figure 7. Plot of the cloud-point pressure versus the weight percentage of the carbohydrate-derivative for AGLU (■); BGLU (●) and BGAL (▲) in supercritical CO_2 at a temperature of 40.0 °C (40).
(Reproduced from reference 40. Copyright 2002 American Chemical Society.)

At the cloud-point pressure, phase separation begins between $scCO_2$ and the sugar ester. Upon lowering the pressure, the material reappears in the solid state. No cloud-point measurements were made above 30 wt% due to limitations arising from the volume of the view cell and the rapid swelling of the sample in the cases of AGLU and BGLU. Considering this cloud-point data and the data presented in Figure 6, it is apparent that the mixtures of AGLU and BGLU show complete miscibility at relatively low pressures with the 3-phase line being shifted to extremely low pressures. It is interesting to note that in Figure 7 there are substantial differences in the cloud points of these three stereoisomers. For the three peracetylated sugars shown in Figure 5 an analysis of the minimum energy structures calculated using density functional theory (DFT) shows an increase in the number of intramolecular C-H···O contacts going from AGLU < BGLU < BGAL (*40*). This increase is correlated with an increase of the cloud point pressure for these compounds suggesting that the intramolecular C-H···O contacts, involving the acetate carbonyl groups and methyl hydrogens, interfere with the ability of CO_2 to form a full solvation sphere. It is important to point out that this is a hypothesis and that one must fully consider the lattice energy as well as the enthalpic and entropic interactions in determination of solvation structures. Nevertheless the understanding of the stereochemical aspects revealed here may provide important insights toward the design of larger CO_2-philic molecules since there is a definite dependence on the configuration of the individual isomers. Additional experiments were performed on two disaccharides, sucrose octaacetate (linked glucose and fructose) and cellobiose octaacetate (linked glucose and glucose). The former was highly soluble in $scCO_2$ (40.0°C) with a 30 wt% solution having a cloud point of 95.1 bar and the latter being completely insoluble. These experiments were extended to a polysaccharide (peracetylated-β-cyclodextrin) in which a 0.5 wt% solution was solubilized in $scCO_2$ (40.0°C) with a cloud point of 193.1 bar. This system had solubilities that were limited to a few weight percent in contrast to the mono- and disaccharide systems. However, the compound is rich in the number of possible applications opening up a new realm of nontoxic systems for use in CO_2.

5. Summary

Based on our computational studies, we proposed that acetylation of polyhydroxy systems such as carbohydrates can be used as a viable method for the preparation of environmentally benign CO_2-philes. The ideas put forth regarding cooperative C-H···O interactions could lead to other CO_2-philes in which CO_2 concurrently acts as a Lewis acid and a Lewis base. The high solubility of peracetylated carbohydrates has several important consequences, namely, the first demonstration of renewable, environmentally benign, nontoxic hydrocarbon-based CO_2-philic systems. The future for these systems appears bright with regard to enabling green chemistry applications using CO_2 in a wide range of processes including food, pharmaceutical, carbohydrate synthesis and materials processing. Use of CO_2 as a solvent is ideal for food and pharmaceutical processes not only due to the non-toxicity of the solvent, but also due to the low critical temperature of CO_2 (31.1°C) of CO_2. Among the carbohydrate systems it is of special importance to mention the use of cyclodextrins which form complexes and have many applications including the sustained release and degradation protection of pharmaceuticals.

References

1. *Technology Vision 2020, The U.S. Chemical Industry.* Multi-organization report administered by the American Chemical Society, Department of Government Relations and Science Policy, December, **1996.**
2. P. T. Anastas, J. C. Warner, *Green Chemistry: Theory and Practice*, Oxford University Press Inc., New York, 1998.
3. Poliakoff, M.; Anastas, P. T. *Nature* **2001**, *413*, 257.
4. Corma, A.; Nemeth, L. T.; Renz, M.; Valencia, S. *Nature* **2001**, *412*, 423-425.
5. DeSimone, J. M. *Science* **2002**, *297*, 799-803.
6. Cross, R. A.; Kalra, B. *Science* **2002**, *297*, 803-807.
7. Poliakoff, M.; Fitzpatrick, J. M.; Farren, T. R.; Anastas, P. T. *Science* **2002**, *297*, 807-810.
8. Sheldon, R. A. *CHEMTECH*, March **1994**, 38-47.
9. Sheldon, R. A. *J. Chem. Tech. Biotechnology* **1997**, *68*, 381-388.
10. Laintz, K.E.; Wai, C. M.; Yonker, C. R.; Smith, R. D. *J. Supercrit. Fluids* **1991**, *4*, 194-198.
11. DeSimone, J. M.; Guan, Z.; Elsbernd, C. S. *Science* **1992**, *267*, 945-947.
12. M. A. McHugh, V. J. Krukonis, *Supercritical fluid extractions: principles principles and practice*, 2nd ed. Butterworth-Heinerman: Boston, MA, 1994.
13. Wells, S. L.; DeSimone, J. M. *Angew. Chem. Int. Ed.* **2001**, *40*, 518-527.
14. Rindfleisch, F.; DiNoia, T. P.; McHugh, M. A. *J. Phys. Chem.* **1996**, *100*, 15581-15587.
15. Beyer, C.; Oellrich, L. R.; McHugh, M. A. *Chem. Eng. & Tech.* 2000, *23*, 592.
16. Sarbu, T.; Styranec, T., Beckman, E. J. *Nature* **2000**, *405*, 165-168.
17. Eckert, C. A.; Knuston, B. L; Debendetti, P. G. *Nature* **1996**, *373*, 313-318.
18. Johnston, K. P.; Harrison, K. L.; Clarke, M. J.; Howdle, S. M.; Heitz, M. P.; Bright, F. V.; Carlier, C.; Randolph, T. W. *Science* **1996**, *271*, 624-626.
19. Ji, M.; Chen, X.; Wai, C. M.; Fulton, J. L. *J. Am. Chem. Soc.* **1999**, *121*, 2631-2632.

20. Ohde, H.; Wai, C. M.; Kim, H.; Kim, J.; Ohde, M. *J. Am. Chem. Soc.* **2002**, *124*, 4540-4541.

21. Subramaniam, B.; Rajewski, R. A.; Snavely, K. *J. Pharm. Sci.* **1997**, *86*, 885-889.

22. Mohamed, R. S.; Debendetti, P. G.; Prud'homme, R. K. *AIChe. J.* **1989**, *35*, 325-328.

23. McClain, J. B.; Londono, D.; Combes, J. R.; Romack, T. J.; Canelas, D. A.; Betts, D. E.; Samulski, E. T.; Wignal, G.; DeSimone, J. M. *Science* **1996**, *274*, 2049-2052.

24. P. G. Jessop, W. Leitner, Eds., *Chemical Synthesis Using Supercritical Fluids*, Wiley-VCH, Weinheim, Germany 1999.

25. Oakes, R. S.; Clifford, A. A.; Rayner, C. M. *J. Chem. Soc., Perkin Trans. 1* **2001**, 917-941.

26. Consani, K. A.; Smith, R. D. *J. Supercrit. Fluids* **1990**, *3*, 51-65.

27. Ellis, D. A.; Mabury, S. A.; Martin, J. W.; Muir, D. C. G. *Nature* **2001**, *412*, 321-324.

28. Kazarian, S. G.; Vincent, M. F.; Bright, F. V.; Liotta, C. L., Eckert, C. A. *J. Am. Chem. Soc.* **1996**, *118*, 1729-1736.

29. Nelson, M. R.; Borkman, R. F. *J. Phys. Chem. A* **1998**, *102*, 7860.

30. Meredith, J. C.; Johnston, K. P.; Seminario, J. M.; Kazarian, S.G.; Eckert, C. A. *J. Phys. Chem.* **1996**, *100*, 10837-10848.

31. Illies, A. J.; McKee, M. L.; Schlegel, H. B. *J. Phys. Chem.* **1987**, *91*, 3489-3494.

32. Nesbitt, D. J. *Chem. Rev.* **1988**, *88*, 843-870.

33. Jucks, K. W.; Huang, Z. S.; Miller, R. E.; Fraser, G. T., Pine, A. S.; Lafferty, W. J. *J. Chem. Phys.* **1988**, *88*, 2185-2195.

34. Beckman, E. J.; Enick, R. M.; Huang, Z. H., Design of Hydrocarbon CO_2-philic polymers, Abstracts Pap. ACS, 223: 315-IEC Part 1 APR 7 2002.

35. Kauffman, J. F. *J. Phys. Chem. A* **2001**, *105*, 3433-3442.

36. *Gaussian 98* (Revision A.7), M. J. Frisch, G. W. Trucks, H. B. Schlegel, G. E. Scuseria, M. A. Robb, J. R. Cheeseman, V. G. Zakrzewski, J. A. Montgomery, R. E. Stratmann, J. C. Burant, S. Dapprich, J. M. Millam, A. D. Daniels, K. N. Kudin, M. C. Strain, O. Farkas, J. Tomasi, V. Barone, M. Cossi, R. Cammi, B. Mennucci, C. Pomelli, C. Adamo, S. Clifford, J. Ochterski, G. A. Petersson, P. Y. Ayala, Q. Cui, K. Morokuma, D. K. Malick, A. D. Rabuck, K. Raghavachari, J. B. Foresman, J. Cioslowski, J. V. Ortiz, B. B. Stefanov, G. Liu, A. Liashenko, P. Piskorz, I. Komaromi, R. Gomperts, R. L. Martin, D. J. Fox, T. Keith, M. A. Al-Laham, C. Y. Peng, A. Nanayakkara, C. Gonzalez, M. Challacombe, P. M. W. Gill, B. G.

Johnson, W. Chen, M. W. Wong, J. L. Andres, M. Head-Gordon, E. S. Replogle and J. A. Pople, Gaussian, Inc., Pittsburgh PA, 1998.

37. Gu, Y.; Kar, T.; Scheiner, S. *J. Am. Chem. Soc.* **1999**, *121*, 9411-9422.

38. Desiraju, G. R. *Acc. Chem. Res.* **1996**, *29*, 441-449.

39. Raveendran, P.; Wallen, S. L. *J. Am. Chem. Soc.* **2002** *124*, 12590-12599.

40. Raveendran, P.; Wallen, S. L. *J. Am. Chem. Soc.* **2002**, *124*, 7275-7276.

Chapter 19

Design and Performance of Surfactants for Carbon Dioxide

Julian Eastoe[1,*], Audrey Dupont[1], Alison Paul[1], David C. Steytler[2,*], and Emily Rumsey[2]

[1]School of Chemistry, University of Bristol, Bristol BS8 1TS, United Kingdom
[2]School of Chemical Sciences, University of East Anglia, Norwich NR4 7TJ, United Kingdom

Phase stability and aggregation structures of various CO_2-active surfactants are reported. Evidence for aggregation in carbon dioxide has been obtained using High-pressure Small-Angle Neutron Scattering (SANS). Custom synthesized double-fluorocarbon chain compounds are shown to be effective at stabilizing water-in-CO_2 microemulsion droplets. However, branched, methylated hydrocarbon surfactants tend to form reversed micelles, which are alone ineffective at dispersing water.

Over the past years, concerns have been raised, as the emission of volatile organic compounds (VOC's) and effluent waste streams represent an environmental threat. Fortunately, CO_2 is one of the few solvents that is not regulated as a VOC by the US Environmental Protection Agency (EPA). Recently, effort has been made to promote the use of supercritical CO_2 (sc-CO_2) and to develop suitable formulations for a whole range of industrial processes. For these applications, CO_2 has a number of attractive properties: low cost, environmental friendliness, it is non-flammable and non-toxic for food and pharmaceutical grade uses. Furthermore, as a supercritical fluid, its critical pressure and temperature (respectively 73.8 bar and 31.1 °C) are readily accessible so that solvent quality can be fine-tuned by T-P variation. Therefore sc-CO_2 offers great opportunities for applications in a variety of domains including extractions (1) and polymer processing (2). More recent applications include the production of silver nanoparticles (3-6), the preparation of nanocrystals (7) and of carbon nanotubes (8).

CO_2-active surfactants have been developed to boost solubility levels through formation of micelles and water-in-CO_2 (w/c) microemulsions. Preliminary studies by Consani and Smith (9) showed that most conventional surfactants are insoluble in CO_2. They studied over 130 commercially available surfactants, among which only a few nonionics exhibited reasonable solubility in CO_2. On the other hand (for sound reasons (10, 11)), certain fluorocarbon (12, 13), and to a lesser extent silicone surfactants (14-17) are known to dissolve in CO_2. However, such compounds often require specialized synthesis and therefore remain expensive. Although supercritical conditions are useful in fractionation processes, the cost and engineering requirement for scale-up of practical applications of CO_2 are much simplified when liquid CO_2 is used at its vapor pressure. The ability to stabilize microemulsions under this condition is therefore an important feature of surfactant design. Being mindful of the high costs, and environmental issues relating to fluoro-surfactants, considerable efforts are also being made to design CO_2-active hydrocarbons.

This article provides a review of relevant literature (§1), a summary of the behavior of di-fluorocarbon chain sulfosuccinates and phosphates (§2), and finally, section 3 covers aggregation of various hydrocarbon surfactants in CO_2.

1. CO_2-active Surfactants

Following the finding by Iezzi *et al.* (10) that fluorocarbons and CO_2 are compatible, Hoefling *et al.* (11-12, 14-15) designed the first effective surfactants for CO_2. These surfactants were made of low cohesive energy density tails; by insertion of groups with low solubility parameters and low polarizability, such as fluoroalkyls (11), fluoroethers (12) and silicones (14, 15). The effect of tail branching was also found to boost solubility and double-tailed surfactants were noted to perform better. To date, amongst others described in section 2 below, fluoroacrylates (18), fluoroethers (12, 19) and a few fluoroalkyls (11, 20) have been shown to be compatible with CO_2.

Low Molecular Weight Fluoroalkyls

Harrison et al. reported the first w/c microemulsion in 1994 (20). A hybrid surfactant, namely F7H7, made of respectively one hydrocarbon and one fluorocarbon chain attached onto the same sulfate head group, was able to stabilize a w/c microemulsion at 35 and 262 bar. For a surfactant concentration of 1.9 wt %, water up to a w = 32 value ([water]/surfactant]) could be dispersed. A spherical micellar structure was confirmed by small-angle neutron scattering (SANS) experiments (21). This surfactant was later the subject of dynamic molecular simulations (22, 23). The calculations were consistent with the SANS data and high diffusivity was predicted, highlighting this important feature of low-density and low-viscosity supercritical fluids (SCF).

In 1997, Eastoe et al. launched a study of fluorinated anionic sulfosuccinate surfactants, which are analogous to the hydrocarbon Aerosol-OT (sodium bis-2-ethylhexyl sulfosuccinate, AOT) (13, 24-26, 27 a). Example molecular structures are shown on Figure 1. Furthermore, fluoroalkylphosphates have been assessed for CO_2-activity (27 b, c). A detailed presentation of these systems is given in section 2 below. In 2001, Erkey and co-workers also showed interest in these very promising sulfosuccinate surfactants (28, 29). Small-angle X-ray scattering (SAXS) was used to characterize di-HCF$_4$ at 0.1 M in CO_2 at 27 °C and 345 bar for w between 0 and 20 (29).

Fluoroethers: PFPE-based surfactants

In 1993, perfluoropolyether (PFPE) carboxylates, with average molecular weights between 2,500 and 7,500, were reported to be soluble in liquid CO_2 (19). However, these high MW polymers were not effective at stabilizing w/c microemulsions. Later, Johnson et al. formed w/c microemulsions with an ammonium carboxylate PFPE (PFPE-COO$^-$NH$_4^+$) surfactant of only 740 MW (30). Success with these surfactants was attributed to the chemical structure itself. PFPE constitutes an extremely CO_2-philic tail group, accentuated by the presence of pendant fluoromethyl groups, which tend to increase the volume at the interface on the CO_2 side and thus favor curvature around the water.

Spectroscopic techniques, such as FTIR spectroscopy and UV-vis (31), were used to give evidence for formation of reverse micelles with bulk like water in the droplet cores, according to the model proposed by Zinsili (32). Electron paramagnetic resonance (EPR) provided further evidence that PFPE-based surfactants form stable reverse micelles in CO_2, and the presence of microemulsified water (31). More recently, standard NMR (^{1}H and ^{19}F) and also

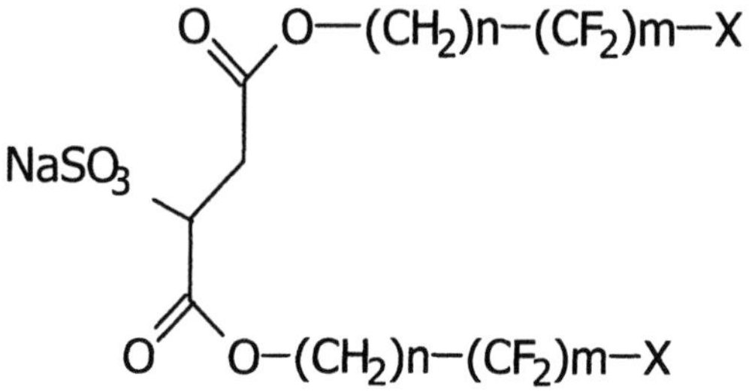

$$O-(CH_2)n-(CF_2)m-X$$

NaSO₃

$$O-(CH_2)n-(CF_2)m-X$$

di-HCF4	n=1	m=4	X=H
di-HCF6	n=1	m=6	X=H
di-CF3	n=1	m=3	X=F
di-CF4	n=1	m=4	X=F
di-CF6	n=1	m=6	X=F
di-CF4H	n=2	m=4	X=F
di-CF6H	n=2	m=6	X=F

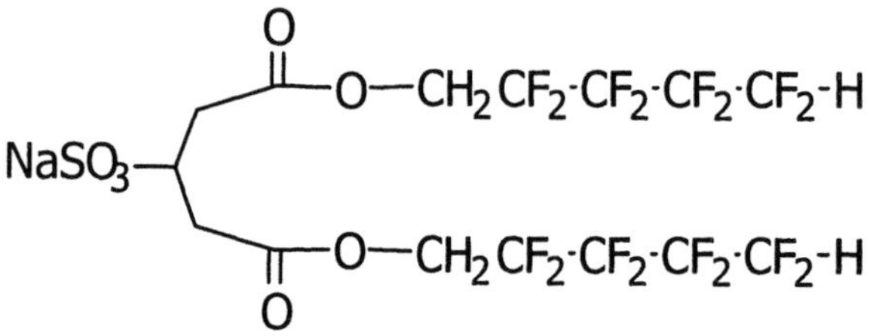

NaSO₃

$$O-CH_2CF_2\cdot CF_2\cdot CF_2\cdot CF_2\cdot H$$

$$O-CH_2CF_2\cdot CF_2\cdot CF_2\cdot CF_2\cdot H$$

di-HCF4GLU

*Figure 1. Partially fluorinated sulfosuccinate surfactants for use in CO_2. Reproduced with permission from Langmuir, **2002**, 18, 3014-3017. Copyright 2002 Am. Chem. Soc.*

rotating frame-imaging (RFI) NMR were used to demonstrate micelle formation in w/c microemulsions with 1.8-wt % PFPE-based surfactant (33). Scattering techniques provide the most definite proof of micellar aggregation. Zielinski *et al.* (34) employed SANS to study the droplet structures in these systems. Conductivity measurements (35) and SANS (36) were also used to study droplet interactions at high volume fraction ϕ in w/c microemulsions formed with a PFPE-COO⁻NH$_4^+$ surfactant (MW = 672). Scattering data were successfully fitted by Schultz distribution of polydisperse spheres (see footnote 37). A range of PFPE-COO⁻NH$_4^+$ surfactants were also shown to form w/c emulsions consisting of equal amount of CO_2 and brine (38-40).

The use of cationic surfactants would be of practical interest in applications on negatively charged surfaces. Recently a polyperfluoroether trimethylammonium acetate (PFPE-TMAA), with an average MW of 1124, formed w/c microemulsions up to w = 32 (41). The results compared favorably with those for anionic PFPE-COO⁻NH$_4^+$ (42).

Fluoroacrylates: PFOA Block Copolymers

In 1992, the fluorinated acrylate polymer, poly (1,1-dihydroperfluoro-octylacrylate) or PFOA, was obtained from homopolymerization in *sc*-CO_2 at 59.4 °C and 207 bars, with MW 270,000 (18). SANS investigations of dilute solutions of PFOA in CO_2 (43), over a wide range of temperatures and pressures, provided clear evidence for favorable interaction between PFOA and CO_2. PFOA grafted with 5 kg mol⁻¹ polyethylene oxide (PEO) segment (PFOA-*g*-PEO) was shown to be CO_2compatible (44, 45).

Block copolymers of polystyrene and PFOA, PS-*b*-PFOA, in CO_2 were also extensively studied by SAXS, SANS (46, 47), high-pressure, high-resolution (HPHR) NMR (48) and pulsed magnetic field gradient (PFG) NMR (49). Neutron scattering data (46) showed that the micellar size increases with the PFOA mass, when holding MW of the PS component constant. The effect of CO_2 density (0.27 to 0.84 g.cm⁻³) at constant temperature (60 °C), and of temperature (29.5 to 76.2 °C) at constant CO_2 density (0.733 g. cm⁻³) was studied on PFOA/CO_2 systems by HPHR NMR (48).

Transitions may occur in block copolymer/CO_2 systems, due to change in temperature and pressure. Subsequently a critical micelle density (CMD), analogous to the critical micellisation concentration (CMC) was found. The existence of a CMD was first argued by McClain *et al.* (47), who observed a transition from aggregates to unimers for the PS-*b*-PFOA/CO_2 system, as solvent density was increased. Further evidence for the CMD was provided by Triolo *et al.* (50-52) and Zhou and Chu (53, 54).

Silicones: Emergence of PDMS

For applications in CO_2, silicones are generally considered less effective than their fluorinated counterparts. Poly(dimethyl siloxane) (PDMS) solubility in CO_2 was first reported in 1996; at a level of 4 %wt, PDMS (Mn ~ 13k) is soluble in CO_2 at 35 °C and 277 bar (55). Block copolymer surfactants consisting of CO_2-philic PDMS and CO_2-phobic ionizable poly(methacrylic acid) (PMA) or poly(acrylic acid) (PAA) were used to form w/c and c/w emulsions (17). These PDMS-based block copolymers exhibited remarkable "ambidextrous" behavior to stabilize PPMA particles both in CO_2 on the one hand, and in water on the other hand (56, 57). Steric stabilization is imparted in CO_2 by PDMS, which is significantly more soluble in a CO_2/MMA mixture than in pure CO_2 (58).

Development of Hydrocarbon Systems: High Hopes

Efforts have been made to develop hydrocarbon systems for CO_2, as they could present significant advantages over high-cost fluorocarbon or siloxane counterparts. Recent advances are covered in section 3 of this article. Solubility of hydrocarbon materials in CO_2 may be achieved by the addition of a polar co-solvent to CO_2 to improve solvent polarity. For instance, AOT was shown by Ihara *et al.* to be completely soluble in CO_2 with ethanol as a co-solvent (59). Along similar lines, Hutton *et al.* in 1999, formed w/c microemulsions with 0.03 M AOT and 15 mol % ethanol or 10 mol % pentanol (60, 61).

Nonionic surfactants are also known to exhibit quite high solubilities in *sc*-CO_2. Nonionic polyethers, C_8E_5 (62) and $C_{12}E_4$ (63), were extensively studied in CO_2 with and without the addition of pentanol as co-solvent. Water solubility up to w = 12 was observed for C_8E_5 with 10 wt % pentanol as a co-solvent. This $C_{12}E_4$ alone dispersed no water, however the addition of pentanol enhanced notably its solubility. Related nonionic surfactants were shown to exhibit solubility in CO_2 (64-66). In section 3 of this chapter a new SANS study of this mixture, and other nonionic surfactants in CO_2, are reported.

Another approach to use AOT as a surfactant for CO_2 microemulsion consists in mixing AOT with a PFPE-based surfactant. This may seem quite disconcerting at first sight, as fluorocarbons and hydrocarbons are notoriously known to be immiscible. However Fulton *et al.* (67); reported microemulsion formation by mixing 15 mM of AOT and 30 mM of PFPE-PO_4 up to w = 12. These systems could be successfully used as micro reactors to synthesize metallic silver (3, 4) and copper nanoparticles (5) and to carry out catalytic hydrogenations (5). Eastoe *et al.* also showed that hydrocarbon surfactants analogous to AOT with branched tails were CO_2 compatible (68). More detail is given in section 3 below.

In 2000, Beckman was once again the pioneer, setting rules for designing hydrocarbon copolymers in CO_2 (69). Different types of monomers were assembled, so that specific interactions with CO_2 were favoured via high flexibility, high free volume, low cohesive energy density and Lewis basicity (via carbonyl groups). Poly(ether-carbonate) was synthesized by copolymerisation of CO_2 with propylene oxide. $(CHO-CO_2)_{25}-(EO)_7-(CHO-CO_2)_{25}$ was soluble in CO_2 at low pressures (below 140 bar) and formed a w/c emulsion, which was stable for hours upon stirring.

Applications: Towards CO_2-"Green" Chemistry

The use of CO_2 as a reaction medium for dispersion (18) or inverse emulsion polymerization (70) has been extensively studied. Due to its excellent transport properties (low viscosity and high diffusivity), CO_2 is usually a beneficial solvent for reactions. For instance, the polymerization of methyl methacrylate is now well established via various methods (71-75).

Based on the polarity difference between CO_2 and the interior of the micelles, w/c microemulsions have found many applications as extraction media. Furthermore, by modifying pressure and temperature, solvent quality may be changed and it becomes, therefore, possible to exert a real control over the extraction process; uptake of solutes inside micelles may be varied. This may be of use for separations/extractions involving bio-chemicals and proteins. In conventional solvents their separation from the reaction medium can be quite complicated, involving tedious processes such as fluid-fluid extraction, decantation, chromatography column, filtration, precipitation. Use of supercritical fluid technology with extraction in reverse micelles seems advantageous for proteins (e.g. 19, 76). This process was also used for the extraction of metals (77-79) and more recently of copper from a filter paper surface (1).

Finally, w/c and c/w PFPE based emulsions have been used for the synthesis of porous materials, which are the skeletal replica of the emulsions after removal of the internal phase. W/c microemulsions allowed for macroporous polyacrylate monoliths to be produced (80-82). Conversely, c/w emulsions may be used for the preparation of well-defined porous hydrophilic polymers (83).

2. Fluorinated Anionic Surfactants

In spite of the various potential applications little is known on the important issue of how to optimize surfactant chemical structure for CO_2 dispersions. This is partly due to a lack of suitable, well-characterized compounds, and that one of the most studied surfactants is a technical grade product perfluorpolyether (PFPE) consisting of a molecular weight distribution (e.g. 19, 30, 31, 34-36, 38). However, fluorinated analogues of Aerosol-OT (sodium bis-2-ethylhexyl sulfosuccinate, AOT) provide a suitable platform for delineating structure-performance relationships in CO_2 (chemical structures and nomenclature given on Figure 1).

Air-water Surface Tensions

Surface tension measurements were used to check surfactant purity, as well as determine critical micelle concentrations (cmc's) and limiting surface tensions at the cmc γ_{cmc}'s. The results are listed in Table 1. Methods and techniques described elsewhere (25, 84) were used to obtain representative equilibrium adsorption isotherms. This tension γ_{cmc} is characteristic of the efficiency of a given surfactant, and it is strongly linked to the hydrophobic chain chemical structure (85).

Water-in-CO_2 Microemulsion Stability

Of twelve different linear chain fluoro-succinates investigated, nine formed w/c microemulsions with water content w=10 ([water]/[surfactant]); these were di-HCF4, di-HCF6, di-CF3, diCF4, diCF6, diCF4H, diCF6H, di-CF4GLU and the cobalt salt Co-HCF4. As the phase transition pressure (P_{trans}) is approached from the high-pressure transparent region, these systems quickly develop opacity around ±5 bars of P_{trans}. Neither of the shortest chain compounds, di-CF2 and di-HCF2, or the long chain di-HCF8 formed stable microemulsions. On the other hand di-CF4, di-CF6 and di-CF4H were very effective, and gave w=10 microemulsions at CO_2 bottle pressure (57 bar) at 15°C. This indicates a strong chemical specificity and is consistent with an optimum chain length for w/c formation. Issues of phase boundary reproducibility, effects of (i) surfactant chain length, (ii) fluorination level and (iii) headgroup/counterion type have been dealt with in detail elsewhere (24-28).

Correlation Between Cloud Points and Aqueous Surface Tensions

Figure 2 demonstrates an interesting correlation between the limiting surface tension γ_{cmc} and the corresponding w/c phase boundary pressure (for w=10 systems at 0.05 mol dm^{-3} and 15°C), for eight different related surfactants (27a). As shown, there is a distinction between the behavior of H-CF$_2$- tipped surfactants (higher γ_{cmc} and P_{trans}) and that for the class of F-CF$_2$ compounds, which on the whole perform best in CO_2.

Table 1. Aqueous Phase Critical Micelle Concentrations (cmc's), Limiting Surface Tensions γ_{cmc}'s and Microemulsion Stability Pressures for Fluorinated Surfactants.

	$cmc / (mmol\ dm^{-3})$	$\gamma_{cmc} / (mN\ m^{-1})$	P_{trans} / bar
di-HCF4	12	26.8	193
di-HCF6	0.8	24.1	163
di-CF3	12	17.8	124
di-CF4	2	17.7	70
di-CF6	0.1	15.5	77
di-CF4H	0.7	15.6	89
di-CF6H	0.05	22.0	139
di-HCF4GLU	11	25.8	181

NOTE: Temperatures; di-HCF4 25°C, di-HCF6 40°C, di-CF3 25°C, di-CF4 30°C, di-CF6 40°C, di-CF4H 25°C, di-CF6H 40°C, di-HCF4GLU 25°C. The phase transition pressures P_{trans} water-in CO_2 microemulsions are for systems with [surf] = 0.05 mol dm^{-3} and w=10 at 25 °C (w=[water]/[surfactant]).

SOURCE: Reproduced with permission from *Langmuir*, **2002**, *18*, 3014-3018. Copyright 2002 Am. Chem. Soc.

Although there have been previous reports of w/c formation with fluoro-surfactants of this type (24, 26, 28, 29), new results on the widest range of compounds available to date are reported here. Gratifyingly, for di-HCF4, there is good agreement between the new results presented here and those of Erkey. For the longer chain-length surfactant (di-CF6H, us, or di-HCF7 in 28) the experimental conditions used here are different and so direct comparisons cannot be made. An important finding is highlighted for the first time: a clear correlation between surface tension of an aqueous solution and the performance of the compound in a w/c microemulsion. This represents a new paradigm for designing highly efficient CO_2-philes.

Water-in-CO_2 Microemulsion Structure

Microemulsion structures were investigated by high-pressure small-angle neutron scattering (SANS). Engineering considerations limit the maximum pressure of the SANS cell to about 600 bar. Therefore, the experimental conditions were chosen as T= 15°C and P = 500 bar, so as to strike a compromise between this and the need to minimize attractive interactions, which increase on approach to P_c. The scattering was consistent with the added D_2O

being dispersed as microemulsion droplets, and fitted radii for three different surfactants are plotted in Figure 3 as a function of water content w. These surfactants exhibit almost identical behavior, at least within the experimental error on R_c (± 2 Å). Although the phase stability depends strongly on surfactant type, there seems to be no obvious effect on internal droplet structure or droplet radius.

The effective molecular area at the water-CO_2 interface, A_h, may be estimated using eq 1 which is valid for spherical droplets.

$$A_h = \frac{3v_{D2O}\, w}{R_c} \tag{1}$$

In eq 1 v_{D2O} is the volume of a water molecule. Hence, A_h can be estimated as 115 ± 5 Å^2, and the intercept is consistent with a radius for the polar core of a "dry" micelle of around 11 Å. This head group area A_h can also be calculated from high Q SANS data with the Porod equation, as described elsewhere (13, 86). These results indicate that fluoro-surfactants at the water-CO_2 interface adopt a lower packing density than a related hydrocarbon surfactant (AOT) at analogous water-alkane interfaces (86).

Microemulsion Formation with Fluorinated Phosphate Surfactants

Based on the success of these fluoro-sulfosuccinates described above di-fluorocarbon phosphates have also been investigated. In terms of synthesis and raw materials costs these surfactants have significant advantages over the sulfosuccinates. Surfactants of this kind have also been studied by DeSimone *et al.* (27 c), and the synthesis and purification are described elsewhere (27b, c). Detailed SANS experiments are described in these papers (27b, c), and it is clear that surfactants of this kind stabilize aqueous nano-droplets. Hence, anionics other than sulfosuccinates may be employed in water-in-CO_2 microemulsions. Significantly, one of these compounds (di-HCF6-P, see ref 27 b) stabilizes microemulsions in liquid CO_2 at vapor pressure; a potentially useful result that may be of importance in facilitating applications.

3. Aggregation of Hydrocarbon Surfactants in CO_2

Owing to the high cost of fluorous chemicals efforts have been made to obtain CO_2-soluble hydrocarbon polymers and surfactants (68, 69). With readily available hydrocarbon surfactants there is some indirect spectroscopic evidence for aggregation in CO_2, but mostly in the presence of high levels of short-medium chain alcohols (60-62). The next section summarizes recent SANS

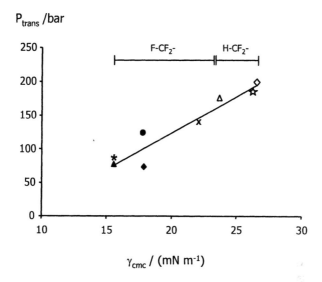

Figure 2. Correlation between P_{trans} in w/c systems and γ_{cmc} at the air-water interface. Phase transition pressures were measured at 25 °C for w = 10 microemulsions and 0.05 mol dm^{-3} surfactant concentration. Line is a guide to the eye. di-CF6 (▲) di-CF4H (*), diCF3 (•), di-CF4 (♦), di-HCF6 (Δ), di-CF6H (×),di-HCF4 (◊), di-HCF4GLU (✳). Reproduced with permission from Langmuir, 2002, 18, 3014-3017. Copyright 2002 Am. Chem. Soc.

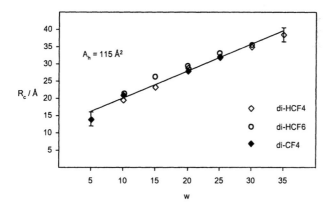

Figure 3. Variation of water-in-CO$_2$ microemulsion droplet radius with water content for three surfactants at T = 15 °C and P = 500 bar, and [surf] = 0.05 mol dm^{-3}. Reproduced from reference 13 by permission of the Royal Society of Chemistry.

evidence for micellisation of custom-made, and commercially available, hydrocarbon surfactants in CO_2.

Aerosol-OT analogues

It is known that γ_{cmc} decreases with the extent of methylation in the hydrophobic chain tip region (86, 87). The highly methyl-branched compounds A and B shown on Figure 4a showed the low γ_{cmc}; 27 and 28 mN m^{-1} respectively compared with 31 mN m^{-1} for AOT (86). Of course, the general structure of these hydrocarbons is closely related to that of the fluorocarbons, which were so effective. The idea was to use this sulfosuccinate backbone, and to substitute hydrocarbon chains, which are as CO_2-philic as possible, in place of the fluorocarbon tails. These two compounds both dissolve in stirred sc-CO_2: at 33°C and 250 bar at up to 0.1-wt%. The solid was initially insoluble, however after around 10 minutes surfactant began to melt, and after 20 minutes a single-phase transparent solution was formed. Under identical conditions AOT did not dissolve at all.

SANS curves obtained after subtraction of the cell+CO_2 background are shown in Figure 4b. Data fitting with the model in ref. 37 was consistent with "dry" reversed micelles, as observed with these surfactants in n-heptane solvent (86). UV-vis dye solubilisation measurements shown in Figure 4c also provide evidence for aggregation. If reversed micelles were present the positively charged chromophore, dimidium bromide (λ_{max} ~ 550 nm), should be incorporated owing to favorable interactions with the surfactant anionic head groups. No dimidium bromide absorbance could be detected in pure CO_2, consistent with an insoluble dye. Further, no uptake was seen with added Aerosol-OT (0.025 mol dm^{-3}). However, with the CO_2-soluble surfactant B as dispersant a spectrum characteristic of dimidium bromide in reversed micelles was obtained (Figure 4c).

Although the common anionic hydrocarbon surfactant Aerosol-OT does not aggregate in carbon dioxide, two related compounds, each possessing a high degree of chain tip methylation, do form reversed micelles in sc-CO_2. These principles may be helpful for designing a wider range of low-cost "CO_2-philic" hydrocarbon surfactants, especially for stabilizing bio- and food-compatible water-in-CO_2 emulsions and microemulsions.

Commercial non-ionics

Micellisation of various hydrocarbon surfactants in sc-CO_2 was also investigated by high pressure SANS. It can be emphasized that this study widely contrasts with other works where an indirect spectroscopic method is used to provide evidence for aggregation (e.g. 62-66). These surfactants, shown in Figure 5, were selected because of their low cost and CO_2-compatible chain tips, which are highly methylated as for the AOT analogues described above. Triton surfactants are commercially available, and are extensively used in industrial (88, 89) and pharmaceutical formulations (90, 91) and in biochemical research (92).

On Figure 5 example high-pressure SANS data confirm the presence of spheroidal aggregates in sc-CO_2. This provides further evidence for CO_2 compatibility of these branched-chained, t-butyl-tipped nonionic surfactants.

Table 2. Fitted Form Factor and Guinier Radii for Triton Surfactants.

Triton Surfactant	Fitted R ± 1/(Å)	Guinier R ± 1/(Å)
TX-45	16	13
TX-100R	18	16
TX-100	14	11

NOTE: SANS data were recorded a 40°C and 500 bar.

Water Uptake

It would be of great interest if water could be dispersed in the polar cores of the systems described above. Water was added to the binary systems (W=$[H_2O]$/[TX]=5), with an additional amount to account for the small water solubility in CO_2. In some cases added water caused the precipitation of the surfactant. Near-Infrared (NIR) spectra obtained for the Triton surfactants, and interpreted as in ref. 13, indicated no clear evidence of water solubilisation in the micelles and the signals were consistent with water mainly located in the bulk CO_2. Further attempts with brine (0.05 and 0.10 mol dm^{-3}) proved to be no more successful.

Surfactant-Alcohol-Water Mixtures

There have been reports (e.g. 60-63) claiming that reversed microemulsions can be induced to form by employing medium chain length alcohols as a "co-surfactant". The physical evidence comes mainly from indirect spectroscopic methods (λ_{max} shifts of probe dyes etc.). To improve understanding about these potentially important systems, SANS experiments were made. Figure 6 shows a Guinier plot (37) for I(Q) data obtained with a system suggested by Howdle *et .al* (62). Note the added alcohol is soluble in both CO_2 and water (weakly) hence, the main role is somewhat unclear since pentanol can act both as a co-solvent and co-surfactant. Since D_2O was used the effective contrast should be at the CO_2-D_2O interface. It is encouraging to see that the calculated radius of ~ 29Å is somewhat larger than that found for a water-free, CO_2-soluble reversed

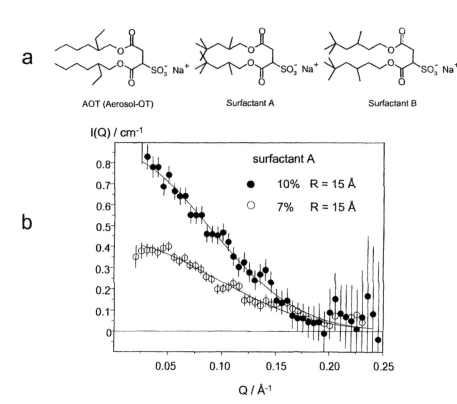

*Figure 4. (a) Surfactants AOT (Aerosol-OT) Sodium bis (2-ethyl-1-hexyl)sulfosuccinate, surfactant A (sodium bis (2,4,4-trimethyl-1-pentyl)sulfosuccinate) and surfactant B (sodium bis (3,5,5-trimethyl-1-hexyl) sulfosuccinate. (b) SANS data obtained after subtracting the cell+sc-CO_2 background for surfactant A as a function of concentration at 0.15 (●) and 0.10 (○) mol dm^{-3}. T = 33 °C and 500 bar. The fits are to a polydisperse sphere model with R^{av} = 14 ± 1 Å and σ/R^{av} = 0.20 (footnote 37). (c) UV-vis spectrum of dimidium bromide dispersed in sc-CO_2 with reversed micelles of surfactant B at 40 °C and 500 bar. The surfactant concentration is 0.025 mol dm^{-3}. Reproduced with permission from Journal of the American Society, **2001**, 123, 988-989. Copyright 2001 Am. Chem. Soc.*

C

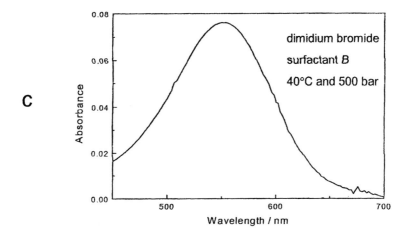

Figure 4. *Continued.*

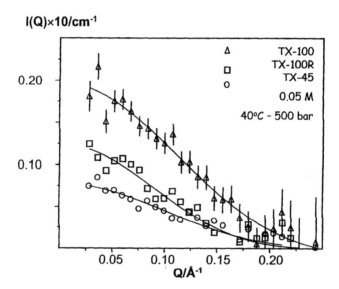

Figure 5. Molecular structures of the Triton surfactants. SANS data obtained after subtraction of the cell and sc-CO₂ background for Triton surfactants at 0.05 mol dm⁻³ and at 40 °C and 500 bars. Example error bars are shown for TX-100 only. The lines are model fits; fitted parameters are given in Table 2.

micelle with similar non-ionic surfactants (Table 2). It should be noted that a high level of surfactant and alcohol (26%) is needed to dissolve only a modest amount of water (4%). As stated in the previous section non-ionic surfactants alone do not stabilize "wet" reversed micelles, but in the presence of high alcohol levels water solubility appears to be significantly improved. Hence, it may be possible to formulate microemulsions in CO_2-rich solvents using commercially available surfactants and alcohols.

Conclusions and Outlook

The benefits of employing surfactants in CO_2 are clear. However, achieving low-cost effective CO_2-philes still represents a major challenge, which continues to occupy a number of research groups worldwide. Owing to high CO_2 compatibility various fluorocarbon surfactants can now be used to stabilize water-in-CO_2 microemulsions. Of these custom synthesized molecules, certain ones (di-CF4 and di-HCF6-P) have been shown to be highly efficient, since they are able to stabilize dispersions at relatively low pressure, just above the vapor pressure of CO_2. A valid, question concerns the economics of fluorocarbon surfactants for applications in CO_2. At current market prices and using scientific (not bulk) suppliers the raw costs can be estimated at \$0.6/g, \$2.4/g and \$50/g for di-HCF4, di-CF4H and di-CF4 respectively (not including human resource costs). For any practical purposes di-CF4H (n=2) would be the preferred option, owing to cost, in spite of the ~20 bar increase in P_{trans} over the more expensive di-CF4. The use of added electrolyte is another strategy that can be employed to reduce phase stability pressures (93), so those relatively cheap active components such as di-HCF4 and di-HCF6 could be used.

A clear correlation has been observed between limiting surface tension γ_{cmc} and surfactant performance in water-in-CO_2 microemulsions, as measured by the phase transition pressure P_{trans}. These results have important implications for the rational design of CO_2-philic surfactants. Studies of aqueous solutions are relatively easy to carry out, and surface tension measurements can be used to screen target compounds expected to exhibit enhanced activity in CO_2. Therefore, potential surfactant candidates can be identified before making time-consuming phase stability measurements in high-pressure CO_2.

Building on the success of the fluoro-sulfosuccinates, hydrocarbon analogues, bearing t-butyl chain termini, were developed as possible candidates for CO_2-active compounds (68). The trends in γ_{cmc} within a surfactant series noted for fluoro-succinates were used to as a guide to design these molecules. This strategy switches a CO_2-insoluble compound, Aerosol-OT, into one which micellises in carbon dioxide. These ideas were extrapolated to commercially available Triton non-ionics, which also micellise in CO_2. Unfortunately, neither micelles of the hydrocarbon succinates, nor the Tritons, are able to disperse water. However, limited water solubility can be achieved if medium chain alcohol is mixed with a non-ionic surfactant. It should be noted that a high level of surfactant and alcohol (26%) is needed to dissolve only a modest amount of

In [I(Q)/cm⁻¹]

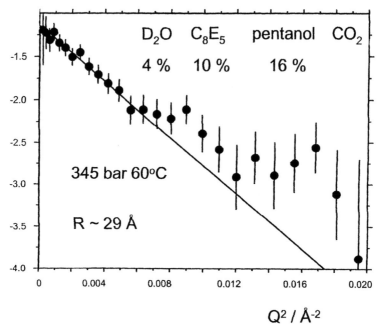

Figure 6. Guinier plot of SANS data for a dispersion of D_2O (4%), nonionic surfactant C_8E_5 (10%), pentanol (16%) and CO_2 at 60 °C and 345 bar. The gradient at low Q is consistent with water clusters of ~ 29 Å.

water (4%). As stated in the previous section non-ionic surfactants alone do not stabilized "wet" reversed micelles, but in the presence of high alcohol levels water solubility appears to be significantly improved. Hence, it may be possible to formulate microemulsions in CO_2-rich solvents using commercially available surfactants and alcohols, as an alternative to custom designing costly CO_2-philes.

Acknowledgements

Acknowledgment is made to the Donors of The Petroleum Research Fund, administered by the American Chemical Society, for partial travel support of JE. This work was funded under EPSRC grants GR/L05532 and GR/L25653. The EPSRC (AD, AP and ER) is thanked for studentship support. We thank CLRC for allocating beam time at ISIS and contributions towards consumables and travel. Dr. J Holmes (University College Cork, Ireland) and Prof. K.P. Johnston (University of Texas, Austin) are thanked for stimulating discussions.

References

1. Campbell, M. L., Apocada, D. L., Yates, M. Z., McCleskey, T. M., Birnbaum, E. R., *Langmuir* **2001**, 17, 5458.
2. Shiho, H., DeSimone, J. M., *Macromolecules* **2001**, *34*, 1198.
3. Ohde, H.; Rodriguez, J. M.; Ye, X.-R.; Wau, C.M. *Chem. Comm.* **2000**, 2353.
4. Ohde, H.; Hunt, F.; Wai, C. M. *Chem. Mater.* **2001**, *13*, 4130.
5. Ohde, H.; Wai, C. M.; Kim, H.; Kim, J.; Ohde, M. *J. Am. Chem. Soc.* **2002**, *124*, 4540.
6. Sun, Y-P., Atorngitjawat, P., Meziani, M. J., *Langmuir* **2001**, *17*, 5707.
7. Shah, P. S., Hussain, S., Johnston, K. P., Korgel, B. A., *J. Phys. Chem. B* **2001**, *105*, 9433.
8. Motiei, M., Rosenfeld Hacohen Y., Calderon-Moreno J., Gedanken, A., *J. Am. Chem. Soc.* **2001**, *123*, 8624.
9. Consani, K. A.; Smith, R. D. *J. Supercrit. Fluids* **1990**, *3*, 51.
10. Iezzi, A. In *Supercritical Fluid Science and Technology.* Johnston, K.P.; Penninger J.M.L. Eds.; ACS Symposium Series 406; American Chemical Society: Washington, DC, 1989;. pp 122.
11. Hoefling, T. A.; Enick, R. M.; Beckman, E. J. *J. Phys. Chem.* **1991**, *95*, 7127.
12. Newman, D. A.; Hoefling, T. A.; Beitle, R. R.; Beckman, E. J.; Enick, R. M. *J. Supercrit. Fluids*, **1993**, *6*, 205.
13. Eastoe, J.; Downer, A.; Paul, A.; Steytler, D. C.; Rumsey, E.; Penfold, J.; Heenan, R. K. *Phys. Chem. Chem. Phys.* **2000**, *2*, 5235.
14. Hoefling, T. A.; Newman, D. A.; Enick, R. M.; Beckman, E. J. *J. Supercrit. Fluids* **1993**, *6*, 165.
15. Hoefling, T. A.; Beitle, R. R.; Enick, R. M.; Beckman, E. J. *Fluid Phase Equilib.* **1993**, *83*, 203.
16. Fink, R., Beckman, E. J., *J. Supercrit. Fluids* **2000**, *18*, 101.
17. Psathas, P. A., da Rocha, S. R. P., Lee, C. T., Johnston, K. P., Lim, K. T., Weber, S., *Ind. Eng. Chem. Res.* **2000**, *39*, 2655.
18. DeSimone, J. M.; Guan, Z.; Eisbernd, C. S. *Science* **1992**, *25*, 945
19. Johnston, K. P.; Harrison, K. L.; Clarke, M. J.; Howdle, S. M.; Heitz, M. P.; Bright, F. V.; Carlier, C.; Randolph, T. W. *Science* **1996**, *271*, 624.
20. Harrison, K.; Goveas J.; Johnston, K. P. *Langmuir* **1994**, *10*, 3535.
21. Eastoe, J.; Bayazit, Z.; Martel, S.; Steytler, D. C.; Heenan, R. K. *Langmuir* **1996**, *12*, 1423.

304

22. Salaniwal, S.; Cui, S.; Cummings, P. T.; Cochran, H. D. *Langmuir* **1999**, *15*, 5188.
23. Salaniwal, S.; Cui, S.; Cochran, H. D.; Cummings, P. T. *Langmuir* **2000**, *39*, 4543.
24. Eastoe, J.; Cazelles, B. M. H.; Steytler, D. C.; Holmes, J. D.; Pitt, A. R.; Wear, T. J. *Langmuir* **1997**, *13*, 6980.
25. Eastoe, J.; Downer, A.; Pitt, A. R.; Simister, E. A.; Penfold, J. *Langmuir* **1999**, *15*, 7591.
26. Eastoe, J.; Downer, A. M.; Paul, A.; Steytler, D. C.; Rumsey, E. *Progr. Colloid Polym. Sci.* **2000**, *115*, 214.
27. a)Steytler, D. C.; Rumsey, E.; Eastoe, J.; Paul, A.; Downer A.. *Langmuir*, **2002**, *18*, 3014. b) Steytler, D. C.; Rumsey, E.; Thorpe, M.; Eastoe, J.; Paul, A.; Heenan, R. K. Langmuir, **2001**, *17*, 7948. c) Keiper, J. S.; Simhan, R.; DeSimone, J. M.; Wignall, G. D.; Melnichenko, Y. B.; Frielinghaus, H. *J. Am. Chem. Soc.* **2002**; *124*; 1834.
28. Liu, Z.; Erkey, C. *Langmuir* **2001**, *17*, 274.
29. Dong, X.; Erkey, C.; Dai, H.-J.; Li, H.-C.; Cochran, H. D.; Lin, J. S. *Ind. Eng. Chem. Res.* **2002**, *41*, 1038.
30. Heitz, M. P.; Carlier, C.; deGrazia, J.; Harrison, K. L.; Johnston, K. P.; Randolph, T. W.; Bright, F. V. *J. Phys. Chem. B* **1997**, *101*, 6707.
31. Clarke, M. J.; Harrison, K. L.; Johnston, K. P.; Howdle, S. M. *J. Am. Chem. Soc.* **1997**, *119*, 6399.
32. Zinsili, P. E. *J. Phys. Chem.* **1979**, *83*, 3223.
33. Fremgen, D. E.; Smotkin, E. S.; Gerald II, R. E.; Klingler, R. J.; Rathke, J. W. *J. Supercrit. Fluids* **2001**, *19*, 287.
34. Zielinski, R. G.; Kline, S. R.; Kaler, E. W.; Rosov, N. *Langmuir* **1997**, *13*, 3934.
35. Lee Jr., C. T.; Bhargava, P.; Johnston, K. P. *J. Phys. Chem. B* **2000**, *104*, 4448.
36. Lee Jr., C. T.; Johnston, K. P.; Dai, H. J.; Cochran, H. D.; Melnichenko, Y. B.; Wignall, G. D. *J. Phys. Chem B* **2001**, *105*, 3540.
37. From a range of possible models a Schultz distribution of spherical particles is often found to give the best fits and physically reasonable parameters. This scattering law may be written

$$I(Q) = \left(\frac{\phi(\rho_{micelle} - \rho_{CO2})^2}{\sum_i V_i X(R_i)} \right) \sum_i \left[V^2_{\ i} P(Q, R_i) X(R_i) \right]$$

where ϕ, R and V are the particle volume fraction, radius and volume, and ρ_i denotes a scattering length density. The spherical form factor is $P(Q, R_i)$, and $X(R_i)$ is the Schultz function, which is characterized by an average radius R^{av} and RMS deviation $\sigma = R^{av}/(Z + 1)^{1/2}$, where Z is a width parameter. For CO_2 the neutron scattering length density may be taken as

ρ_{CO2} = (mass density x 2.498 x 10^{10} cm^{-2}, as described in reference 43) hence effects of P and T on ρ_{CO2} can be taken into account. Calculated scale factors are typically ± 15% of those expected owing to sample compositions, indicating a physically reasonable model. With samples which are close to a phase boundary P_{trans} it is often necessary to introduce a structure factor, and the attractive Ornstein-Zernicke model $S(Q, \zeta)$ is often employed. This $S(Q)$ (or $S(Q)_{att}$) is entirely effective, accounting for additional scattering at low Q. The O-Z function describes a decaying particle distribution with ξ a correlation length, and the strength of interactions are related to $S(0)$ via the isothermal compressibility

$$S(Q,\xi) = 1 + \left[\frac{S(0)}{1+(Q\xi)^2} \right]$$

Sample compositions and scattering length densities are known quantities, hence the adjustable parameters are R_c^{av} and σ / R_c^{av}, and if used ζ and $S(0)$. Uncertainties in R_c^{av}, σ / R_c^{av} and ζ may be taken as ± 2 Å, ± 0.02 and ± 10 Å respectively. We use the multi-model FISH program for our SANS analysis (see reference 13). For dilute non-interacting systems (absence of any obvious $S(Q)$) estimates for the micelle radii R can also be obtained using the Guinier law which is valid at low Q ($QR<1$)

$$\ln[I(Q)] = \ln[I(0)] - \frac{(QR_g)^2}{3} \text{ ,with } R = \sqrt{5R_g/3}$$

In the above R_g is a radius of gyration and $I(0)$ is an intensity factor related to concentration and contrast.

38. Lee Jr., C. T.; Psathas, P. A.; Johnston, K. P.; deGrazia, J.; Randolph, T. W. *Langmuir* **1999**, *15*, 6781.
39. daRocha, S. R. P; Johnston, K. P. *Langmuir* **2000**, *16*, 3690.
40. Johnston, K. P. *Curr. Opin. Colloid Interface Sci.* **2001**, *5*, 351.
41. Lee Jr., C. T.; Psathas, P. A.; Ziegler, K. J.; Johnston, K. P.; Dai, H. J.; Cochran, H. D.; Melnichenko, Y. B.; Wignall, G. D. *J. Phys. Chem. B* **2000**, *104*, 11094.
42. daRocha, S. R. P.; Harrison, K. L.; Johnston, K. P. *Langmuir* **1999**, *15*, 419-428.
43. McClain, J. B.; Londono, D.; Combes, J. R.; Romack, T. J.; Canelas, D. A.; Betts, D. E.; Wignall, G. D.; Samulski, E. T.; DeSimone, J. M. *J. Am. Chem. Soc.* **1996**, *118*, 917.
44. Fulton, J. L.; Pfund, D. M.; McCLain, J. B.; Romack, T. J.; Maury, E. E.; Combes, J. R.; Samulski, E. T.; DeSimone, J. M.; Capel, M. *Langmuir* **1995**, *11*, 4241.

306

45. Chillura-Martino, D.; Triolo, R.; McClain, J. B.; Combes, J. R.; Betts, D. E.; Canelas, D. A.; DeSimone, J.M.; Samulski, E. T.; Cochran, H. D.; Londono, J. D.; Wignall, G. D. *J. Mol. Struct.* **1996**, *3*, 383.
46. McCLain, J. B.; Betts, D. E.; Canelas, D. A.; Samulski, E. T.; DeSimone, J. M.; Londono, J. D.; Cochran, H. D.; Wignall, G. D.; Chillura-Martino, D.; Triolo, R. *Science* **1996**, *274*, 2049.
47. Londono, J. D.; Dharmapurikar, R.; Cochran, H. D.; Wignall, G. D.; McClain, J. B.; Betts, D. E.; Canelas, D. A.; DeSimone, J. M.; Samulski, E. T.; Chillura-Martino, D.; Triolo, R. *J. Appl. Crystallogr.* **1997**, *30*, 690.
48. Dardin, A.; Cain, J. B.; DeSimone, J. M.; Johnson Jr., C. S.; Samulski, E. T. *Macromolecules* **1997**, *30*, 3593.
49. Cain, J. B.; Zhang, K; Betts, D. E.; DeSimone, J. M.; Johnson Jr, C. S. *J. Am. Chem. Soc.* **1998**, *120*, 9390.
50. Triolo, F.; Triolo, A.; Triolo, R.; Londono, J. D.; Wignall, G. D.; McClain, J. B.; Betts, D. E.; Wells, S.; Samulski, E. T.; DeSimone, J. M. *Langmuir*, **2000**, *16*, 416.
51. Triolo, R.; Triolo, A.; Triolo, F.; Steytler, D. C.; Lewis, C. A.; Heenan, R. K.; Wignall, G. D.; DeSimone, J. M. *Phys. Rev. E* **2000**, *61*, 4640.
52. Triolo, A; Triolo, F.; Celso, F. Lo; Betts, D. E.; McClain, J. B.; DeSimone, J. M.; Wignall, G. D.; Triolo, R. *Phys. Rev. E* **2000**, *62*, 5839.
53. Zhou, S.; Chu, B. *Macromolecules* **1998**, *31*,. 5300.
54. Zhou, S.; Chu, B. *Macromolecules* **1998**, *31*, 7746.
55. O'Neil, M. L.; Cao, Q.; Fang, M.; Johnston, K. P.; Wilkinson, S. P.; Smith, C. D.; Kerschner, J. L.; Jureller, S. H. *Ind. Eng. Chem. Res.* **1996**, *37*, 3067.
56. Yates, M. Z.; Li, G.; Shim, J. J.; Maniar, S.; Johnston, K. P.; Lim, K. T.; Webber, S. E. *Macromolecules* **1999**, *32*, 1018.
57. Li, G.; Yates, M. Z.; Johnston, K. P.; Lim, K. T.; Webber, S. E. *Macromolecules* **2000**, *33*, 1606-2.
58. O'Neil, M. L.; Yates, M. Z.; Johnston, K. P.; Smith, C. D.; Wilkinson, S. P. *Macromolecules.* **1998**, *31*, 2838.
59. Ihara, T.; Suzuki, N.; Maeda, T.; Sagara, K.; Hobo, T. *Chem. Pharm. Bull.* **1995**, *43*, 626.
60. Hutton, B. H.; Perera, J. M.; Grieser, F.; Stevens, G. W. *Coll. Surf. A-Physicochem. Eng*, **1999**, *146*, 227.
61. Hutton, B. H.; Perera, J. M.; Grieser, F.; Stevens, G. W. *Coll. Surf. A-Physicochem. Eng.* **2001**, *189*, 177.
62. McFann, G. J.; Johnston, K. P.; Howdle, S.M. *AIChE J.* **1994**, *40*, 543-555.

307

63. Lui, J.; Han, B.; Li, G.; Liu, Z.; He, J; Yang, G. *Fluid Phase Equilib.* **2001**, *187-188*, 247.
64. Dimitrov, K.; Boyadzhiev, L.; Tufeu, R.; Cansell, F.; Barth, D. *J; Supercrit. Fluids* **1998**, *14*, 41.
65. Liu, J.; Han, B.; Li, G.; Zhang, X.; He, J.; Liu, Z. *Langmuir* **2001**, *17*, 8040.
66. Liu, J.; Han, B.; Wang, Z.; Zhang, J.; Li, G.; Yang, G. *Langmuir* **2002**, *18*, 3086.
67. Ji, M.; Chen, X.; Wai, C.; Fulton, J. L. *J. Am. Chem. Soc.* **1999**, *121*, 2631.
68. Eastoe, J.; Paul, A.; Nave, S.; Steytler, D.C.; Robinson, B.H.; Rumsey, E.; Thorpe, M.; Heenan, R.K. *J.Am.Chem.Soc.* **2001**, *123*, 988.
69. Sarbu, T.; Styranec, T.; Beckman, E. J. *Nature* **2000**, *405*, 165.
70. Adamsky, F. A.; Beckman, E. J. *Macromolecules* **1994**, *27*, 312.
71. Nikitin, L. N.; Said-Galiyev, E. E.; Vinokur, R. A.; Kholkhlov, A. R.; Gallyamov, M. O.; Schaumburg, K. *Macromolecules* **2002**, *35*, 934.
72. Mizumoto, T.; Sigumura, N.; Moritani, M.; Yasuda, H.; Sato, Y.; Masuoka, H. *Macromolecules* **2001**, *34*, 5200.
73. Giles, M. R.; Howdle, S. M. *Eur. Pol. J.* **2001**, *37*, 1347.
74. Giles, M. R.; Hay, J. N.;Howdle, S. M.; Winder, R. J. *Polymer* **2000**, *41*, 6715.
75. Hems, W. P.; Yong, T.-M.; vanNunen, J. L. M.; Cooper, A. I.; Holmes, A. B.; Griffin, D. A. *J. Mater. Chem.* **1999**, *9*, 1403.
76. Ghenciu, E. G.; Russell, A. J.; Beckman, E. J.; Steele, L.; Becker, N. T. *Biotech. Bioeng.* **1998**, *58*, 572.
77. Yazdi, A.; Beckman, E. J. *Mater. Res. Soc. Symp. Proc.* (Materials and Processes for Environmental Protection) **1994**, 344, 211.
78. Li, J.; Beckman, E. J. *Ind. Eng. Chem. Res.* **1998**, 37, 4768.
79. Yates, M. Z.; Apocada, D. L.; Campbell, M. L.; McCleskey, T. M. *Chem. Commun.* **2001**, 25.
80. Cooper, A. I.; Holmes, A. B. *Adv. Mater.* **1999**, *11*, 1270.
81. Cooper, A. I.; Wood, C. D.; Holmes, A. B. *Ind. Eng. Chem. Res.* **2000**, *39*, 4741.
82. Wood, C. D.; Cooper, A. I. *Macromolecules* **2000**, *34*, 5.
83. Butler, R.; Davies, C. M.; Cooper, A. I. *Adv. Mater.* **2001**, *13*, 1459.
84. Nave, S.; Eastoe, J.; Penfold, J. *Langmuir* **2000**, *16*, 8733.
85. Hudlicky, M.; Pavlath, A.E. *Chemistry of Organic Fluorine Compounds 2*, ACS Monograph 187; American Chemical Society: Washington, DC, 1995, Chapter 6.
86. Nave, S.; Eastoe, J.; Heenan, R. K.; Steytler, D. C.; Grillo, I. *Langmuir* **2000**, *16*, 8741.

87. Pitt, A. R.; Morley, S. D.; Burbidge, N. J.; Quickenden, E. L. *Coll. Surf. A-Physicochem. Eng* **1996**, 114, 321.
88. Kim, E. J.; Hahn, S. H. *Materials Science and Engineering A-Structural Materials Properties Microstructure and Processing* **2001**, 24, 303.
89. Fu, X.; Qutubuddin, S. *Coll. Surf. A-Physicochem. Eng.* **2001**, *179*, 65.
90. Liu, Z.; Bendayan, R.; Wu, X. Y. *J. Pharm. Pharmacol.* **2001**, *53*, 779.
91. Nyholm, T.; Slotte, J. P. *Langmuir* **2001**, *17*, 4724.
92. Hagerstrand, H.; Bobacka, J.; Bobrowska-Hagerstrand, M.; Kralj-Iglic, V.; Fosnaric, M.; Iglic, A. *Cellular & Mol. Biol. Lett.* **2001**, *6*, 161.
93. Paul, A, *Ph.D. thesis* University of Bristol, Bristol UK 2001.

Chapter 20

Nanoparticle Formation in Rapid Expansion of Water-in-Carbon Dioxide Microemulsion into Liquid Solvent

Mohammed J. Meziani[1], Pankaj Pathak[1], Lawrence F. Allard[2], and Ya-Ping Sun[1,*]

[1]Department of Chemistry and Center for Advanced Engineering Fibers and Films, Howard L. Hunter Chemistry Laboratory, Clemson University, Clemson, SC 29634–0973
[2]High Temperature Materials Laboratory, Oak Ridge National Laboratory, Oak Ridge, TN 37831–6062

Nanocrystalline metal (silver and copper) and metal sulfide (silver sulfide, cadmium sulfide, and lead sulfide) particles were prepared via RESOLV (Rapid Expansion of a Supercritical Solution into a Liquid SOLVent) with water-in-carbon dioxide microemulsion as solvent for the rapid expansion. The nanoparticles were characterized using UV/vis absorption, X-ray powder diffraction, and transmission electron microscopy methods. The results of the different nanoparticles are compared and discussed in reference to those of the same nanoparticles produced via RESOLV with the use of conventional supercritical solvents.

309

Introduction

The use of supercritical carbon dioxide (CO_2) in the preparation of nanoscale materials has received much recent attention. There are several obvious advantages of supercritical CO_2 over conventional organic solvents. CO_2 is abundant and environmentally friendly as a solvent system. The high diffusivity in supercritical CO_2 allows the acceleration of some chemical reactions. The dependence of solubility on the fluid density over a wide range also enables manipulation and control of solute precipitation. A significant disadvantage of CO_2 as a solvent, on the other hand, is the generally low solubility for most solutes. Cosolvents are often used to improve the solubility. However, in order to dissolve hydrophilic compounds, a commonly employed approach is to add surfactant to CO_2 to form water-in-CO_2 microemulsion (1-4). In particular, perfluorinated surfactant has been used to stabilize water-in-CO_2 microemulsion in the preparation of metal and semiconductor nanoparticles (5-11). For example, Wai, Fulton and coworkers synthesized silver and silver halide nanoparticles in the microemulsion of water in supercritical CO_2 with a fluorinated surfactant (5, 6). The nanoparticles were collected via rapid expansion and characterized using optical and electron microscopy techniques. The estimated sizes of the nanoparticles were 5-15 nm with a relatively broad size distribution. The same group also evaluated the use of two different reducing agents, sodium cyanoborohydride and *N,N,N',N'*-tetramethyl-*p*-phenylenediamine, with water-in-CO_2 microemulsion in the preparation of silver and copper nanoparticles (11). Similarly, Johnston, Korgel and coworkers used the water-in-CO_2 microemulsion to synthesize nanoscale cadmium sulfide particles (7). These authors reported that a higher water-to-surfactant molar ratio for the microemulsion resulted in larger nanoparticles (7). Johnston, Korgel and coworkers also prepared silver nanocrystals that were coated with fluorinated ligands for the dispersion in CO_2 at moderate temperatures and pressures (8). In a related approach, they synthesized fluorocarbon-coated silver, iridium, and platinum nanocrystals of 2 -12 nm in diameter in supercritical CO_2 by arrested precipitation from soluble organometallic precursors (12). An advantage of this approach is in the recovery of the nanoparticles from microemulsion without irreversible aggregation.

In our laboratory, we have modified the supercritical fluid processing method known as RESS (Rapid Expansion of Supercritical Solution) (13-18) by expanding the supercritical solution into a liquid solvent, or RESOLV (Rapid Expansion of a Supercritical Solution into a Liquid SOLVent), to produce nanoscale semiconductor and metal particles (1, 9, 19-22). For the solubility of metal salts, supercritical ammonia, THF, and acetone were used in the rapid expansion at relatively higher temperatures. The nanoparticles thus obtained were small (less than 10 nm), with relatively narrow size distributions. In an effort to replace the organic solvents with CO_2-based systems for RESOLV at ambient temperatures, we used a water-in-CO_2

microemulsion to dissolve silver salt, similar to what was reported by Wai, Fulton and coworkers (*5, 6, 11*) and Johnston, Korgel and coworkers (*7, 8*). However, instead of *in situ* chemical reduction, the microemulsion with silver salt was rapidly expanded into a room-temperature solution for chemical reduction under ambient conditions, yielding silver nanoparticles (*9*). Here we report the use of the same water-in-CO_2 microemulsion in the preparation of other nanoparticles via RESOLV. The results are compared with those from the preparation of silver nanoparticles.

Experimental Section

Materials

Silver nitrate ($AgNO_3$), copper nitrate ($Cu(NO_3)_2$), cadmium nitrate ($Cd(NO_3)_2$), lead nitrate ($Pb(NO_3)_2$), sodium borohydride ($NaBH_4$), and sodium sulfide (Na_2S) were purchased from Aldrich. Spectroscopy- or HPLC-grade organic solvents were used as received. Poly(N-vinyl-2-pyrrolidone) (PVP) of average molecular weight $M_W \approx 360,000$ was obtained from Sigma and used without further purification. SCF-grade CO_2 was purchased from Air Products.

Perfluoropolyether carboxylic acid (PFPE-COOH, $M_W \approx 655$) was obtained from Ausimont. The conversion to the ammonium salt (perfluoropolyether ammonium carboxylate, PFPE-NH$_4$) was accomplished via the neutralization reaction of PFPE-COOH with aqueous ammonium hydroxide, followed by the removal of water and excess ammonia under vacuum at 65 °C. The conversion was monitored and confirmed by results from NMR measurements.

Measurements

The apparatus for the preparation of nanoparticles via RESOLV is illustrated in Figure 1. It consisted of a syringe pump for generating and maintaining pressure during the rapid expansion process and a gauge for monitoring the system pressure. The heating unit consisted of a cylindrical solid copper block of high heat capacity in a tube furnace. The copper block was wrapped with stainless steel tubing coil and inserted tightly into a stainless steel tube to ensure close contacts between the tubing coil and copper block for efficient heat transfer. The expansion nozzle was a fused silica capillary hosted in a stainless steel tubing, which was inserted into the collection chamber containing a room-temperature solution.

UV/vis absorption spectra were recorded on a Shimadzu UV-3100 spectrophotometer. Powder X-ray diffraction measurements were carried out on a Scintag XDS-2000 powder diffraction system. Transmission electron microscopy (TEM) images were obtained on Hitachi 7000 and Hitachi HF-2000 TEM systems.

312

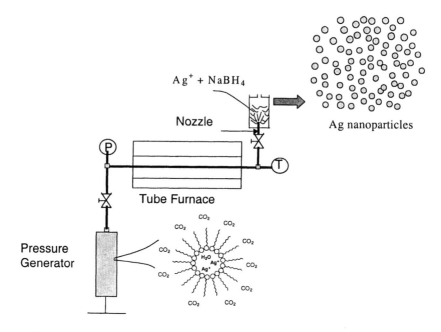

Figure 1. The experimental setup for RESOLV with the microemulsion as solvent for rapid expansion.

Nanoparticle Preparation via RESOLV

In a typical experiment for copper (Cu) nanoparticles, a water-in-CO_2 microemulsion was prepared by adding an aqueous $Cu(NO_3)_2$ solution (0.2 M, 0.25 mL) and then the surfactant PFPE-NH_4 to the syringe pump to achieve the desired water-to-surfactant molar ratio (Wo) of 10. The syringe pump was loaded with CO_2 to a final pressure of 2,000 psia. The mixture in the syringe pump was stirred for 2 h to facilitate the formation of reverse micelles of an aqueous $Cu(NO_3)_2$ core. The PFPE-NH_4-stabilized microemulsion of aqueous $Cu(NO_3)_2$ in CO_2 was rapidly expanded through a 50-microns fused silica capillary nozzle into a room-temperature solution of $NaBH_4$ (0.026 M) in ethanol. The maintaining of an inert environment was achieved via purging nitrogen gas during the rapid expansion. The room-temperature solution also contained PVP polymer ($M_W \approx 360,000$) as a protection agent (5 mg/mL) for the produced Cu nanoparticles.

The same procedure and similar experimental parameters were used in the preparation of other metal and semiconductor nanoparticles.

Results and Discussion

Water-in-CO_2 microemulsion was used to dissolve metal salts in the production of nanoparticles via RESOLV. In order to evaluate the solubility of $Cu(NO_3)_2$, for example, the same microemulsion as that used in the rapid expansion was prepared in a high-pressure optical cell. With $Cu(NO_3)_2$ in the water phase, which exhibited the distinctive blue color of aqueous Cu^{2+} (*10*), the microemulsion appeared homogenous. According to the observed absorbance (the band centered at ~ 740 nm), the $Cu(NO_3)_2$ salt was completely dissolved in the PFPE-NH_4-stabilized water-in-CO_2 microemulsion. The other metal salts were similarly soluble, resulting in microemulsions that appeared equally homogeneous.

The formation of silver (Ag) nanoparticles in the rapid expansion of PFPE-NH_4-stabilized aqueous $AgNO_3$-in-CO_2 microemulsion into a room-temperature reductant solution has been reported (*9*). In order to use Ag as a reference in this study, the same preparation of Ag nanoparticles via RESOLV was repeated.

The results were found to be reproducible and similar to what had been reported. For example, the Ag nanoparticles obtained by rapidly expanding a microemulsion of aqueous $AgNO_3$-in-CO_2 (Wo = 10) into a room-temperature solution of $NaBH_4$ in ethanol exhibit the characteristic plasmon absorption band at ~ 430 nm. Shown in Figure 2 is a typical TEM image of the Ag nanoparticles thus

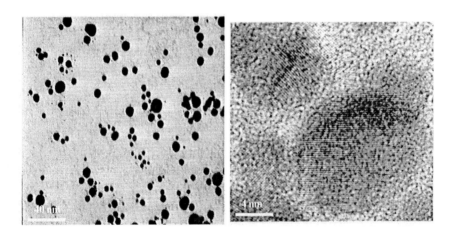

Figure 2. A typical TEM image of the Ag nanoparticles obtained via RESOLV with the microemulsion of Wo equal to 10 (left). The high-resolution image (right) shows the lattice fringes in individual Ag nanoparticles.

produced. A statistical analysis of the particles yields an average size of 9.5 nm in diameter and a size distribution standard deviation of 3.8 nm, which agree well with those from previous experiments (9).

Similar to Ag, the Cu nanoparticles obtained from RESOLV with the microemulsion as solvent for the rapid expansion formed a stable suspension in ethanol under PVP polymer protection, which was essentially indistinguishable from a typical homogeneous solution. The nanoparticles were air-sensitive in both the suspension and the solid state to form copper oxide in the presence of oxygen. Thus, an inert atmosphere had to be maintained via purging nitrogen gas to prevent the nanoparticles from oxidation. Shown in Figure 3 is the absorption spectrum of

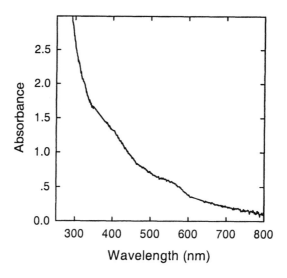

Figure 3. The UV-vis absorption spectrum of the Cu nanoparticles (suspended in ethanol) obtained via RESOLV with the microemulsion of Wo equal to 10.

the Cu nanoparticles in ethanol. The shoulder at 570 nm is characteristic of nanoscale Cu, attributed to the plasma excitation in copper colloids (23-25). According to a known simulation concerning the absorption spectra of spherical Cu particles (25, 26), the Cu nanoparticles responsible for the absorption spectrum in Figure 3 should be less than 12 nm in diameter.

The nanoparticle sizes were determined more accurately by using TEM, which is under high vacuum. For the TEM analysis, the suspended Cu nanoparticles were deposited on a collodion film-coated copper grid, followed by the removal of solvent. Shown in Figure 4 is a typical TEM image of the Cu nanoparticles. A statistical analysis yields an average particle size of 9 nm and a size distribution standard deviation of 3 nm. The results confirm the estimation on the basis of the absorption spectrum that the Cu nanoparticles should be less than 12 nm.

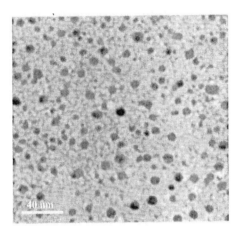

Figure 4. A TEM image of the Cu nanoparticles obtained via RESOLV with the microemulsion of Wo equal to 10.

The Cu nanoparticles were identified by X-ray powder diffraction analysis under nitrogen. The solid-state sample for the analysis was prepared by removing the solvent ethanol from the suspension, followed by washing the remaining solids with acetone (to remove some of the surfactant) and water. The X-ray powder diffraction pattern of the sample thus obtained is shown in Figure 5. It matches well with that of the bulk Cu in the JCPDS library. The broadening in the diffraction peaks may be attributed to the nanoscale nature of the Cu particles. Thus, the average size of the Cu nanoparticles is also estimated from the peak broadening in terms of the Debye-Scherrer equation (27).

$$D = K\lambda/\beta \cos \theta \qquad (1)$$

where D is the average particle diameter in Å, λ is the X-ray wavelength, β is the corrected peak broadening (FWHM), and θ is the diffraction angle. K is a constant dependent on the particle shape and the way in which D and β are defined. Since it

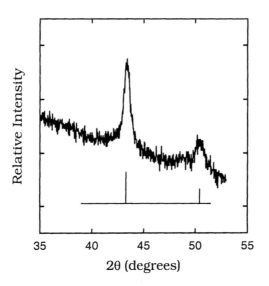

2θ (degrees)

Figure 5. The X-ray powder diffraction peaks of the Cu nanoparticles obtained via RESOLV with the microemulsion of Wo equal to 10.

is generally close to unity, it is set at unity for simplicity. The average particle size thus calculated is 10 nm, similar to the value obtained from the TEM analysis.

The results presented above show that Cu nanoparticles can be produced via RESOLV with PFPE-NH$_4$-stabilized water-in-CO$_2$ microemulsion as solvent for the rapid expansion. The formation of Cu nanoparticles is apparently similar to that of Ag nanoparticles under comparable experimental conditions (9). The same approach is also applicable to the synthesis of nanoscale metal sulfides, including silver sulfide (Ag$_2$S), cadmium sulfide (CdS), and lead sulfide (PbS) nanoparticles.

For Ag$_2$S nanoparticles, the same PFPE-NH$_4$-stabilized aqueous AgNO$_3$-in-CO$_2$ microemulsion (Wo = 10) was used in the rapid expansion. A room-temperature solution of Na$_2$S in ethanol was at the receiving end. The Ag$_2$S nanoparticles produced in RESOLV were again protected with PVP polymer in the room-temperature solution, resulting in a solution-like stable suspension. The absorption spectrum of the Ag$_2$S nanoparticles in ethanol suspension is compared with that of the Ag nanoparticles in Figure 6. The spectrum is a featureless curve, which is responsible for the yellow color of the suspension. The Ag$_2$S nanoparticles produced via RESOLV with microemulsion are relatively small, as shown in the

TEM image in Figure 7. The average particle size and size distribution standard deviation are 5.3 nm and 0.8 nm, respectively. Also shown in Figure 7 is a high-resolution TEM image of a Ag_2S nanoparticle, which appears to be a single crystal. The interreticular spacing is 0.26 nm, corresponding to that of the (-121) lattice planes.

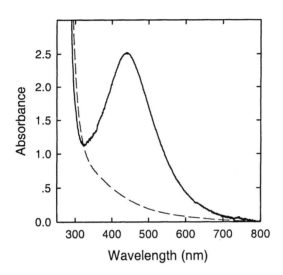

Figure 6. The UV/vis absorption spectrum of the Ag_2S nanoparticles obtained via RESOLV with the microemulsion of Wo equal to 10 (– – –) is compared with that of the Ag nanoparticles prepared under similar experimental conditions (————).

The nanoparticle sample in the solid state was also characterized by X-ray powder diffraction analysis. The diffraction pattern again agrees well with that of bulk Ag_2S (monoclinic) in the JCPDS library (Figure 8).

The same receiving Na_2S solution was used in the preparation of CdS and PbS nanoparticles. The metal salts $Cd(NO_3)_2$ and $Pb(NO_3)_2$ were again dissolved in the aqueous core of the reverse micelles in CO_2. In RESOLV, the $PFPE-NH_4$-stabilized aqueous $Cd(NO_3)_2$- or $Pb(NO_3)_2$-in-CO_2 was rapidly expanded into the room-temperature Na_2S solution to form CdS or PbS nanoparticles. UV/vis absorption

318

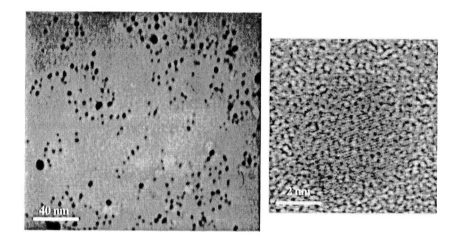

Figure 7. A typical TEM image of the Ag_2S nanoparticles obtained via RESOLV with the microemulsion of Wo equal to 10 (left). The high-resolution image (right) shows a single-crystal Ag_2S nanoparticle.

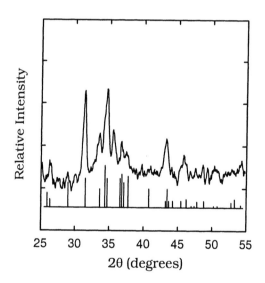

Figure 8. The X-ray powder diffraction peaks of the Ag_2S nanoparticles obtained via RESOLV with the microemulsion of Wo equal to 10.

spectra of the CdS and PbS nanoparticles in stable ethanol suspensions (under PVP polymer protection) are shown in Figure 9. The spectra are rather broad, without the distinctive shoulders as in the spectra of the CdS and PbS nanoparticles produced via RESOLV with supercritical ammonia as the solvent (*19*). The lack of significant structures in the absorption spectra might be a result of relatively broader nanoparticle size distributions. According to TEM analyses, the average particle size and size distribution standard deviation are 4.7 nm and 0.9 nm, respectively, for CdS and 4.3 nm and 0.7 nm, respectively, for PbS. These results are comparable with those of the Ag_2S nanoparticles produced under similar experimental conditions.

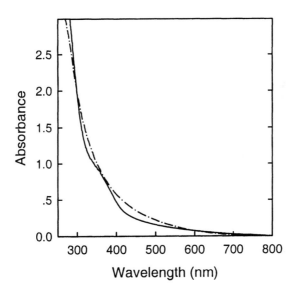

Figure 9. The UV/vis absorption spectra of the CdS (———) and PbS (— — —) nanoparticles obtained via RESOLV with the microemulsion of Wo equal to 10.

The high-resolution TEM images show that the CdS and PbS nanoparticles are mostly single crystals (Figure 10). The lattice spacings are 0.33 nm and 0.20 nm for CdS, likely corresponding to those of (111) and (220) planes in the cubic structure. For PbS, The lattice spacings are 0.33 nm, 0.29 nm, and 0.21 nm, consistent with those of the (111), (200), and (220) planes, respectively.

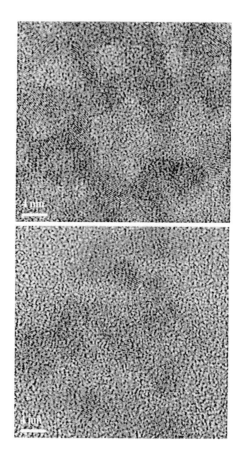

Figure 10. High-resolution TEM images of the CdS (top) and PbS (bottom) nanoparticles.

According to Zielinski *et al.*, the PFPE-NH$_4$-stabilized water-in-CO$_2$ microemulsion of Wo equal to 11 contains water droplets of ~ 4 nm in average diameter, and the droplets and droplet structure are little affected by experimental parameters such as the system pressure (*28*). Thus, the reverse micellar core in the microemulsion used in this study (Wo = 10) should have an average diameter close to 4 nm. Since there is evidence that the average size of the nanoparticles produced via RESOLV is dependent on the size of the pre-expansion reverse micelles in CO$_2$ (*9*), it may be more than just a coincidence that the metal sulfide nanoparticles are of average sizes comparable to that of the pre-expansion water core (Table 1). The

Table I. A Summary of the Parameters of the Nanoparticle

Nanoparticles	X-ray Identification	TEM Size (nm)	σ (nm)
Cu	Cubic	9.0	3.1
Ag	Cubic	9.5	3.8
Ag_2S	Monoclinic	5.2	0.9
CdS	Primarily cubic	4.7	0.9
PbS	Cubic	4.3	0.7

larger Ag and Cu nanoparticles are probably associated with the post-expansion agglomeration of the water cores and/or the initially formed smaller nanoparticles.

Conclusion

In the preparation of nanoparticles via RESOLV, the size distributions are generally broader with the water-in-CO_2 microemulsion as solvent for the rapid expansion than those with conventional supercritical solvents such as ammonia. The difference is less significant for the metal sulfide nanoparticles. However, an advantage of the microemulsion is the tunable size of the reverse micellar core (via changing the Wo value), which makes it possible to vary the average size of the produced nanoparticles (9). New strategies for an optimization of the processing conditions to narrow the nanoparticle size distribution are under exploration, and the results are encouraging. For example, the rapid expansion of the microemulsion containing a silver salt into a basic room-temperature aqueous $NaBH_4$ solution was found to produce Ag nanoparticles of a relatively narrow size distribution (9). Investigations toward gaining more mechanistic insights are in progress.

Acknowledgment. We thank P. Atorngitjawat, Y. Lin, D. Zweifel, and D. Hill for experimental assistance. Financial support from DOE (DE-FG02-00ER45859) and the Center for Advanced Engineering Fibers and Films (NSF-ERC at Clemson University) is gratefully acknowledged. We also acknowledge the sponsorship by the Assistant Secretary for Energy Efficiency and Renewable Energy, Office of Transportation Technologies, as part of the High Temperature Materials Laboratory User Program, Oak Ridge National Laboratory, managed by UT-Battelle, LLC, for the Department of Energy (DE-AC05-00OR22725).

322

References

1. Sun, Y.-P.; Rollins, H. W.; Bandara, J.; Meziani, M. J.; Bunker, C. E. in *Supercritical Fluid Technology in Materials Science and Engineering: Synthesis, Properties, and Applications*; Sun, Y.-P., Ed.; Marcel Dekker: New York, 2002, p. 491.
2. Fox, K. C. *Science* **1994**, *265*, 321.
3. McClain, J. B.; Betts, D. E.; Canelas, D. A.; Samulski, E. T.; DeSimone, J. M.; Londono, J. D.; Cochran, H. D.; Wignall, G. D.; Chillura-Martino, D.; Triolo, R. *Science* **1996**, *274*, 2049.
4. Johnston, K. P.; Harrison, K. L.; Clarke, M. J.; Howdle, S. M.; Heitze, M. P.; Bright, F. V.; Carlier, C.; Randolph, T. W. *Science* **1996**, *271*, 624.
5. Ji, M.; Chen, X. Y.; Wai, C. M.; Fulton, J. L. *J. Am. Chem. Soc.* **1999**, *121*, 2631.
6. Ohde, H.; Rodriguez, J. M.; Ye, X. R.; Wai, C. M. *Chem. Comm.* **2000**, *23*, 2353.
7. Holmes, J. D.; Bhargava, P. A.; Korgel, B. A.; Johnston, K. P. *Langmuir* **1999**, *15*, 6613.
8. Shah, P. S.; Holmes, J. D.; Doty, R. C.; Johnston, K. P.; Korgel, B. A. *J. Am. Chem. Soc.* **2000**, *122*, 4245.
9. Sun, Y.-P.; Atorngitjawat, P.; Meziani, M. J. *Langmuir* **2001**, *17*, 5707.
10. Meziani, M. J.; Sun, Y.-P. *Langmuir* **2002**, *18*, 3787.
11. Ohde, H.; Hunt, F.; Wai, C. M. *Chem. Mater.* **2001**, *13*, 4130.
12. Shah, P. S.; Husain, S.; Johnston, K. P.; Korgel, B. A. *J. Phys. Chem. B* **2001**, *105*, 9433.
13. (a) Petersen, R. C.; Matson, D. W.; Smith, R. D. *J. Am. Chem. Soc.* **1986**, *108*, 2100. (b) Matson, D. W.; Norton, K. A.; Smith, R. D. *Chemtech.* **1989**, *19*, 480.
14. Eckert, C. A.; Knutson, B. L.; Debenedetti, P. G. *Nature* **1996**, *383*, 313.
15. Tom, J. W.; Debenedetti, P. G. *J. Aerosol Sci.* **1991**, *22*, 555.
16. Mawson, S.; Johnston, K. P.; Combes, J. R.; Desimone, J. M. *Macromolecules* **1995**, *28*, 3182.
17. Lele, A. K.; Shine, A. D. *Ind. Eng. Chem. Res.* **1994**, *33*, 1476.
18. Weber, M.; Thies, M. C. in *Supercritical Fluid Technology in Materials Science and Engineering: Synthesis, Properties, and Applications*; Sun, Y.-P., Ed., Marcel Dekker: New York, 2002, p. 387.
19. (a) Sun, Y.-P.; Rollins, H. W. *Chem. Phys. Lett.* **1998**, *288*, 585. (b) Sun, Y.-P.; Guduru, R.; Lin, F.; Whiteside, T. *Ind. Eng. Chem. Res.* **2000**, *39*, 4663.
20. Sun, Y.-P.; Rollins, H. W.; Guduru, R. *Chem. Mater.* **1999**, *11*, 7.
21. Sun, Y.-P.; Riggs, J. E.; Rollins, H. W.; Guduru, R. *J. Phys. Chem. B* **1999**, *103*, 77.

22. Meziani, M. J.; Rollins, H. W.; Allard, L. F.; Sun, Y.-P. *J. Phys. Chem. B* **2002**, *106*, 11178.
23. Savinova, E. R.; Chuvilin, A. L.; Parmon, V. N. *J. Mol. Cat.* **1988**, *48*, 217.
24. Creighton, J. A.; Eadon, D. G. *J. Chem. Soc., Faraday Trans.* **1991**, *87*, 3881.
25. Lisiecki, I.; Pileni, M. P. *J. Am. Chem. Soc.* **1993**, *115*, 3887.
26. Mie, G. *Ann. Phys.* **1908**, *25*, 377.
27. Klug, H. P.; Alexander, L. E. *X-Ray Diffraction Procedures*; John Wiley & Sons: New York, 1959.
28. Zielinski, R. G.; Kline, S. R.; Kaler, E. W.; Rosov, N. *Langmuir* **1997**, *13*, 3934.

Chapter 21

Fine Particle Pharmaceutical Manufacturing Using Dense Carbon Dioxide Mixed with Aqueous or Alcoholic Solutions

Edward T. S. Huang[1,*], Hung-yi Chang[1,2], C. D. Liang[1], and R. E. Sievers[1,3]

[1]Center for Pharmaceutical Biotechnology, Department of Chemistry & Biochemistry, and CIRES, 214 UCB, University of Colorado, Boulder, CO 80309–0214
[2]Chemical Engineering Department, National Cheng-Kung University, Taiwan
[3]Aktiv-Dry LLC, 655 Northstar Court, Boulder, CO 80304

Abstract

This paper describes a newly patented CAN-BD (Carbon dioxide Assisted Nebulization with a Bubble Dryer®) process, utilizing dense CO_2 to micronize solutes to fine particles in a diameter range of 0.6 to 5 µm. The potential applications of this process are in thin film deposition, fine powder generation, and drug delivery.

The fine particles are generated by (a) intimately mixing dense CO_2 (at super- or sub-critical conditions) and a liquid solution (containing a dissolved solute of interest) in a small volume tee at about 83 bar and room temperature, (b) expanding this mixture through a 10 cm long capillary tube flow restrictor (with inner diameters of 50, 74 or 100 µm) into a drying chamber at atmospheric pressure to generate an aerosol, and (c) drying the aerosol plume with preheated air or nitrogen gas at temperatures between 10 and 65°C to form dry powders. Fine dry powders of disaccharide sugars, proteins, water-soluble and alcohol-soluble drugs have been generated with a lab CAN-BD unit (using a glass drying chamber with a volume of one to two liters) at a liquid flow rate of 0.3 to 2 mL/min. In a scaled up prototype unit (utilizing a 170-liter drying chamber with a thin stainless steel wall), aqueous solutions with 10% solute have successfully been nebulized and dried at a liquid flow rate of 20 mL/min.

This paper presents experimental results of nebulizing aqueous solutions of mannitol and myo-inositol utilizing a lab CAN-BD unit. The effect of certain operating parameters on particle characteristics has been investigated. The particle size (a) decreases with reduction in solute concentration, and (b) decreases with increase in the ratio of dense CO_2 to aqueous solution flow rates.

Supercritical Fluid Technology for Particle Generation

A fluid is said to be in a supercritical state when it exists above its critical pressure (P_C) and critical temperature (T_C). A supercritical fluid (SCF) exhibits significant solvent strength when it is compressed to liquid-like densities (1). On a plot of reduced density ($\rho_R = \rho/\rho_C$) vs. reduced pressure ($P_R = P/P_C$) with reduced temperature ($T_R = T/T_C$) as a parameter, the reduced fluid density can increase from gas-like density (~0.1) to liquid-like density (>2.0) within a reduced temperature range of 0.9-1.2, as the fluid is compressed beyond a reduced pressure greater than 1.0. Near the critical point, the slope of a constant temperature (T_R) curve would be almost infinite, which implies that a slight change in pressure can cause a tremendous change in density. Since fluid density is closely related to fluid solvent power, a small change in pressure and temperature near the critical point can easily regulate the solvent power of a SCF. The most commonly used SCF is carbon dioxide. This fluid is a popular supercritical solvent, due to its inert nature as well as its mild critical conditions (P_C of 73.8 bar and T_C of 31.1°C).

In the past two decades, a lot of work has been done on applications of SCF technology in the fields of extraction, materials and particle engineering (1, 2). A brief review of fine particle generation processes using supercritical (SC) CO_2 will be presented in the following. For detailed information, the readers are referred to literature articles (3, 4, 5). Particle generation processes can be classified as follows:

(A)   Solid-SCF system: Due to the significant solvent power of SC CO_2, this solvent can dissolve various intermediate and high molecular weight solutes. The process taking advantage of this property is called Rapid Expansion from Supersaturated Solution, RESS (6). It involves a binary solid-SC CO_2 system, in which the solid to be micronized is dissolved in SC CO_2 at a pressure greater than the P_C of CO_2 (73.8 bar). This single-phase mixture is then expanded through a nozzle or a flow restrictor from a supercritical pressure to a low pressure. Pressure reduction causes CO_2 to lose its solvent power, which results in precipitating solids as micron-size particles. Utilization of this process is limited because not many solids have high solubility in SC CO_2. Sometimes a co-solvent (e.g., alcohol) can be added to increase solubility, but such a method is not suited for water-soluble pharmaceuticals.

(B)   Solid-SCF-Solvent system: The types of particle generation in this category normally use SC CO_2 as an anti-solvent. The other two components in the ternary system are a solvent and a solid (or solute). For this type of SCF process to work, (a) the solvent needs to be "fully" miscible with the SC CO_2 and (b) the solid needs to be soluble in the solvent but not significantly in CO_2. Since water is only partially miscible with CO_2 (i.e., about 2 mole % CO_2 in water at 100 bar), water cannot be used in this process. Several organic solvents (e.g., alcohol, acetone and DSMO, etc.) are fully or nearly fully miscible with SC CO_2 and are used frequently in this application. Under supercritical conditions, SC CO_2 is mixed with the liquid organic solution. Since SC CO_2, acts as an anti-solvent, and is miscible with the organic solvent, it extracts and swells the solvent, and super-saturation of the solution results in crystallization (or precipitation) of solid. The types of SCF processes belonging to this category are,

(a)   GAS (gas Anti-Solvent) or SAS (Supercritical fluid Anti-Solvent) (7, 8)

(b)   ASES (Aerosol Solvent Extraction System) (9, 10)

(c)   PCA (Precipitation with Compressed fluid Anti-solvent) (11)

(d)   SEDS (Solution Enhanced Dispersion by Supercritical fluids) (12)

One drawback of the anti-solvent processes is the requirement of the use of an organic solvent. This may lead to a problem of residual organic solvent left in the final dry powder product. Furthermore, some solutes are not stable in organic solvents. Another disadvantage is the requirement of high-pressure autoclaves, since anti-solvent crystallization must occur under supercritical conditions of CO_2. This requirement makes it difficult to use the process in scaled-up applications (13). In an attempt to extend applications of these processes to the water-soluble solids, it is necessary to achieve "full" miscibility between SC CO_2 and the aqueous solution. This has been accomplished by adding an organic co-solvent (e.g., ethanol) to the system (14). Hence, the spray nozzle of the SEDS process was modified to handle three incoming streams, i.e., SC CO_2, a co-solvent, and an aqueous solution containing the solid of interest. However, this method also ends up with an organic solvent in the system, so the problems mentioned above still remain.

There are other 3-component processes, i.e., PGSS (15) and CAN-BD (16, 17). PGSS stands for Particle Generation from Supercritical Solution. This process is similar in principle to RESS. The SC CO_2 is dissolved in a molten solute or a liquid solution with suspensions of interest, and the resulting solution under high pressures is fed through a nozzle (or a flow restrictor) to effect a rapid expansion at ambient conditions. This process does not rely on the solubility of a solute in SC CO_2, but rather on the solubility of SC CO_2 in a liquid solution. Lastly, the CAN-BD process to be described in the following is unique and differs from all the above processes described. In CAN-BD, dense CO_2 and the liquid solution to be nebulized are intimately mixed in a small volume tee to form microbubbles and microdroplets. The amount of CO_2 in this mixture, which facilitates aerosolization and drying as the aerosol plume is formed in a drying chamber, is more important than the solubility change requirement of all the supercritical fluid processes described above.

Principle of the CAN-BD Process

The CAN-BD process is a patented nebulization process (16,17). The experimental setup of the process has been described elsewhere (19 - 21). A schematic of the CAN-BD apparatus is shown in Figure 1. The unique features of this process are (a) simultaneous micronizing and drying, (b) drying at low temperatures, 10 to 65°C, (c) no organic solvent required when nebulizing an aqueous solution, (d) no high-pressure autoclave required, and (e) a continuous process, which can easily be scaled-up.

Solvents used in the CAN-BD process can be water, alcohol or water-alcohol mixtures, or other solvent with which CO_2 is compatible. Solute concentrations typically vary from 0.1 to 10 % by weight. Liquid CO_2 at room temperature is compressed with a syringe pump to a pressure above its critical pressure (e.g., 83 bar). At a constant pressure (e.g., 83 bar), the dense CO_2 and the liquid solution are intimately mixed in a small volume tee at sub or super-critical conditions of CO_2 (e.g., near room temperature and 83 bar) to form an emulsion. The nebulization is usually effected with a liquid flow rate of 0.3 mL/min and a dense CO_2 flow rate of 1 to 3 mL/min. The resulting emulsion is expanded through a ~10 cm long capillary tube flow restrictor into a two-liter glass-drying chamber at atmospheric pressure to generate aerosols of microbubbles and microdroplets. This set of experiments utilized three restrictors with inner diameters of 50, 74, and 100 μm, subsequently larger diameter restrictors have been used with much larger drying chambers.

Pre-heated air or nitrogen gas is passed through the drying chamber at 20 liters/min to provide heat for aerosol drying. The drying of the aerosols can be achieved with the drying chamber temperature maintained between 30 to 65°C if water is used as a solvent. If the solvent is alcohol, the drying can be achieved at temperatures much lower than 30°C. Finally, the dry powders are collected on a filter membrane (with a pore size of 0.45 μm), which is attached to the bottom of the drying chamber. The process involves continuous flow of a solution and dense CO_2 through 1/16 inch OD stainless steel tubing to generate dry powder. (No high-pressure autoclave is required for the CAN-BD process).

These are illustrative conditions and a wide variety of operating parameters have been tested. For example, lab-scale units have been used to generate fine powders of sugars (trehalose, lactose, sucrose), proteins (trypsinogen, lactate dehydrogenase, ovalbumin), water-soluble drugs (tobramycin sulfate, albuterol sulfate, and cromolyn sodium) and alcohol-soluble drugs (naproxen, amphotericin B, and budesonide) (19 – 26). For the CAN-BD treatment of the alcohol-soluble solutes, the drying temperature can be as low as 5°C, since alcohol with its higher vapor pressure is much easier to evaporate than water. In a scaled up prototype CAN-BD unit (utilizing a 170-liter drying chamber with a thin stainless steel wall), a 10% aqueous solution of NaCl or mannitol has successfully been nebulized at a liquid solution flow rate of 20 mL/min (22).

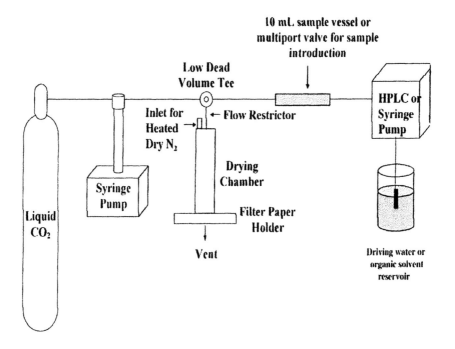

Figure 1   An experimental CAN-BD apparatus

Materials and Analyses

Mannitol (99%) was supplied by Pfanstiehl Lab (Waukegan, Illinois), myo-inositol (99%) by Sigma. Carbon dioxide (99%) and nitrogen (99%) were supplied by Air Gas. Samples of micronized powders were analyzed by a scanning electron microscope (ISI, model SX-30). The mean aerodynamic particle size distribution was measured using the Model 3225 Aerosizer^R DSP, which uses a laser time of flight principle.

Results and Discussions

In the CAN-BD process, the characteristics of the powders generated (e.g., particle size and morphology, etc.) can be affected by the process operating parameters, i.e., drying temperature, CO_2 pressure, solute concentration, flow rates of dense CO_2 and aqueous solution, etc. Parametric studies were conducted to determine the particle characteristics that are most sensitive to process parameters.

A. Parametric study using mannitol:

The results of a parametric study conducted with aqueous solutions of mannitol are summarized in Table I. The base micronization test cases (Runs 1 and 2) were carried out using a 9.5 cm long flow restrictor with an inner diameter (ID) of 74 µm. The aqueous solution containing 10% mannitol was pumped to the mixing tee at a rate of 0.3 mL/min, while dense CO_2 under a constant pressure of 83 bar was pumped into the second port of the mixing tee at a rate of about 2 mL/min. A preheated nitrogen gas stream was delivered into the drying chamber at 20 liters/min to maintain the temperature in the chamber at 50°C.

Runs 1 and 2 (Table I) were the base cases. The data show an average aerodynamic diameter of the particles to be 1.5 µm, with the diameters of 95% of the particles less than 3.5 µm, and the diameters of 5% of the particles less than 0.8 µm. Figure 2a shows the SEM image of the dry powders generated in Run 1. The remaining experiments were carried out by changing one parameter at a time from those in the base cases.

When the drying temperature was varied, no significant change in particle size was observed between 85 and 50°C (Runs 3 and 1). The mean particle size increased to 2.08 µm when the temperature was lowered to 30°C (Run 4). This lower drying temperature probably caused the microbubbles and microdroplets to dry slower and have more opportunity to aggregate, so this could be the cause for the larger particle size. Yet this average particle size of 2 µm is still within

Table I.   Parametric Study at 83 bar and 30-85°C (mannitol in water)

Run no.	Conc. (%)	I.D. (μm)	Temp (C)	Liq. (mL/min)	CO_2 (mL/min)	CO_2/liq.	Particle size (μm) >5%	Avg. D	<95%
Base case (74 μm ID flow restrictor; Liquid at 0.3 mL/min; CO_2 at 83 bar; N_2 at 20 L/min)									
Run 1	10	74	50	0.3	2.2	7.3	0.80	1.53	3.54
Run 2	10	74	50	0.3	2.2	7.3	0.80	1.48	3.31
Effect of drying temperature									
Run 3	10	74	85	0.3	2.1	7.0	0.81	1.48	3.45
Run 1	10	74	50	0.3	2.2	7.3	-	1.53	-
Run 4	10	74	30	0.3	2.5	8.3	0.90	2.08	5.09
Effect of solute concentration									
Run 1	10	74	50	0.3	2.2	7.3	-	1.53	-
Run 5	1	74	50	0.3	2.0	6.7	0.69	1.19	2.11
Run 6	0.2	74	50	0.3	2.0	6.7	0.60	0.98	1.61
Effect of CO_2 flow rate									
Run 7	10	50	50	0.3	0.8	2.7	0.84	1.79	5.65
Run 1	10	74	50	0.3	2.2	7.3	-	1.53	-
Run 8	10	100	50	0.3	13.0	43.3	0.65	1.24	2.90
Run 9	1	100	50	0.3	13.0	26.0	0.60	1.02	1.82

the size range required for effective pulmonary delivery; optimum particle size for the pulmonary drug delivery is 1 to 3 μm (27).

Runs 1, 5 and 6 show the effect of reducing solute concentrations. Average particle size decreased from 1.53, to 1.19, to 0.98 μm, when the solution concentration was reduced from 10%, to 1%, to 0.2%, respectively. The SEM image of fine particles generated from 0.2% mannitol solution is shown in Figure 2b. The aerodynamic particle size distribution (number weighted) is shown in Figure 3. The effect of solute concentration on resulting particle size is obvious by comparing Figures 2a and 2b.

The effect of changing the dense CO_2 flow rate was also investigated. This was accomplished by increasing the flow restrictor ID from 50, to 74, to 100 μm, while maintaining the aqueous flow rate at 0.3 mL/min and CO_2 pressure at 83 bar during the tests. In doing so, the dense CO_2 rate increased from 0.8 to 13 mL/min, and the average particle size decreased from 1.79 to 1.24 μm (Runs 7, 1 and 8). Run 9 was conducted to show the combined effect of low concentration and high CO_2 flow rate. As the solution concentration was reduced from 10% to 1%, the particle size was further lowered from 1.24 μm (Run 8) to 1.02 μm. In subsequent experiments, restrictor ID's measuring 381 μm have been tested.

B. Parametric study using myo-inositol:

A second parametric study was conducted using aqueous solutions of myo-inositol. The results of the study are summarized in Table II. The base micronization cases (Runs 1, 2 and 3) were carried out using a 9.5 cm long flow restrictor with an ID of 74 μm. The aqueous solution containing 5% myo-inositol was pumped to the mixing tee at a rate of 0.3 mL/min, while dense CO_2 under a constant pressure of 83 bar was pumped into the second port of the mixing tee at a rate of about 2 mL/min. Preheated nitrogen gas was delivered into the drying chamber at 20 liters/min to maintain the temperature of the chamber at 65°C.

Runs 1, 2 and 3 (Table II) were replicate base cases with measured average aerodynamic diameters of particles of 1.52, 1.45 and 1.64 μm, respectively. A mean diameter of 1.54 μm was calculated from the three runs. Runs 4, 1, 5, 6 and 7 show that decreasing solute concentrations (from 10% to 0.1 %) reduced particle size (from 1.74 to 0.99 μm). This is consistent with the trend observed for the mannitol (see triangles and diamonds in Figure 6). The CO_2 pressure effect was investigated by compressing dense CO_2 from 83 to 124 and then to 207 bar. Runs 1, 9 and 8 show that average particle size decreased from 1.52 to 1.27 μm as the pressure was increased to 207 bar. The effect of dense CO_2 flow

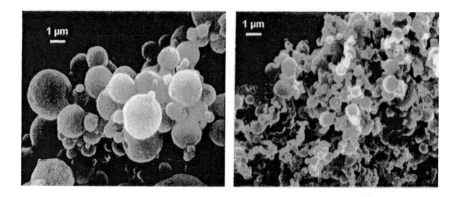

Figure 2a (left) SEM of particles from CAN-BD of a 10 % mannitol
aqueous solution (Run 1 from Table 1).
Figure 2b (right) SEM of particles from CAN-BD of a 0.2 % mannitol
aqueous solution (Run 6 from Table 1)

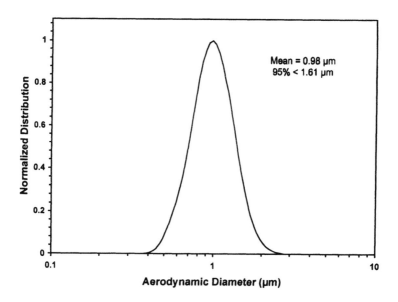

Figure 3   Aerodynamic diameters of particles from CAN-BD of a
0.2 % mannitol aqueous solution (Run 6 from Table 1).

Table II. Parametric Study at 65°C and 83-207 bar (myo-inositol in water)

Run no.	Conc. (%)	I.D. (μm)	Temp (C)	P (bar)	Liq. (mL/min)	CO_2 (mL/min)	CO_2/liq.	Particle size (μm) >5%	Particle size (μm) Avg. D	Particle size (μm) <95%
Base case (74 μm ID flow restrictor; Liquid at 0.3 mL/min; CO_2 at 83 bar; N_2 at 20 L/min)										
Run 1	5	74	65	83	0.3	2.7	9.0	0.76	1.52	3.69
Run 2	5	74	65	83	0.3	2.2	7.3	0.84	1.45	2.99
Run 3	5	74	65	83	0.3	1.9	6.3	0.88	1.64	3.63
Effect of solute concentration										
Run 4	10	74	65	83	0.3	3.1	10.3	0.93	1.74	4.44
Run 1	5	74	65	83	0.3	2.7	9.0	-	1.52	-
Run 5	2	74	65	83	0.3	3.5	11.7	0.81	1.46	3.21
Run 6	1	74	65	83	0.3	3.2	10.7	0.74	1.29	2.35
Run 7	0.1	74	65	83	0.3	1.4	4.7	0.62	0.99	1.64
Effect of CO_2 pressure										
Run 8	5	74	65	207	0.3	7.5	25.0	0.71	1.27	2.36
Run 9	5	74	65	124	0.3	5.3	17.7	0.81	1.54	3.69
Run 1	5	74	65	83	0.3	2.7	9.0	-	1.52	-
Effect of CO_2 flow rate										
Run 10	5	50	65	83	0.3	0.5	1.7	0.85	1.73	4.48
Run 1	5	74	65	83	0.3	2.7	9.0	-	1.52	-
Run 11	5	100	65	83	0.3	14.0	46.7	0.61	1.46	4.51

rate was also investigated. The flow restrictor ID was increased from 50, 74 to 100 µm, while maintaining CO_2 pressure at 83 bar and the aqueous flow rate 0.3 mL/min. In doing so, the dense CO_2 rate increased from 0.5 to 14 mL/min, and the particle size decreased from 1.73 to 1.46 µm (Runs 10, 1 and 11). The SEM images in Figures 4a and 4b show particle size reduction when solution concentration was reduced from 5 to 0.1 %. Figure 5 shows the particle size distribution of the powders generated from a 0.1% myo-inositol solution.

Figure 7 shows a plot of particle size vs. the ratio of CO_2 to aqueous solution mass flow rates. The volumetric flow rates presented in Tables I and II were converted to mass flow rates utilizing the densities of water and dense CO_2. It is interesting to observe that, for the two sets of the data of myo-inositol (i.e., the effects of CO_2 pressure as well as CO_2 flow rate), the figure shows a correlation of particle size vs. mass flow rate ratio (the triangles in Figure 7). Particles reduce in size as the CO_2 to solution flow rate ratio increases. The data of mannitol (the effect of CO_2 flow rate), as represented by the diamonds, also fall in line with the data of myo-inositol.

The anti-solvent type supercritical precipitation processes described earlier, e.g., GAS, are highly dependent on the solubility (or miscibility) between CO_2 and solvent. In the case of CAN-BD, the particle size appears to be affected more by the CO_2 to water ratio, and much less by the solubility of CO_2 in the water. We plan to do further work in the future to substantiate the correlation shown in Figure 7 by varying other pertinent process parameters.

From the above two sets of experimental data, we concluded that the particle size is affected most significantly by two parameters, i.e., the solute concentration and the ratio of CO_2 to liquid solution flow rates. CAN-BD, while a supercritical or near-critical fluid technology, has more in common with conventional spray drying than with precipitation processes from supercritical solutions, in which changes in solubility predominate. The CAN-BD process can be viewed as spray drying with CO_2-assisted nebulization to produce smaller microbubbles and microdroplets than can be obtained with nozzles now used in spray drying. In other experiments, commercial scale spray dryers have been employed successfully with CAN-BD nebulizing nozzles installed in place of jet nebulizer or rotating disk methods of forming droplets.

Conclusions

(1) A new nebulization process (CAN-BD) can generate fine dry particles with mean diameters of substantially less then 5 µm.

Figure 4a (left) SEM of particles from CAN-BD of a 5 % myo-inositol
aqueous solution (Run 1 from Table 2).
Figure 4b (right) SEM of particles from CAN-BD of a 0.1 % myo-inositol
aqueous solution (Run 7 from Table 2)

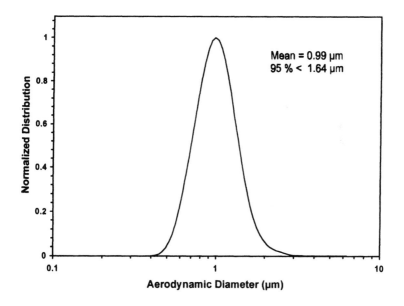

Figure 5   Aerodynamic diameters of particles from CAN-BD of a
0.1 % myo-inositol aqueous solution (Run 7 from Table 2).

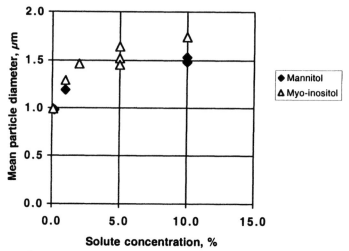

Figure 6 Solute concentration vs. particle size

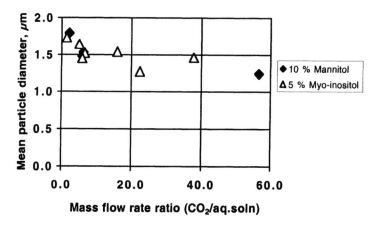

Figure 7 CO₂/aq. soln ratio vs. particle size
(for constant solute concentration)

337

(2) Parametric studies utilizing mannitol and myo-inositol indicate that smaller particles are obtained as the solute concentration decreases or the ratio of dense CO_2 to liquid solution flow rates increases.

(3) Micronization and drying occur simultaneously, resulting in simplification of process and equipment design.

(4) The process can be carried out at low temperatures (e.g., 40 °C), which enables processing of temperature-sensitive solutes such as drugs, vaccines, antibiotics, and proteins.

(5) Drying requires only seconds, and no organic solvent is required in the process. However, if the solute is soluble and stable in an organic solvent, organic solutions can also be nebulized by CAN-BD.

(6) No high-pressure autoclaves are required.

(7) The continuous CAN-BD process has recently been scaled-up to operate in large commercial dryers.

Acknowledgements

Dr. S.P. Cape, Dr. B.P Quinn and J.A. Villa helped review the manuscript. The authors acknowledge the support of the Colorado Tobacco Research Program (Award number 1R-031) for this work.

References

[1] McHugh, M.A.; Krukonis, V.J. *Supercritical Fluid Extraction – Principles and Pratice*; ISBN 0-7506-9244-8; Butterworth-Heinemann: Boston, **1994**.

[2] *Extraction of Natural Products using Near-critical Solvents*; King, M.B.; Bott, T.R., Eds.; ISBN 0-7514-0069-6; Blackie Academic & Professional: London, **1993**.

[3] Jung, J.; Perrut, M. *J. Supercritical Fluids.* **2001**, *20*, 179-219.

[4] Reverchon, E.; Perrut, M. *Proceedings of 7th Meeting on Supercritical Fluids*, Antibes, France. **2000**, 3-20.

[5] Subramanium, B.; Rajewski, R.A.; Snavely, K. *J. Pharm. Sci.* **1997**, *86 (8)*, 885-890.

[6] Matson, D.W.; Peterson, R.C.; Smith, R.D. *Advances in Ceramics.* **1987**, *21*, 109.

[7] Gallagher, R.M.; Coffey, M.P.; Krukonis, V.J.; Hillstrom, W.W. *J. Supercritical Fluids.* **1992**, *5*, 130-142.

[8] Debenedetti, P.G.; Tom, J.W.; Yeo, S.D.; Lim, G.B. *J. Control Release.* **1993**, *24*, 27.

[9] Bleich, J.B.; Muller, B.W.; Wassmus, W. *Int. J. Pharm.* **1993**, *97*, 111.

[10] Foster, N.R.; Bezanehtak, M.; Charoenchaitrakool, M.; Combes, G.; Dehghani, F.; Sze, L.; Thiering, R.; Warwick, B.; Bustami, R.; Chan, H.K.

338

Proceedings of 5th International Symposium on Supercritical Fluids, Atlanta, GA, USA (4/8/2000).

[11] Dixon, D.J.; Johnston, K.P. *Am. Inst. Chem. Eng. J.* **1991.** *37,* 1441.

[12] Hanna, M. H.; York, P. *Proceedings of the Respiratory Drug Delivery Conference V.* Arizona, **1996.**

[13] Thiering, R.; Dehghani, F.; Foster, N.R. *J. Supercritical Fluids.* **2001,** *21,* 159-177.

[14] Palakodaty, S.; York, P. *Pharm. Res.* **1999,** *16,* 976-985.

[15] Weidner, E.; Knez, Z.; Novak, Z. *Proceedings on 3rd Meeting on Supercritical Fluids,* Tome 3, Strasbourg, Germany. **1994,** 229-234.

[16] Sievers, R.E.; Karst, U. U.S. Patent 5,639,441, **1997.**

[17] Sievers, R.E.; Karst, U. European Patent 0677332 B1, **2002.**

[18] Xu, C.Y.; Sievers, R.E.; Karst, U.; Watkins, B.A.; Karbiwnyk, C.M.; Andersen, W.C.; Schaefer, J.D.; Stoldt, C.R. *Chapter 18 in Green Chemistry: Frontiers in Benign Chemical Synthesis and Processes;* Anastas, P.T. and Williamson, T.C., Eds.; Oxford University Press, Oxford. **1998,** 313-335.

[19] Sievers, R.E.; Milewski, P.D.; Sellers, S.P.; Miles, B.A.; Korte, B.J.; Kusek, K.D.; Clark, G.S.; Mioskowski, B.; Villa, J.A. *Ind. Eng. Chem. Res.* **2000,** *39,* 4831-4836.

[20] Sellers, S.P.; Clark, G.S.; Sievers, R.E.; Carpenter, J.F. *J. Pharm. Sci.* **2001,** *90,* 785-797.

[21] Sievers, R.E.; Huang, E.T.S.; Villa, J.A.; Kawamoto, J.K.; Evans, M.M.; Brauer, P.R. *Pure Appl. Chem.* **2001,** *73(8),* 1299-1303.

[22] Sievers, R.E.; Huang, E.T.S.; Villa, J.A.; Walsh, T.R. *Proceedings of 8th Meeting on Supercritical Fluids,* Bordeaux, France. **2002,** 73-75.

[23] Villa, J.A.; Sievers, R.E.; Huang, E.T.S.; Sellers, S.P.; Carpenter, J.F.; Matsuura, J. *Proceedings of 7th Meeting on Supercritical Fluids,* Antibes, France. **2000,** 83-88.

[24] Sievers, R.E.; Huang, E.T.S.; Villa, J.A.; Engling, G.; Brauer, P.R. *J. of Supercritical Fluids,* **2003** (in press).

[25] Sievers, R.E.; Huang, E.T.S.; Villa, J.A. *J. Aerosol Medicine,* **2001,** *14,* 390.

[26] Sievers, R.E.; Huang, E.T.S.; Villa, J.A.; Walsh, T.R.; Meresman, H.V.; Liang, C.D.; Cape, S.P. *Proceedings to Respiratory Drug Delivery VIII,* Tucson, Arizona. **2002,** 675-677.

[27] Corkery, K. Inhalable *Respiratory Care.* **2000,** *45(7),* 831-835.

Chapter 22

Nanocrystal Synthesis and Stabilization in Supercritical Solvents

Parag S. Shah, Keith P. Johnston[*], and Brian A. Korgel[*]

Department of Chemical Engineering and Texas Materials Institute Center for Nano- and Molecular Science and Technology, The University of Texas, Austin, TX 78712-1062

Sterically stabilized nanocrystals in both sc-ethane and sc-CO_2 follow lower critical solution temperature (LCST) phase behavior, with increased dispersibility at higher solvent densities. Perfluorodecanethiol ligands allow for nanocrystal dispersibility in pure CO_2. A single phase arrested growth technique was developed to synthesize fluorinated ligand capped nanocrystals in sc-CO_2. These particles were well passivated with the stabilizing ligand, protecting them from irreversible aggregation. It was found that the nanocrystal size depended on solvent density: at conditions of adequate steric stabilization—i.e., high solvent density and long ligands—the particles grew much smaller and more monodisperse than at poor solvent conditions.

Nanocrystals ranging from 20 to 100 Å in diameter represent a unique class of colloids, capable of exhibiting a variety of size-dependent optical and electronic properties that could be used in a variety of technologies, including coatings, environmental, chemical processing, medical, electronic, and sensing combined with ~0.20 M organic (chloroform) solution of tetraoctylammonium bromide, phase transfer catalyst. The organic phase was collected and the metal ions were reduced using an aqueous sodium borohydride (~0.44 M) solution in the presence of dodecanethiol. The organic phase, rich in nanocrystals, was subsequently collected and washed with ethanol to remove the phase transfer catalyst and any reaction byproducts, and then resuspended in chloroform or hexane (13).

Silver nanocrystals coated with 1H,1H,2H,2H-Perfluorodecanethiol were synthesized at room temperature using a single-phase arrested growth method in a polar solvating medium (18). Initially an aqueous silver nitrate solution (0.03 M) was combined with an acetonic solution of tetraoctylammonium bromide (0.20 M) to form a cloudy single phase mixture. To that was added an aqueous solution of sodium borohydride (0.44 M) in the presence of the fluorinated thiol. The capped silver nanocrystals flocculated in the water/acetone solution due to inadequate steric stabilization in the mixed solvent, resulting in large brown clumps of nanocrystals. This precipitate was separated and redispersed in pure acetone. The nanocrystals could be washed using water as the anti-solvent and subsequently redispersed in polar solvents such as acetone and ethanol or in fluorinated solvents such as Freon or fluorinert.

Nanocrystal synthesis in sc-CO_2 was conducted in a high pressure variable volume view cell (9,19). Initially, the front of the cell was loaded with the precursor, silver acetylacetonate (Ag(acac)), and pressurized with liquid CO_2 until reaching the desired precursor concentration. The system was then brought to reaction conditions by pressurizing the back of the cell by adding CO_2, and heating the cell using heating tape. Once reaction conditions were reached, the fluorinated thiol (perfluorodecanethiol or perfluorooctanethiol) and hydrogen were added to the cell through two injection loops attached in series to the front of the cell. Varying the injection loop length and the pressure of the hydrogen, the ratio of hydrogen and thiol to precursor could be explicitly controlled. Typical reaction times were 3 hours after which the cell was cooled and depressurized by venting the back of the cell. CO_2 was slowly vented from the front of the cell, leaving only the nanocrystals, excess thiol and reaction by-products. The nanocrystals were collected from the cell using acetone and cleaned using heptane as the anti-solvent. The nanocrystals could then be redispersed in acetone, sc-CO_2 or fluorinated solvents (Fluorinert or Freon).

Nanocrystal Characterization

High-resolution transmission electron microscopy (TEM) images were obtained using a JEOL 2010F microscope with 1.4 Å point-to-point resolution equipped with a GATAN digital photography system for imaging and operating with a 200 kV accelerating voltage. Low-resolution TEM images were acquired on a Phillips EM280 microscope with a 4.5 Å point-to-point resolution operating with an 80 kV accelerating voltage. For imaging, the nanocrystals were deposited from solution onto 200 mesh carbon coated copper TEM grids. Average particle size and polydispersity were determined using *Scion Image for Windows* software.

UV-visible absorbance spectra were measured using either a high-pressure optical cell equipped with sapphire windows on opposing sides with an optical path length of 0.9 cm with a 1.75 cm aperture or a variable volume view cell equipped with sapphire windows on opposing sides with an optical path length of 3 cm. The cell was loaded with nanocrystals by evaporating a drop of nanocrystal dispersion into the cell. Temperature and pressure were controlled by using CO_2 back pressure and wrapping the cell in heating tape. The absorbance measurements were acquired on a Beckman DU-40 UV-VIS Spectrophotometer at 1 nm intervals from 200–800 nm.

Results and Discussion

Nanocrystal Stabilization

Nanocrystals are novel colloids made up of crystalline cores surrounded by adsorbed ligands as shown in Figure 1. Since colloid stability is provided by the adsorbed ligands, understanding ligand solvation is very important. As particles collide, well solvated ligands repel each other through osmotic and elastic forces. In order to stabilize the nanocrystal dispersion, these forces must overcome the long range attraction between particles due to van der Waals forces.

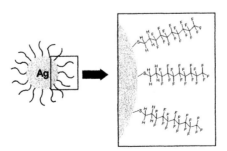

Figure 1. Schematic of perfluorodecanethiol capped silver nanocrystal.

For colloids in SCFs, the strength of the steric repulsion is most closely related to the density of the solvent (20,13,21). The cohesive energy density of the solvent is higher at increased solvent density, resulting in better ligand-solvent interactions (15). In SCFs, the critical flocculation density (CFD) defines the solvent conditions where the dispersion is no longer stable. Monte Carlo simulations have shown that the CFD in SCFs corresponds to the density where the stabilizer ligand is soluble at all concentrations, defined as the upper critical solution density (UCSD) (22,23). However, it is important to note that the simulation results in Ref. 22-23 do not account for colloidal core-core attractions, which are relatively strong for metal nanocrystals. In nanocrystal dispersions the CFD occurs when the steric repulsion can no longer overcome the interparticle van der Waals attraction. For polydisperse systems, the CFD can be blurred, due to the fact that larger particles have stronger core-core attractions than smaller particles (13).

Short alkanes such as decane and dodecane are completely miscible in sc-CO_2 at moderate pressures and temperatures. Based on the earlier simulations, it would seem that nanocrystals capped with dodecanethiol should disperse in sc-CO_2 at moderate conditions; however, this is not the case. Gold and silver nanocrystals capped with dodecanethiol ligands were exposed to sc-CO_2 at pressures as high as 483 bar and temperatures up to 80°C without any visible dispersibility. This indicates that ligands with better CO_2 compatibility need to be developed to allow for effective dispersibility. The dodecanethiol capped nanocrystals were dispersible in sc-ethane, as the solvent-ligand interactions were much more compatible.

Figure 2 shows the UV-visible absorbance spectra of dodecanethiol-coated silver nanocrystals dispersed in ethane at various densities (13). The absorbance peaks correspond to the lowest order surface mode resulting from a plasma oscillation with uniform polarization across the volume of the silver particle (24). The absorbance spectra in Figure 2a were obtained at 35°C by sequentially lowering the system pressure, beginning from the highest ethane density and allowing the system to equilibrate at each reported pressure. Figure 2b shows the absorbance spectra at constant pressure of 414 bar as the sample temperature was raised from 25°C to 55°C, while allowing the system to equilibrate at each reported temperature. In both sets of spectra, the total absorbance decreases with decreasing ethane density, as an increasing proportion of particles precipitate from solution. The same trend of improved nanocrystal dispersibility with increased pressure occurs at 45°C. Additionally, dodecanethiol coated gold nanocrystals show the same LCST type behavior. Although the absorbance spectra from small gold nanocrystals ($d=\sim18$ Å) was visible at pressures as low as 136 bar at 25°C, a distinct plasmon peak at ~510 nm, indicative of larger gold nanocrystals ($d=\sim42$ Å), was only observed at the highest pressure of 414 bar. At 35°C, the absorbance peak at 510 nm could not be detected, even at the highest pressure studied of 414 bar, indicating that the ethane density was too low to disperse the larger gold nanocrystals at this temperature. These data clearly show that reduced ethane density decreases the dispersibility of dodecanethiol-coated nanocrystals.

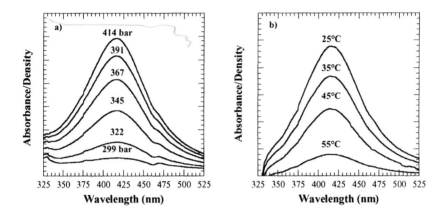

Figure 2. Absorbance spectra for silver nanocrystals in sc-ethane – a) 35°C and various pressures; b) 414 bar and various temperatures. (Reproduced from reference 13. Copyright 2002 American Chemical Society.)

The lower critical solution temperature (LCST) phase behavior exhibited by the nanocrystals is often found for low molecular weight solutes in supercritical fluids (25,26) and also for polymers dissolved in SCFs, and results from compressibility differences between the polymer and the solvent (15). As the temperature increases or the pressure decreases, the solvent prefers to leave the solute to increase its volume and entropy. The same mechanism that governs phase separation in supercritical fluids also drives flocculation of two surfaces with steric stabilizers, as has been shown with theory (22) and simulation (23).

TEM was used to measure the size distribution of the dispersed nanocrystals as a function of ethane density. Figure 3 shows the pressure-dependence of the silver nanocrystal size distribution in sc-ethane at 35°C. Nanocrystals dispersed at 414 bar were collected and examined by TEM. The pressure was lowered to 317 bar and a second sample was collected. The as-synthesized nanocrystal dispersion is relatively polydisperse in size. After suspending the nanocrystals in ethane at 414 bar (Figure 3b) the histogram still exhibits a broad distribution with a similar size range compared to the original preparation, however, it shifts slightly to favor the smaller nanocrystals less than 40 Å in diameter, and the average diameter drops from 53 Å to 37 Å. Nanocrystals collected at 317 bar

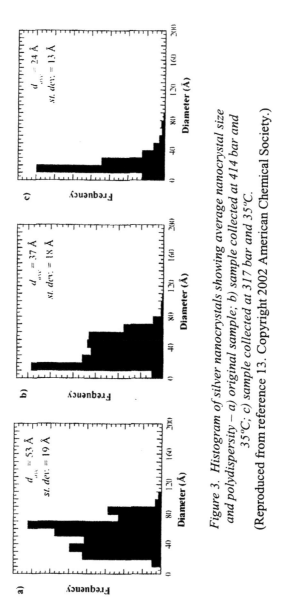

Figure 3. Histogram of silver nanocrystals showing average nanocrystal size and polydispersity – a) original sample; b) sample collected at 414 bar and 35°C; c) sample collected at 317 bar and 35°C.
(Reproduced from reference 13. Copyright 2002 American Chemical Society.)

exhibit a significantly smaller size with a narrower size distribution. The average diameter of dispersed dodecanethiol-capped silver nanocrystals decreases considerably with ethane density, from 53 Å prior to dispersion in ethane, to 37 Å at 414 bar, and 24 Å at 317 bar.

Nanocrystals exhibit strongly size-dependent core-core attractions. A decrease in pressure from 414 bar to 317 bar results in a small decrease in ligand solvation, which drops the average nanocrystal size by 13 Å. As seen in Figure 3c, at 35°C particles larger than 60 Å in diameter do not disperse in ethane at 317 bar. At 414 bar, Figure 3b, the cutoff size increases to ~80 Å. Higher solvent densities provide enhanced ligand tail solvation necessary to suspend larger particles.

Finally, the precipitation and redispersion of the silver nanocrystals was found to be nearly reversible. After precipitating the largest nanocrystals of a polydisperse dispersion by lowering the system pressure from 414 bar to 276 bar, and then repressurizing to 414 bar, 90% of the silver nanocrystals redispersed. Reversible nanocrystal flocculation has potential value in fine-tuning size-dependent separations with minor variations in pressure. Reversible solvation conditions are difficult to achieve using a conventional anti-solvent approach.

Unlike hydrocarbon solvents, sc-CO_2 has a low polarizability per volume which gives rise to weak van der Waals forces. Sterically stabilized systems in sc-CO_2 require "CO_2-philic" groups with low cohesive energy densities to provide favorable energetic interactions with the solvent. The first sterically stabilized nanocrystal dispersion in sc-CO_2 was achieved by substituting a partially fluorinated alkanethiol for the typical dodecanethiol, shown in Figure 1 (18). Perfluorodecanethiol capped nanocrystals were dispersible in liquid and sc-CO_2, as well as polar solvents such as acetone and in fluorinated solvents such as Freon or fluorinert. O'Neill et al. showed that polymer solubility in CO_2 correlates with the surface tension of the polymer, γ, which is a measure of the cohesive energy density (15). Lower surface tensions indicate a lower cohesive energy density and thus a greater solubility in CO_2. The surface tension of dodecane is 24.5 mN/m, while for perfluorodecane it is 14.0 mN/m. The stability of the nanocrystals in CO_2 is consistent with the low cohesive energy density of perfluorodecane. The dispersibility in polar solvents may be explained by the large dipole moment of a CH_2—CF_2 group (see Figure 1), which produces substantial dipole-dipole interactions with polar solvents. It is possible that dipole-quadrupole interactions could further enhance dispersibility in sc-CO_2.

Figure 4 shows the absorbance spectra of perfluorodecanethiol capped nanocrystals in sc-CO_2 at 80°C. The same trend as for the dodecanethiol capped nanocrystals in ethane are seen and dispersibility is lost as the solvent density is lowered. The CFD is determined to be ~276 bar as the plasmon peak absorbance disappears altogether, at lower pressures. Compared to the perfluorodecanethiol ($C_8F_{17}C_2H_4SH$) ligand, perfluorooctanethiol ligands ($C_6F_{13}C_2H_4SH$) were unable to disperse nanocrystals in CO_2, although they worked effectively in both fluorinated and polar solvents. Perfluorooctanethiol, which has 20% fewer fluorinated carbons than perfluorodecanethiol, is unable to

346

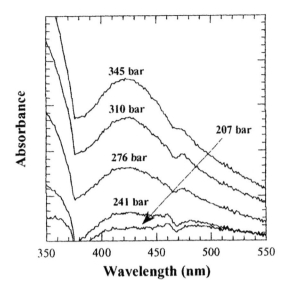

Figure 4. *UV-visible absorbance spectra of perfluorodecanethiol capped silver nanocrystals dispersed in sc-CO_2 at 80°C.* (Reproduced from reference 19. Copyright 2002 American Chemical Society.)

provide the steric repulsion needed in sc-CO_2, where core-core van der Waals attractions are typically stronger than conventional solvents.

Nanocrystal Synthesis

Silver nanocrystals were synthesized in sc-CO_2 by reducing Ag(acac) with hydrogen at a variety of precursor and thiol concentrations and pressures. The nanocrystals were generally well-passivated by the ligands (both perfluorodecanethiol and perfluorooctanethiol). They redisperse in acetone and fluorinated solvents such as Freon and fluorinert. Figure 5 shows representative TEM images of perfluorodecanethiol coated silver nanocrystals synthesized in sc-CO_2 at high reaction densities. The reasonably monodisperse nanocrystals form small regions of ordered close-packing. The average edge-to-edge separation distance between neighboring nanocrystals is ~21 Å, significantly larger than what is found for hydrocarbon-stabilized nanocrystals (decanethiol gives ~12.5 Å (27)), due to the increased stiffness of the fluorinated ligand. The lattice spacing in the high-resolution TEM images in Figure 5 reveal lattice spacing of 2.3 Å corresponding to the d_{111} spacing in silver.

The moments of the size distribution, μ_1 and μ_3, can be used to determine the relative extent of Ag condensation and particle coagulation on the growth mechanism. By measuring the arithmetic mean radius, $r_1 = \sum r_i / N_x$, cube-mean radius, $r_3 = \sqrt[3]{\sum r_i^3 / N_x}$, and harmonic mean radius, $r_h = N_x / \sum (1/r_i)$, where N_t is the total number of particles, of the samples, the moments of the size

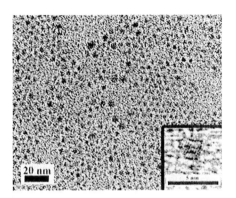

Figure 5. Representative TEM images of silver nanocrystals capped with perfluorodecanethiol synthesized in sc-CO₂. (Reproduced from reference 19. Copyright 2002 American Chemical Society.)

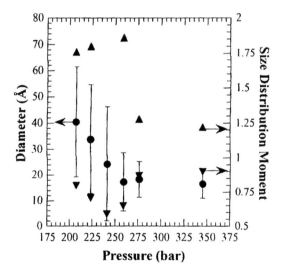

Figure 6. Average nanocrystal diameter (●) and size distribution moments, μ_1 (▲) and μ_3 (▼), of perfluorodecanethiol-coated silver nanocrystals synthesized at 80°C with a precursor concentration of 3.5 mM and thiol:precursor ratio of 2.5 as a function of solvent pressure. The error bars represent the standard deviation of the samples. (Reproduced from reference 19. Copyright 2002 American Chemical Society.)

distribution, $\mu_1 = r_3/r_h$ and $\mu_3 = r_1/r_3$ can be determined. Nanocrystals synthesized through condensation of free atoms, or small oligomers, onto growing metal cores are relatively monodisperse with $\mu_1 = \mu_3 = 1$; whereas, coagulative growth results in broad size distributions with $\mu_1 > 1.25$ and $\mu_3 < 0.905$ (28).

Our most significant finding is the distinct dependence of the nanocrystal size and polydispersity on the solvent density during particle formation. Well-passivated silver nanocrystals could be synthesized at a wide range of solvent pressures, from 209 to 345 bar. However, they do not *redisperse* at pressures below 276 bar (at 80°C)—as observed in the UV-visible absorbance spectra in Figure 4. Figure 6 shows the size and polydispersity of perfluorodecanethiol capped nanocrystals synthesized at 80°C with a 3.5 mM precursor concentration and a thiol:precursor ratio of 2.5. Above 276 bar, the particles exhibit an average diameter of ~20 Å and the synthesis pressure does not noticeably affect the nanocrystal size or polydispersity. However, below 276 bar, both the nanocrystal size and polydispersity increase with decreasing pressure. Nanocrystals synthesized below the dispersibility limit of 276 bar are still well-passivated by the ligands and redisperse in acetone, fluorinated solvents, and sc-CO_2 at 80°C and pressures greater than 276 bar. The size distribution moments indicate that particles grow by coagulation at pressures below 276 bar. Above 276 bar, μ_1 and μ_3 approach 1, revealing that growth occurs through a combination of both coagulation and condensation. These results indicate that high solvent density slows particle aggregation, which largely prevents coagulation and precipitates particles in the small size range. At lower solvent density, where ligand solvation is poor, particles grow primarily by coagulation.

Along with reaction density, the precursor concentration can also significantly affect the nanocrystal size and size distribution. Figure 7 plots the particle size and polydispersity for nanocrystals capped with perfluorodecanethiol synthesized under both poor and good solvation conditions. Under poor solvent conditions, low solvent density or short ligands (perfluorooctanethiol), both the size and polydispersity are affected by the precursor concentration. Under good solvent conditions, high solvent density and long ligands (perfluorodecanethiol), only the size distribution appears affected with higher precursor concentration leading to broader size distributions. Increasing the precursor concentration causes μ_1 and μ_3 to deviate increasingly from 1, signifying that these conditions favor coagulation. Higher precursor concentrations increase the initial concentration of silver nuclei, which increases the collision rate between nuclei leading to an increased coagulation effect. Additionally, at low solvent densities the increased precursor concentration results in larger nanocrystals as the number of collisions increases, however, these collisions are limited at higher densities and the average nanocrystal size remains small. Nanocrystals synthesized at either low or high solvent density redispersed in acetone, fluorinated solvents such as Freon and fluorinert, and sc-CO_2. Coagulative growth dominates nanocrystal formation

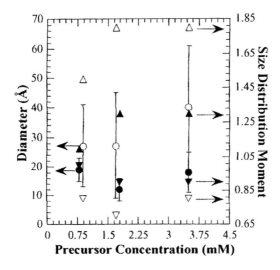

Figure 7. Average diameter (●) and size distribution moments, μ_1 (▲) and μ_3 (▼), of perfluorodecanethiol stabilized silver nanocrystals grown at 80°C with a thiol:precursor ratio of 2.5:1, as a function of precursor concentration. The closed symbols correspond to 276 bar while the open symbols correspond to 207 bar. The error bars represent the standard deviation of the samples. (Reproduced from reference 19. Copyright 2002 American Chemical Society.)

using the shorter perfluorooctanethiol ligands under all solvent conditions (data not shown) (9). In addition, the nanocrystal size increases with increasing precursor concentration, indicating poor solvent conditions as discussed above. The shorter ligand does not provide the adequate steric repulsion to prevent coagulative growth, as expected since the ligand cannot disperse nanocrystals in sc-CO$_2$.

Analysis of the growth mechanism reveals that low solvent density or short steric stabilizers increases coagulation and leads to larger nanocrystals. To better understand this phenomenon it is important to consider the relevant conditions for coagulative growth. Coagulative nanocrystal growth requires collisions between particle clusters along with close approach of the particle surface, to allow for metal fusion (29). At very early times in the particle formation process, nanocrystals will be only partially coated with ligands and particle coagulation will compete with ligand adsorption. However, sterically stabilized nanocrystals undergo many collisions without coagulation and particle growth. Additionally, fully passivated nanocrystals can be flocculated using anti-solvents and redispersed without any change in particle size. This means that sufficient ligand binding density is needed to prevent irreversible coagulation of nanocrystals.

As a general guideline, the metal cores fuse if they approach to within 5 Å of each other (29). Since the interparticle separation for perfluorodecanethiol-coated nanocrystals in the *absence* of solvent is 21 Å, well-passivated nanocrystals are protected from fusion and coagulative growth under all solvent conditions. *Partially* passivated nanocrystals, however, are capable of fusing into larger clusters if the cores can approach to within 5 Å upon collision. Early in the growth process, the ligand surface coverage is low and particle coagulation competes with ligand adsorption. Therefore, to prevent coagulative nanocrystal growth, the solvent must enable the ligands to fully extend and prevent cluster coagulation between partially capped cores until capping becomes sufficiently high. At high solvent densities, the ligands provide a strong steric repulsion, opposing the van der Waals core-core attraction and preventing excessive coagulation during the growth stage. In contrast, at lower solvent densities, the ligands provide only a weak repulsion at best and are unable to prevent coagulative collisions between cores. Although high ligand surface coverage inhibits nanocrystal growth at all solvent conditions, growth due to coagulation is arrested at much lower coverages earlier in the growth process under good solvent conditions. This leads to smaller nanocrystals with tighter size distributions as evidenced in Figures 6 and 7.

Conclusions

Both sc-ethane and sc-CO_2 provide density tunable dispersibility for nanocrystals. Partially fluorinated ligands enabled the first example of a sterically stabilized nanocrystal dispersion in pure CO_2. The nanocrystals show LCST phase behavior with increased dispersibility at higher solvent densities. Additionally, arrested precipitation to synthesize nanocrystals in sc-CO_2 has been developed. The technique yields chemically robust nanocrystals that are fully passivated with fluorinated ligands allowing for collection and redispersion of the particles without any change in size or polydispersity. The nanocrystal size produced depends on both the solvent density and length of the ligand, with smaller less polydisperse particles formed at conditions of adequate steric stabilization.

Acknowledgments

This work is supported in part by the STC Program of the National Science Foundation under Agreement No. CHE-9876674 and the Welch Foundation. B.A.K. also thanks NSF (Agreement No. CTS-9984396) for support through a CAREER Award.

References

1.   Brust, M.; Walker, M.; Bethell, D.; Schiffrin, D. J.; Whyman, R. *J. Chem. Soc., Chem. Commun.* **1994**, *7*, 801-802.

2.   Alivisatos, A. P. *Science* **1996**, *271*, 933-937.
3.   Murray, C. B.; Norris, D. J.; Bawendi, M. G. *J. Am. Chem. Soc.* **1993**, *115*, 8706-8715.
4.   Sun, S.; Murray, C. B. *J. App. Phys.* **1999**, *85*, 4325-4330.
5.   Sun, S.; Murray, C. B.; Weller, D.; Folks, L.; Moser, A. *Science* **2000**, *287*, 1989-1992.
6.   Puntes, V. F.; Krishnan, K. M.; Alivisatos, A. P. *Science* **2001**, *291*, 2115-2117.
7.   Holmes, J. D.; Bhargava, P. A.; Korgel, B. A.; Johnston, K. P. *Langmuir* **1999**, 6613-6615.
8.   Holmes, J. D.; Ziegler, K. J.; Doty, R. C.; Pell, L. E.; Johnston, K. P.; Korgel, B. A. *J. Am. Chem. Soc.* **2001**, *123*, 3743-3748.
9.   Shah, P. S.; Husain, S.; Johnston, K. P.; Korgel, B. A. *J. Phys. Chem. B* **2001**, *105*, 9433-9440.
10. Holmes, J. D.; Johnston, K. P.; Doty, R. C.; Korgel, B. A. *Science* **2000**, *287*, 1471.
11. Ziegler, K. J.; Doty, R. C.; Johnston, K. P.; Korgel, B. A. *J. Am. Chem. Soc.* **2001**, *123*, 7797-7803.
12. Clarke, N. Z.; Waters, C.; Johnson, K. A.; Satherly, J.; Schiffrin, D. J. *Langmuir* **2001**, *17*, 6048-6050.
13. Shah, P. S.; Holmes, J. D.; Johnston, K. P.; Korgel, B. A. *J. Phys. Chem. B* **2002**, *106*, 2545-2551.
14. O'Shea, K.; Kirmse, K.; Fox, M. A.; Johnston, K. P. *J. Phys. Chem.* **1991**, *95*, 7863.
15. O'Neill, M. L.; Cao, Q.; Fang, M.; Johnston, K. P.; Wilkinson, S. P.; Smith, C. D.; Kerschner, J. L.; Jureller, S. H. *Ind. Eng. Chem. Res.* **1998**, *37*, 3067-3079.
16. Johnston, K. P.; Harrison, K. L.; Clarke, M. J.; Howdle, S. M.; Heitz, M. P.; Bright, F. V.; Carlier, C.; Randolph, T. W. *Science* **1996**, *271*, 624.
17. Ohde, H.; Hunt, F.; Wai, C. M. *Chem. Mater.* **2001**, *13*, 4130-4135.
18. Shah, P. S.; Holmes, J. D.; Doty, R. C.; Johnston, K. P.; Korgel, B. A. *J. Am. Chem. Soc.* **2000**, *122*, 4245-4246.
19. Shah, P. S.; Husain, S.; Johnston, K. P.; Korgel, B. A. *J. Phys. Chem. B* **2002**, 106, 12178-12185.
20. Yates, M. Z.; Shah, P. S.; Johnston, K. P.; Lim, K. T.; Webber, S. *J. Colloid Interface Sci.* **2000**, *227*, 176-184.
21. Calvo, L.; Holmes, J. D.; Yates, M. Z.; Johnston, K. P. *J. Supercrit. Fluids* **2000**, *16*, 247-260.
22. Meredith, J. C.; Johnston, K. P. *Macromolecules* **1998**, *31*, 5518-5528.
23. Meredith, J. C.; Sanchez, I. C.; Johnston, K. P.; de Pablo, J. J. *J. Chem. Phys.* **1998**, *109*, 6424-6434.
24. Bohren, C. F.; Huffman, D. R. *Absorption and Scattering of Light by Small Particles*; John Wiley & Sons: New York, 1983.

25. McHugh, M.; Krukonis, V. *Supercritical Fluid Extraction*; Buttersworth Publishers: Boston, 1986.

26. Eckert, C. A.; Knutson, B. L.; Debenedetti, P. G. *Nature* **1996**, *383*, 313-318.

27. Korgel, B. A.; Fullam, S.; Connolly, S.; Fitzmaurice, D. *J. Phys. Chem. B* **1998**, *102*, 8379-8388.

28. Pich, J.; Friedlander, S. K.; Lai, F. S. *Aerosol Science* **1970**, *1*, 115-126.

29. Israelachvili, J. *Intermolecular & Surface Forces*, 2nd ed.; Academic Press, Inc.: San Diego, 1992.

Chapter 23

Supercritical Melt Micronization Using the Particles from Gas Saturated Solution Process

P. Münüklü[1], F. Wubbolts[1], Th. W. De Loos[2], G. J. Witkamp[1], and P. J. Jansens[1]

[1]Laboratory for Process Equipment, Delft University of Technology, Leeghwaterstraat 44, 2628 CA Delft, The Netherlands
[2]Laboratory of Applied Thermodynamics and Phase Equilibria, Delft University of Technology, Julianalaan 136, 2628 BL Delft, The Netherlands

Compressed CO_2 dissolves in melts of fats. Solubility measurements show at higher pressures, a minimum in the liquid-vapour isopleth, which is an indication of type III phase behaviour. Batch particle formation experiments show an influence of CO_2/melt-ratio, feed rate, temperature and pressure on the particle size and structure. This information is crucial for the development of applications and for optimization of the process. Three main shapes can be distinguished, spherical (solid or hollow), distorted and sponge-like particles. The particle sizes are in the range of 5-200µm

The grinding of soft and heat sensitive materials forms a problem that is traditionally tackled by using cryogenic milling in order to avoid product degradation or clogging of partially molten particles. Cryogenic milling uses liquid nitrogen to cool down the material and make it brittle. Because of the high specific nitrogen consumption this is an expensive and energy consuming technique. Nowadays, supercritical fluids are being increasingly used with some important techniques as media for fine particle formation. One of the important techniques is the "particle from gas-saturated solution" process (PGSS), first studied by E.Weidner and Z.Knez (1).

In the PGSS technique, supercritical carbon dioxide is dissolved in a melt at a high pressure. The obtained mixture is depressurized over a nozzle, where a spray of fine droplets is produced. These droplets solidify and form fine particles. The advantage of the PGSS method over milling is that CO_2 is not toxic and inexpensive. The process has many similarities to spray drying (3) where a solvent evaporates from a solution. The difference between spray drying and PGSS is that in spray drying the feed is sprayed into hot drying medium e.g. air. In the PGSS process the feed contains CO_2 and is sprayed into ambient air at room temperature. The CO_2 in this case can be compared to the moisture in spray drying. Due to its high vapor pressure, the CO_2 boils off during the expansion. A similar effect may occur in a certain regime of the spray drying process, where the moisture is evaporating from the droplet while it is boiling. In spray drying hollow particles can be formed through three different mechanisms. One of the mechanisms results in the formation of a semi impervious surface layer at the droplet surface where the droplet puffs out when vapor is formed inside. The second mechanism happens, when the moisture in the droplet evaporates before a complete solid crust is formed. The last mechanism is observed when air entrained in the feed contributes to air voids in the droplet. Only the first two mechanisms might provide useful information about the formation of particles in the PGSS process. More detailed Particle formation in spray drying is given by K. Masters (3).

Until now, no experimental studies have yet been published on supercritical melt micronisation for fats. The aim of this work is to develop and to understand a process for the micronisation of edible fats, polymer or waxes. The requirements for such powders are from manufacturers and consumers different. For example, for a good dispersion of particles in solutions the particles should not be too small or hollow always, because flotation of the particles appears at the surface of the solution. Another application is the wettablity of the powder, where non-spherical particles with large surfaces areas are required.

The spherical particles again can be preferred in order to avoid agglomeration with other ingredients and fast sedimentation of the used substrate in food applications and cosmetics etc. The study is started with a mixture of an edible fat, Rapeseed 70 (RP 70), and CO_2. Further solubility measurements were done to measure the phase behavior of the Rapeseed 70/CO_2 mixture and to determine the operating window. Batch experiments were done to determine the influence of CO_2 concentration, melt temperature, atomization pressure, ambient and nozzle temperature on particle morphology.

Experimental

Set-up for Solubility Measurements

With a Cailletet apparatus (*2*), schematically shown in Figure 1, dew points and bubble points of mixtures of known compositions can be determined visually. The sample of the mixture is confined above mercury in the sealed end of a capillary Pyrex glass tube, which fits in the closing plug of a stainless-steel autoclave filled with mercury. The open end of the tube is immersed in the mercury, which serves as a pressure intermediate between the sample and the hydraulic oil, which is used for pressurizing the system. The sample is stirred by a stainless steel ball and a magnet. The glass capillary tube is kept at the desired temperature by a thermostat with circulating oil or water depending on the needed temperature.

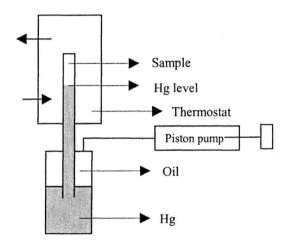

Figure 1: Experimental set-up for solubility measurements

Set-up for Batch Experiments

The experiments where carried out using the test facility shown in Figure 2. After melting at T_{melt}, the fat is poured into the heated pressure vessel (*a*). The CO_2 (*b*) enters the pressure vessel from the bottom until equilibrium was

356

reached at $P_{dissolve}$, (X_{CO2}). CO_2 pressure is built up in the autoclave with the pressure control valve (d). When the desired pressure is reached, the CO_2/fat mixture was pressurized to atomization pressure (P_{atom}) using Helium (j). After the desired atomisation pressure is reached, the stirrer is turned off and valves (e) and (f) are closed. By opening valve (f) in the direction of the nozzle (g), a fine spray of droplets is sprayed in the collection vessel (h). To prevent the particles from being released from the collection vessel into the air, the CO_2 outlet of the collection vessel passes a filter (i). A serie of experiments were carried out where T_{melt}, $P_{dissolve}$, and P_{atom}. were varied.

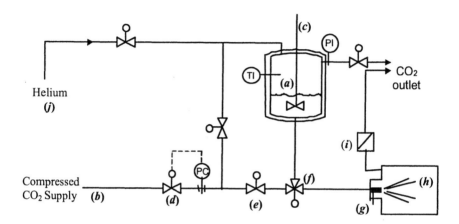

Figure 2: Experimental set-up for batch experiments

The nozzle (5) during expansion is shown in Figure 3. After expansion, re-circulation occurs where the droplets collide and depending on the melt temperature, (T_{melt}), droplet velocity and the vapour pressure of CO_2 different particle morphologies occur.

Figure 3: Spraying nozzle

Results and Discussions

Solubility Measurements

The experimental window of the measurements is shown in Table 1. The results from the solubility measurements are plotted in a P-T diagram, Figure 4, as lines of constant composition (isopleths).The P-x diagrams that were constructed, are shown in Figure 5. It can be seen that the isotherms converge at 30wt% CO_2. This is a consequence of the pressure minimum in the 70wt% fat isopleth shown in Figure 4. An increase in pressure or a decrease in temperature results in a higher solubility of carbon dioxide in the fat melts. As can be seen in Figure 4, the vertical S+L→ (Liquidus curve) phase boundary curves shift to lower temperatures with increasing the CO_2 concentration indicating melting point depression. Solid-Liquid equilibria were measured only for 83wt%, 90wt%, 95wt% and 100wt% of fat.

Table 1- Experimental window of Solubilty Measurements

	Temperature in °C	Pressure in bar	Weight % CO_2
Minimum	52	29	5
Maximum	94	307	30

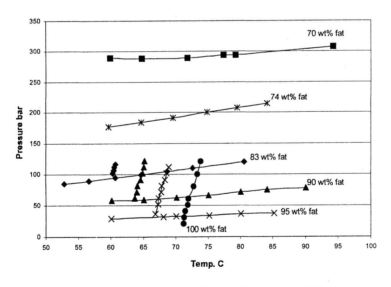

Figure: 4 P-T crossection of isopleths showing bubble-point pressures (L+V→L), horizontal curves, and solid-liquid (S+L→L), vertical curves, for the binary system CO_2/RP 70.

358

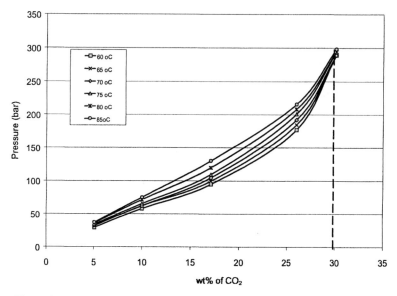

Figure5: *P-x* projection of isopleth showing bubble-point pressures (L+V→L) for the binary system CO_2/RP 70 at temperatures of 60, 65, 70, 75, 80 and 85 °C.

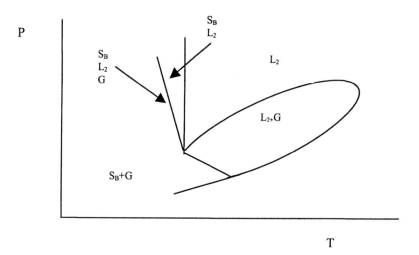

Figure 6 – *P, T* cross-section of a typical Type III system at high rapeseed concentration.

Figure 6 is selected using the classification by Scott & Konijnenburg (7), which describes only six different basic types of phase diagrams for binary systems. In Figure 4 a minimum can be observed at high pressure in the 70wt% fat isopleth, which corresponds to convergence of the isotherms in Figure 5. This is a clear indication of type III phase behavior, which is represented in Figure 6. A liquid-liquid phase split is in general typical for a type III system (9).

The CO_2 / RP70 system has a metastable liquid-liquid phase split with respect to the solid fraction because the metastable region is close to the critical temperature of CO_2. Hence liquid-liquid equilibria were not observed experimentally. Upon expansion, the melt starts in the melting region (L_2) and goes to the solid+gas (*S+G*) region crossing solid + liquid (S_B+L_2) or liquid+gas (*L+G*) regions. These possible pathways may have an effect on the morphology of the obtained product.

Fat Micronisation Experiments

The experimental window of the batch experiments is shown in Table 2.

Table 2 – Experimental window of batch experiments

	T_{melt} in °C	$P_{atom.}$ in bar	$P_{dissolve}$ in bar	CO_2 in w%
Minimum	60	70	0	0
Maximum	80	158	145	21

Size and Shape

In the discussion of the results a distinction is made, with help of the scanning electron microscopy (SEM), between spheres (solid or hollow), distorted and sponge-like (*6*) particles. All the other morphologies are defined by the combination of these three basic shapes e.g. distorted agglomerated hollow spheres. The three shapes are shown in Figure 7.

Effect of atomization pressure

Table 3 shows the effect of the hydraulic atomization on the particle formation. No clear difference in particle morphology was observed for different atomization pressures (*$P_{atom.}$*). The obtained powder consists of agglomerated particles (Fig. 8.b). The particles tend to stick together and are distorted. Further, the size of the particles is for each experiment in the range of 10 to 50 microns. The bulk density of the particles did not change considerably (Fig. 8a).

Contrary to the expectations the atomization pressure has no effect on the size of the primary particles within the experimental window. The relatively low bulk density and the appearance of the punctured particles suggest that the particles are hollow.

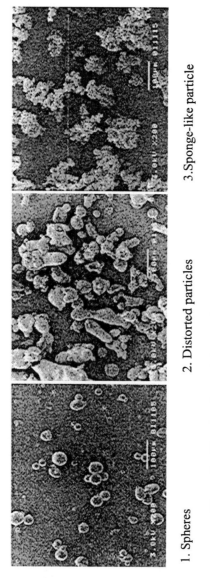

1. Spheres   2. Distorted particles   3.Sponge-like particle

Figure:7 – Three main particle morphologies

Table 3–The effect of the hydraulic atomization on the particle morphology

Exper. Nr.	$P_{dissolve}$ bar	$P_{atom.}$ bar	T_{melt} °C	Bulk-density kg/m^3	Calcul.wt.%CO_2	Particle size (10-50µm)
1	70	70	70	204	11.4	Aggl. hol. Particles
2	70	90	70	218	11.4	Aggl. hol. Particles
3	70	120	70	256	11.4	Aggl. hol. Particles
4	70	145	70	219	11.4	Aggl. hol. particles
5	70	158	70	213	11.4	Aggl. hol. particles

Effect of the dissolved wt% of CO_2 in the melt

In Table 4 the effect of the concentration of CO_2 on the particle morphology is shown. In the order of increasing wt.% carbon dioxide, the structure of the particle changes from solid spheres (0wt% CO_2) through hollow spheres (8-18wt% CO_2) to sponge- like particles (>18wt% CO_2), Figure 9. From experiment 6 to 8 the particles also decrease in size. Together with the change in particle morphology, the bulk density decreases significantly. In the experiments the temperature of the collection vessel was increased from ambient (15-20°C) to 32 –40°C.

Table 4- The effect of the wt% CO_2 on the particle morphology

Exper.nr.	$P_{dissolve}$ bar	$P_{atom.}$ bar	T_{melt} °C	Bulk-density kg/m^3	Calculated wt.% CO_2	Particle size (10-200µm)
11	0	145	70	404	0	Dist. Solid spheres
6	50	145	70	267	8.11	Distort. hol. Spheres
7	65	145	70	263	10.62	Distort. hol. Spheres
8	80	145	70	238	12.94	Distort. hol. Spheres
9	120	145	70	167	18.25	Aggl. hol. Particles
10	145	145	70	118	20.93	Sponge-like particles

Initial temperature of the fat melt at 70 and 140 bar atomization pressure

In Table 5 the effect of the melt temperature on the particle morphology is shown. In all experiments hollow spheres were formed. At 60-65°C melt temperature distorted particles were formed. At 80°C and 140 bar atomization pressure agglomeration occurred. Whereby at 80°C and 70 bar atomization pressure, Figure10, hollow and spherical particles are formed.

362

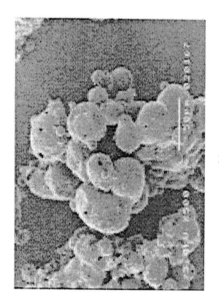

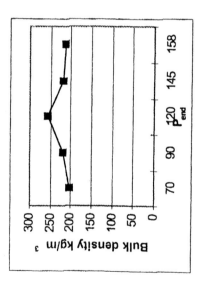

a)

b)

Figure: 8 - a) Bulk density at different end-pressures, b) Exp. nr. 5, agglomerated hollow particles

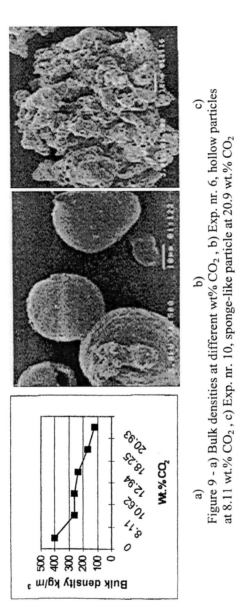

a)

b)

c)

Figure 9 - a) Bulk densities at different wt% CO_2, b) Exp. nr. 6, hollow particles at 8.11 wt.% CO_2, c) Exp. nr. 10, sponge-like particle at 20.9 wt.% CO_2

Table 5 –The effect of the melt temperature on the particle morphology

Exper.nr.	$P_{dissolve}$ bar	$P_{atom.}$ bar	T_{melt} °C	Bulk-density kg/m^3	Calculated wt.% CO_2	Particle size (5-80μm)
12	70	70	60	199	12.76	Distort hol. spheres
13	70	70	65	196	11.93	Distort. hol. spheres
14	70	70	70	240	11.41	Hollow spheres
15	70	70	75	243	11.19	Hollow spheres
16	70	70	80	247	10.13	Hollow spheres
17	70	140	60	186	12.76	Dist. hol. particles
18	70	140	65	234	11.93	Dist. hol. particles
19	70	140	70	255	11.41	Hollow spherical particles
20	70	140	75	254	11.19	Hollow spherical particles
21	70	140	80	264	10.13	Aggl. hol. particles

The results show that the particle formation is strongly dependent on the temperature of the fat melt. Furthermore the results again show that the atomization pressure has little effect on the particle morphology. At a higher temperature of the droplet, more energy has to be removed before solidification occurs, i.e. it takes more time to form a crust. In this way there is more time for the carbon dioxide to evaporate and less or no collapsing or distortion of the particles occurs. If the particles are not fully solidified before they collide, then they stick together which explains the agglomeration in experiment 21, Figure 11, table 2. It has to be noticed that surprisingly the particles at the atomisation pressure of 140 bar are on average larger than the particles formed at the atomization pressure of 70 bar. To gain understanding about the formation mechanism of this fact, more experiments and analysis will follow.

Effect of Nozzle Temperature and Ambient Air

Experiments have not shown a clear influence of nozzle temperature and on bulk density and particle size or morphology. The influence of the ambient air temperature, which has also a cooling effect on the droplet, is an important point, which needs further studies.

Particle Formation mechanism

Figure 12 shows possible mechanisms of particle formation during PGSS operation. Under the influence of CO_2 concentration, melt temperature, atomization pressure and feed rate, these mechanisms predict the formation of completely solid spherical particles, a hollow spherical particles, agglomerated distorted particles or sponge like particles.

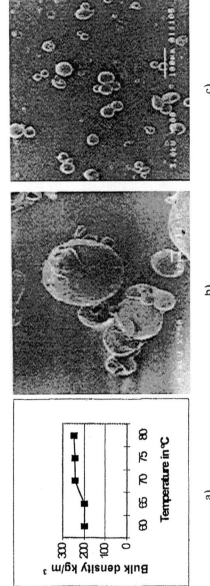

a)

b)

c)

Figure10: – (a) Bulk densities at different temperature (b) Exp. nr. 12, reformed particle formed at T_{melt}: 60 °C (c) Exp. nr. 16, hollow particles formed at T_{melt}: 80 °C and 70 bar.

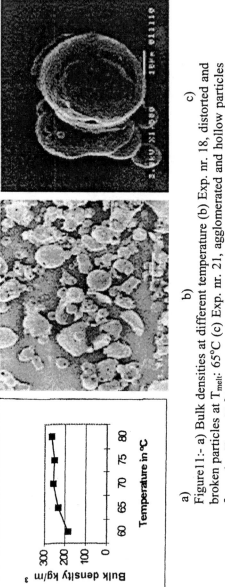

a)

b)

c)

Figure11:- a) Bulk densities at different temperature (b) Exp. nr. 18, distorted and broken particles at T_{melt}: 65°C (c) Exp. nr. 21, agglomerated and hollow particles formed at T_{melt}: 80 °C and 140 bar.

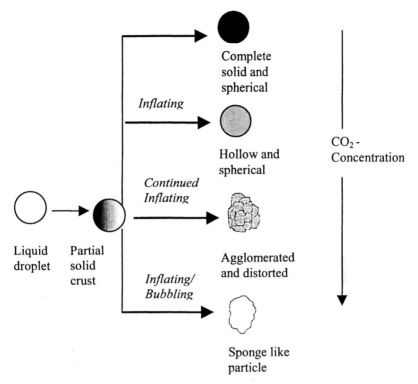

Figure12: Characteristics of particles obtained from the PGSS method with increasing initial CO_2 concentration.

The molten fat / CO_2 mixture expands due to the pressure drop over a nozzle whereby the liquid is broken-up into droplets under the influence of shear. These droplets cool down and start to solidify due to expansion, convective cooling and CO_2 evaporation. The solidification begins at the surface of the droplet. In the absence of CO_2 completely solid particles are formed. In the presence of CO_2 inflation of the droplets occurs and hollow particles with a solid crust are formed. At relatively low melt temperatures, immediately after the droplets are formed, solidification begins and a solid crust is formed before all of the CO_2 can escape. Under such conditions, dependent on the permeability and the flexibility of the crust the droplets/particles will be deformed or even shattered leading to the observed hollow distorted or sponge like morphologies.

At relatively high melt-temperatures, solidification will proceed more slowly, which on the one hand facilitates the formation of spherical (primary) particles and on the other hand promotes agglomeration since the wet droplet/particles are more likely to stick together when they collide. Similarly the effect of the CO_2-concentration in the melt can be explained in terms of competition between the solidification rate of the melt and the escape rate of CO_2. Both temperature and CO_2-concentration also have an indirect effect on the particle formation process in the sense that they have a strong effect on relevant physical properties of the melt (like viscosity, surface tension and conductivity). An increase of the atomisation pressure yielded larger particles. This is not in-line with the prediction of simple atomisation models and will be the subject of further study.

Conclusion

The solubility of carbon dioxide in fat depends on the pressure and temperature. From the solubility measurement data indications for type IIIa phase behavior were observed.

From batch experiments three main particle shapes were obtained depending on the fat melt temperature and the CO_2 concentration. Solid spheres are obtained at 0wt% of CO_2 in the fat melt at 140 bar and 70°C and undistorted hollow spheres can be obtained at a constant CO_2 concentration of 10wt%, and a constant atomization pressure of 70 and 140 bar and a melt temperature which was ranging from 70-80°C. At 21wt% of CO_2 in the fat melt sponge like particles were obtained. Based on the analyses between spray drying and PGSS the formation of solid, hollow and sponge like particle can be explained.

The shape and the size of the formed particles can be manipulated by the temperature of the melt and by the CO_2 concentration. The bulk densities of the formed powder decrease with increasing amount of dissolved carbon dioxide and increase with increasing melt temperature. There is no significant effect of atomization pressure and nozzle temperature on the bulk density.

From the particle formation mechanism it is seen that the major cooling effect of the droplet is in the fast temperature drop and the CO_2 evaporation due to expansion. At low melt temperature, 60°C, 70 bar, a surface crust of the particle is formed because the melt temperature is near the solid point. Because of this fact no spherical particles are formed. At high temperature, 70-80°C, and 70 bar, formation of spherical particle are observed because the droplet has to remove more heat to reach the solid point, which gives more time for the formation of spherical particle. At 80°C and 140 bar agglomeration is obtained because of the high pressure and heat transfer, which leads to high droplet velocities. The flexibility of the droplet surface is variable with changes in the CO_2 concentration and the melt temperature.

References

1. Weidner, E.; Knez, Z., WO patent 95/21688, 1995.
2. De Loos, Th.W; van der Kooi, H.J.; Ott, P.L., J. Chem.Eng.Data **1986**, 31, 166-168.
3. Masters, K. Spray Drying Handbook, 3rd Edition, Godwin, London, GB. 1979.
4. De Loos, Th. W., Supercritical Fluids, Fundamentals for Application, E.Kiran and J.M.H. Levelt-Senger, Series E: Applied Sciences, Kluwer Academic Publisher, Dodrecht, NL, 1994,Vol.273.
5. Duijndam, R.B, Master thesis, Delft University of Technology, Delft, NL, 2002.
6. Svarovsky, L., Powder Testing Guide, Methods of Measuring the Physical Properties of Bulk Powders, Elsevier, London, GB, 1987, Essex, page (81-92).
7. Van Konijnenburg, P.H. and Scott, R.L., Critical lines and phase equilibria in binary Van der Waals mixtures, Phil.Trans. NL, 1980, 298A: 495-540.
8. McHugh, M., Supercritical Fluid Extraction, Principle and practice, 2nd edition, Butterworth-Heinemann, NL, 1993.
9. Bertucco, A., and Vetter, G., Thermodynamic Properties at High Pressure, Chapter 2 in: High Pressure Process technology: Fundamentals and Applications, Elsevier Science, Amsterdam, NL, 2001.

Chapter 24

Synthesis of Nanostructured Sorbent Materials Using Supercritical Fluids

Thomas S. Zemanian, Glen E. Fryxell, Oleksey Ustyugov,
Jerome C. Birnbaum, and Yuehe Lin

Pacific Northwest National Laboratory, Richland, WA 99352

Abstract

There exists a significant need today for high capacity, high efficiency sorbent materials to selectively sequester toxic metal species from groundwater or wastestreams. We have an on-going effort at PNNL to design, synthesize, characterize and evaluate functionalized nanoporous materials for environmental remediation, sensing/detection and device applications. During the course of these studies we have developed novel synthetic methods employing supercritical fluids (SCFs) that provide powerful new synthetic capabilities for molecular self-assembly. The monolayer coatings provided by these SCF methods have higher surface coverage, greater crosslinking, fewer defects and greater stability than those prepared via conventional methods. This manuscript will summarize the SAMMS synthetic strategy, their advantages as sorbent materials and how SCF methodology enhances their preparation and properties.

Mesoporous Ceramics

The surfactant templated synthesis of mesoporous ceramics was first reported in 1992 [1], and since that time there has been a veritable explosion in the number of papers in the area. There has been particular interest in the functionalization of mesoporous ceramics [2-5]. As outlined in Figure 1, the original synthesis employed rod-shaped micelles composed of cationic surfactant molecules as the pore template (more recently, this methodology has been extended to a wide variety of other surfactant systems and reaction conditions). When exposed to routine sol-gel conditions, the cationic micelles undergo an anionic metathesis with silicate anions, resulting in a "glass-coated log" which

370

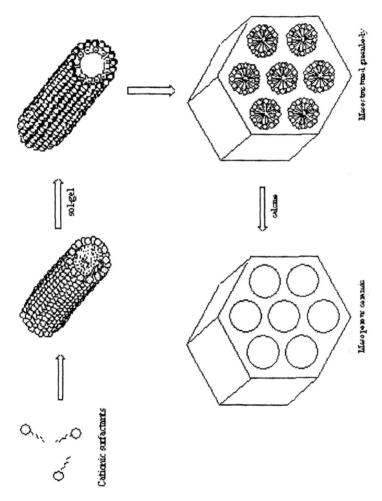

Figure 1. Schematic diagram showing the genesis of MCM-41 from a micelle template, and successive generations of self-assembly.

self-assembles and ultimately precipitates out of solution. This pre-ceramic greenbody is collected and typically subjected to calcination (there have been non-thermal surfactant removal methods developed since the original reports). Calcination first serves to fuse the silicic acids units into a hard ceramic backbone, then subsequently the surfactant molecules are burned out of the mesoporous matrix, exposing the pores. The pore diameter can be systematically varied from about 20Å to about 300Å depending on the reaction conditions and template chosen. The MCM-41 (Mobil Catalytic Material #41) materials generally demonstrate excellent thermal stability, maintaining their pore structure up to about 600°C (some of the larger pore, thicker walled mesoporous materials are stable up to almost 900°C).

The honeycomb morphology of the MCM-41 materials condenses a huge amount of surface area into a very small volume. Typical surface areas for the MCM-41 materials are in the range of 900-1000 m^2/g, and it's not uncommon to see cases of 1200+ m^2/g. In addition, since this morphology is captured by a rigid ceramic backbone, the pore structure is rigidly held open and is not subject to solvent swelling, as commonly encountered with polymeric systems. However, the interface is still that of a ceramic oxide, and while such oxide surfaces do indeed sorb heavy metal ions, such sorption is reversible and fairly weak. A more robust metal-binding strategy is needed for responsible environmental remediation.

Self-Assembled Monolayers on Mesoporous Supports (SAMMS)

Silane-based self-assembled monolayers are a very effective method for imparting specific chemical identity to a ceramic oxide surface [6]. They offer the advantages of easy installation, covalent anchoring to the surface, and facile modification. The appropriate level of interfacial hydration is critical to the success of monolayer formation, and is one of the parameters that differentiates self-assembly from simple silanol capping chemistry [5]. Approximately 2 monolayers of water (roughly 0.17 mmoles/m^2, or 3 microliters/m^2) are needed to hydrolyze the silane to the tris(hydroxy)silane intermediate that is responsible for self-assembly (a full 3 equivalents are not necessary since the condensation reactions between the tris(hydroxy)silane and the surface, and between the silanes themselves, regenerates some of the water). The tris(hydroxy)silane is hydrogen bound to the oxide surface, but is free to migrate horizontally, allowing the van der Waal's interaction between the hydrocarbon chains to drive molecular aggregation, ultimately leading to monolayer formation. Once aggregated, these tris(hydroxy)silanes can then undergo reaction with either a surface silanol or a siloxane bridge to afford a covalent anchor to the oxide surface. This is then followed by condensation chemistry between the silanes themselves, to form crosslinking siloxane bridges between the monomer components. If the monolayer precursor is a trichlorosilane, then the liberated HCl catalyzes these condensation processes; if the precursor is a trialkoxysilane, then these condensation reactions are slower and are commonly driven thermally to get them to proceed at a useful rate.

Deposition of a self-assembled monolayer within MCM-41 requires the addition of water to achieve complete monolayer coverage. This is due to the fact that the final step in the synthesis of MCM-41 is a calcination step (typically at 540°C), and as a result the surface is completely dessicated and severely silanol depleted [5]. The silanol population seems to vary somewhat from lab to lab, but is clearly related to the calcination time and temperature. Failure to include this water limits the surface chemistry to simple silanol capping reactions, and as a result the degree of surface coverage is limited to the number of surface silanols.

By coating the internal pore surfaces of MCM-41 with functionalized self-assembled monolayers it is possible to impart excellent chemical selectivity for specific toxic metallic species (see Figure 2). For example, installation of a thiol-terminated monolayer provides unprecedented mercury sequestering capability [7,8]. Thiol-SAMMS also is effective for removing other "soft" heavy metals, such as Cd, Ag and Au. In fact, the metallated thiols themselves (*e.g.* Hg and Ag) are also excellent sorbents for sequestering "soft" anions (*e.g.* radioiodide) as well [9].

A stereospecific receptor site for tetrahedral oxometallate anions has been realized in the form of octahedral transition metal cation complexes [10]. For example, the Cu (II) tris (ethylenediamine) SAMMS (Cu-EDA SAMMS) have demonstrated excellent affinity for anions such as chromate and arsenate, and have significant potential for the sequestration of pertechnetate. With the global concerns over arsenic in drinking water, these materials can truly save lives.

Actinide species like U (VI), Pu (IV) and Am (III) can be selectively sequestered with SAMMS that are decorated with carbamoylphosphine oxide (CMPO) analog ligands [11]. In this case, selectivity is excellent, with no competition from the ubiquitous alkaline and alkaline earth cations, common transition metals or complexants like EDTA.

Radiocesium is a significant issue as a part of the DOE clean-up effort. Selective sequestration of cesium has been addressed by polymer-bound crown ethers (*e.g.* the SuperLig 644) and the crystalline silicotitanates. While both of these methods are effective, they each have their drawbacks. By lining the pores of SAMMS with ferrocyanide complexes, it is possible to remove all the cesium from waste simulant solutions in a matter of minutes [12]. This sorbent is not

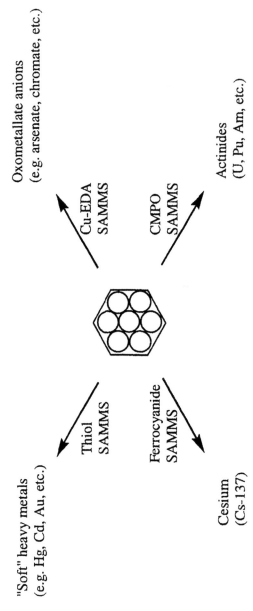

Figure 2. By tailoring the interfacial chemistry of the monolayer, it is possible to selectively sequester a variety of environmentally problematic metal species.

only regenerable, but vitrifiable as well, providing significant versatility to the waste processing flow-sheet.

In each of these cases, the rigid open pores structure of the SAMMS materials allows for rapid sorption kinetics; sorption is typically complete in a matter of minutes as opposed to hours for polymeric sorbents. The binding affinities are high (typical K_d values are in excess of 20,000), so that less than 1% SAMMS by weight can remove more than 99% of the target species from complex mixtures. Since SAMMS is a silica-based sorbent material, once laden with a toxic metal it can easily be incorporated into a vitrification process for permanent disposition, a significant advantage for radionuclide processing and disposal.

What is self-assembly?

Self-assembly is the spontaneous aggregation of molecules, or nanoparticles, into an ordered, organized supramolecular assembly (see Figure 3). This is a contra-entropic process, and as such must have a significant enthalpic driver in order to occur spontaneously. This enthalpic driver has two primary components; the interaction of the head group with the surface, and the van der Waal's interactions between the hydrocarbon chains. It is important to recognize that both of these drivers are not simply energetic payback that occurs every time these molecules are free in solution, but rather is the result of competition between the solvation dynamics of the head group and its interaction with the surface, the competition of the solvation of the hydrocarbon body of the silane molecules and their aggregation. While the final energetic state for the particular molecule/surface system is relatively fixed, the initial solvated state is directly related to the nature of the solvent system employed. The rate and efficiency of self-assembly is dependant on the energy difference between these two states and hence is directly dependant on the solvent system used.

Why are SCF's beneficial for self-assembly?

Supercritical fluids offer many advantages for self-assembly, particularly within a nanoporous matrix. SCFs have modest solvent power that can be tuned by altering the fluid density through manipulation of the reaction temperature and pressure. This tunable solvent power means that the solvent shielding experienced by the fluid-borne silane reagents can be held to a minimum. As a result of the reduced solvent-solute interactions, solute-solute interactions are allowed to predominate, and these are precisely the associative forces that drive self-assembly (along with head group/surface interactions). The low viscosity and high diffusivity of SCFs are widely recognized. These properties enhance mass-transport and delivery of the reagents throughout the nanoporous matrix. In addition, reaction temperatures are not limited by the solvent's boiling point, allowing reaction temperatures to be more closely tailored to the chemistry, rather than being dictated by the volatility of the solvent. The siloxane reagents and reaction by-products are soluble in $SCCO_2$, and in fact the hydrolysis products (typically methanol or ethanol) are excellent adjunct solvents for $SCCO_2$.

Perhaps the most import benefit provided by SCFs however is that any associative chemical reaction (or process) can be accelerated by carrying that reaction out under conditions of high pressure. In an associative process, the molar volume of the products is smaller than that of the starting materials, therefore there is a net negative change in reaction volume as the reaction proceeds. This volume reduction is also reflected in the transition state. Thus, the Δv° and $\Delta v\ddagger$ are both negative. The net result is that the transition state is stabilized relative to the starting materials, so the activation energy of the reaction is lowered, and the reaction rate is faster. In addition, since the energy of the products is lowered relative to that of the starting materials, if the reaction is an equilibrium process, the position of the equilibrium is shifted towards the products.

Formation of SAMMS at Ambient Pressure. The traditional synthesis of 3-mercaptopropyl trimethoxysilane (MPTMS) SAMMS involves suspending the MCM-41 (50Å pores) in toluene, hydrating them, adding a slight excess of MPTMS and heating for 6 hours. At this point the SAMMS has a surface population density of approximately 3.5 silanes/nm^2, and is composed of approximately equal amounts of terminal and internal silanes, with a small amount (*ca.* 10%) isolated silane (see Figure 4(a)) [13]. Additional silane or longer reaction times do not improve this surface coverage or crosslinking level.

However, if the reaction is continued by azeotropically removing the alcohol and toluene/water azeotrope, then the surface coverage is raised to 4.5 silanes/nm^2, the internal silane to terminal silane ratio is approximately 2-to-1, and the isolated silane is eliminated (see Figure 4(b)). It appears that the alcohol and residual water hydrogen bond to the dangling hydroxyl defects and hinders their crosslinking. By removing the alcohol, and by driving the condensation equilibrium through the removal of water, these limitations can be obviated.

For reference, silane-based monolayers prepared on flat wafers and other non-porous substrates have been reported to have a population density of 4.8 silanes/nm^2 [14]. Thus, the toluene preparation of thiols SAMMS seems to provide reasonably good surface coverage of the internal pore surfaces of the mesoporous silica, but there are certainly defects (*e.g.* pinhole defects and dangling hydroxyls) in the monolayer structure, leaving the question, "Is the level of coverage limited by the pore curvature, or is it due to the procedure employed?". As we shall see, it is due to the procedure.

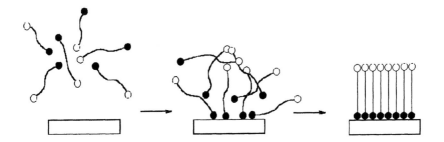

Figure 3. Simplified schematic showing formation of a self-assembled monolayer.

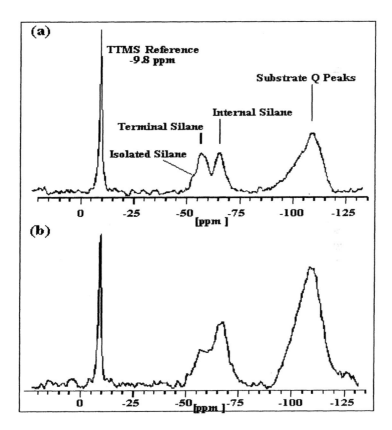

Figure 4. ^{29}Si NMR spectrum of MPTMS SAMMS prepared in refluxing toluene. Spectrum (a) is obtained from a sample after 4 hours of reflux, while spectrum (b) was obtained from a similar sample with the addition of azeotropic removal of water and alcohol from the reaction mixture after the reflux period. (Reproduced from reference 13. Copyright 2001 American Chemical Society.)

Self-assembly in SCFs

Dissolving MPTMS in $SCCO_2$ and exposing it to a hydrated sample of MCM-41 results in very rapid monolayer formation. In less than 5 minutes the deposition is complete [13]. The rapidity of monolayer deposition is due to the enhanced self-assembly possible in $SCCO_2$, along with the pressure accelerated hydrolysis/condensation chemistry of monolayer formation. At this point the surface coverage is complete, with a population density of approximately 6.4 silanes/nm^2 (even in the presence of excess silane) The higher coverage results from the relative lack of pinhole defects. The speciation of silanes at this point is quite similar to the original toluene reflux sample, with approximately equal populations of terminal and internal silanes, along with a small amount of the isolated silanes. While the aggregation and anchoring of the silanes to form the monolayer appears to be quite rapid in $SCCO_2$, the crosslinking chemistry appears to be somewhat slower.

If the $SCCO_2$ SAMMS are exposed to these reaction conditions for a longer period of time, the surface population doesn't change, but the speciation does undergo a slow evolution with time (see Figure 5). First the isolated silane disappears, followed by a gradual decrease in the amount of terminal silane, with concomitant growth of the internal silane. Apparently, the dangling hydroxyl defects appear to "walk" through the monolayer and upon encounting one another they self-annihilate, undergoing condensation to form the desired siloxane bridge. Thermal curing of silane-based self-assembled monolayers is known [14] however this is the first reported case of a pressure-enhanced thermal curing process. The monolayers produced using this SCF methodology are the highest quality (*i.e.* lowest defect density) silane-based monolayers reported to date.

An additional benefit of the enhanced coverage and minimal defect density can be seen in Figure 6. Alkaline conditions allow dissolution of SiO_2, such that even at neutral pH 5% of bare MCM-41 is dissolved after a period of 4 hours. Presumably this is driven by the elevated free energy inherent to high surface area materials. (By analogy, surface tension drives coalescence of small droplets in, say, an aerosol into larger droplets.)

The alkane-based monolayer on the SAMMS protects the underlying SiO_2 from alkaline attack, as can be seen from the data for toluene deposited SAMMS and $SCCO_2$ deposited SAMMS in Figure 6. However, at elevated pH, even the toluene deposited SAMMS are not proof against alkaline assault. The existence of pinhole defects allows attack of the underlying substrate; the caustic solution "gets under the paint" where the "paint" is chipped, and removes the protective monolayer to expose more SiO_2 to attack. At a pH of approximately 9 the toluene deposited SAMMS begin to suffer this fate. The $SCCO_2$ deposited SAMMS, in contrast, are stable to alkaline condition with pH in excess of 11.

This is almost certainly due to the very low defect density of the $SCCO_2$ deposited monolayers; the caustic solution has no breach to attack. Since much of the high level liquid radioactive waste in the US is held at extremely alkaline conditions, this feature might allow processing of the liquid wastes without the need to drastically modify the pH of the stream before sorption of the target ions.

Alternative supports

Aerogels. Aerogels are another high-surface area ceramic oxide support. In this case, the nanostructure is not templated by a specific molecule or molecular aggregate, but rather comes from supercritical fluid replacement of the liquid medium and drying of a sol-gel mixture, resulting in a random, fractal and somewhat strained structure. Aerogels are quite inexpensive to produce and can have extremely high surface area. Unfortunately, they have resisted functionalization since the capillary forces of traditional condensed phase solvents are sufficient to crush the frail walls of the aerogel support. However, when using SCFs as the reaction medium, there are no separate liquid and gas phases, no meniscus, and hence no capillary forces, allowing the surface chemistry to be executed without any damage to the nanoporous support [15].

Note that integration of the spectrum shown (Figure 7) reveals that the surface coverage is very similar to that obtained for MCM-41 (6.2 silanes/nm^2). Note also the relative intensities of the silanes peaks and the Q peaks of the support give an approximate estimate of the relative populations of these components (not strictly true since they have different relaxation rates). Since the support represents a "diluent" as far as sorbent activity is concerned, this provides some degree of insight into the "atomistic efficiency" of these sorbent materials.

Worth noting as well, is the fact that once the monolayer is in place it appears to structurally reinforce the fragile aerogel structure, as the coated materials are capable of withstanding immersion in liquid solutions without structural collapse. Thus, SCF methodology not only represents an efficient method for installing self-assembled monolayers within aerogels, it is in fact the only method known to date for accomplishing this chemistry [15].

Zeolites. SCF's can not only be used to deposit fully dense self-assembled monolayers within mesopores very rapidly, they can also be used to deposit silanes inside pores that would otherwise be too small to admit the silane monomers into using conventional methods. Specifically, we have succeeded in depositing MPTMS inside of 6Å zeolites [16]. The MPTMS molecule is approximately 6Å in length, and the trimethoxysilyl head group is approximately6Å in diameter, making this deposition all the more remarkable. The driving force for this reaction is thought to be two-fold. Firstly, there is a

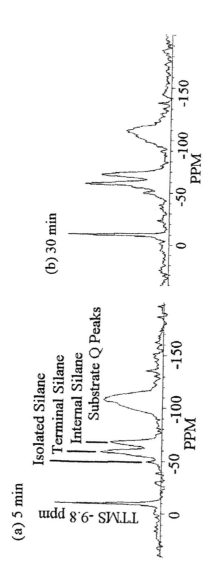

(a) 5 min

(b) 30 min

TTMS -9.8 ppm

Isolated Silane
Terminal Silane
Internal Silane
Substrate Q Peaks

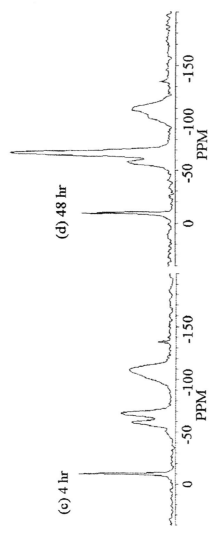

Figure 5. ^{29}Si NMR of MPTMS SAMMS prepared in SCCO2 (7500 psi and 150°C) as a function of treatment time, showing the evolution of silane speciation.

(Reproduced from reference 13. Copyright 2001 American Chemical Society.)

382

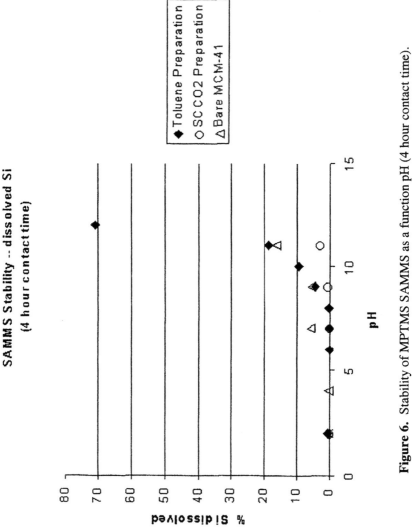

Figure 6. Stability of MPTMS SAMMS as a function pH (4 hour contact time). (Reproduced from reference 13. Copyright 2001 American Chemical Society.)

pressure-driven delivery of the silane to the internal pore surfaces by the high-pressure. Secondly, the absence of a viscous condensed phase solvent bilayer associated with the ceramic oxide surface allows for facile silane transport into the small diameter pores. It is important to recognize that these functionalized zeolites are not coated with fully dense self-assembled monolayers; there is simply not enough room with either 6Å pores or 10Å chambers for these molecules to stand upright and form a well-ordered monolayer. Rather, from the spectroscopic data obtained, it appears that these materials have isolated pairs of silanes, arranged "back-to-back" along the pore axis (see Figure 8). Again, this represents not only an efficient way to chemically modify a zeolite surface, it also represents the only method reported to date for systematically incorporating an organic functional moiety into a 6Å zeolite structure.

Conclusions

SCFs are excellent media for self-assembly. In particular, $SCCO_2$ is an excellent reaction solvent for preparing siloxane-based self-assembled monolayers inside nanoporous ceramic supports. The low viscosity and high diffusivity of SCFs provide rapid mass transport of the silane reagent to the internal pore surfaces of the support. In contrast to common condensed phase solvents, the modest solvent power of SCFs provides minimal hindrance to self-assembly, both in terms of head group/surface interaction and in terms of the chain-chain van der Waals interaction of the monolayer, thereby allowing for facile molecular aggregation during the self-assembly process. The high pressure of the reaction medium accelerates the hydrolysis and condensation chemistry of siloxane deposition, significantly accelerating monolayer formation. Deposition of a self-assembled monolayer within a nanoporous ceramic is an associative process, and as such the reaction rate can be accelerated and the equilibrium can be driven towards the product by carrying out the reaction at elevated pressure. Moreover the SCF deposited monolayer has fewer defects due to a pressure enhanced annealing process that allows the dangling hydroxyl defects to walk through the monolayer and self-annihilate, ultimately resulting in a highly crosslinked structure. The lack of pinhole and crosslinking defects results in a monolayer that is considerably more robust chemically than those prepared via conventional methods.

Similar methodology has also been applied to the preparation of monolayer coated aerogels, which are virtually impossible to make via conventional condensed phase solvent methods. In SCFs, monolayer deposition proceeds smoothly and results in the same high level of crosslinking observed in the MCM-41 materials. Once the monolayer is deposited, it serves to structurally reinforce the aerogel structure sufficiently to where it becomes stable

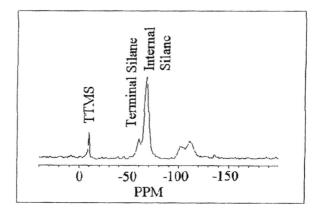

Figure 7. ^{29}Si NMR of MPTMS monolayer coated silica aerogel.

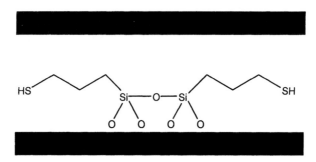

Figure 8. Proposed "back-to-back" structure of silane pairs in zeolite pore.

towards liquid phases and can be employed as a high surface area sorbent material. SCF silanation is such a powerful method that we have even been able to effectively functionalize the internal pore surfaces of 6Å zeolites. In this case, a fully dense, upright monolayer is not formed, but rather the silanes form pairs within the pores, with the alkyl chains directed away from each other. This methodology is the only known way to effectively install organic functionality within a sub-nanometer pore.

Acknowledgement:
This research was supported by the U. S. Department of Energy (DOE) Environmental Managed Science Program. This work was performed at Pacific Northwest National Laboratories, which is operated for the DOE by Battelle Memorial Institute under contract DE AC06-76RLO 1830.

References

1) Kresge, C. T.; Leonowizc, M. E.; Roth, W. J.; Vartuli, J. C.; Beck, J. S. *Nature*, **1992**, *359*, 710-712. Beck, J. S.; Vartuli, J. C.; Roth, W. J.; Leonowizc, M. E.; Kresge, C. T.; Schmitt, K.; D.; Chu, C. T.-W.; Olson, D. H.; Sheppard, E. W.; McCullen, S. B.; Higgins, J. B.; Schlenker, J. L. *J. Am. Chem. Soc.* **1992**, *114*, 10834-10842.

2) Feng, X.; Fryxell, G. E.; Wang, L. Q.; Kim, A. Y.; Liu, J. *Science,* **1997**, *276*, 923-926.

3) Liu, J.; Feng, X.; Fryxell, G. E.; Wang, L. Q.; Kim, A. Y.; Gong, M. *Advanced Materials*, **1998**, *10*, 161-165.

4) Moller, K.; Bein, T. *Chem. Mater.* **1998**, *10*, 2950-2963.

5) "Designing Surface Chemistry in Mesoporous Silica" G. E. Fryxell and Jun Liu in "Adsorption at Silica Surfaces" edited by Eugene Papirer, Marcel Dekker, pp. 665-688, **2000**.

6) A. Ulman in "An Introduction to Organic Thin Films From Langmuir-Blodgett to Self-Assembly" Academic Press, New York, 1991, pp. 245-269.

7) a) "Synthesis and Applications of Functionalized Nanoporous Materials for Specific Adsorption" J. Liu, G. E. Fryxell, S. Mattigod, T. S. Zemanian, Y. Shin, L. Q. Wang, in "Studies in Surface Science and Catalysis" Vol. 129, pp. 729-738, edited by A. Sayari, et al., Elsevier, **2000**. b) Mattigod, S. V.; Feng, X.; Fryxell, G. E.; Liu, J.; Gong, M. *Separation Science and Technology*, **1999**, *34*, 2329-2345. c) Kemner, K. M.; Feng, X.; Liu, J.; Fryxell, G. E.; Wang, L. -Q.; Kim, A. Y.; Gong, M.; Mattigod, S. V. *J. Synchotron Rad.*, **1999**, *6*, 633-635.

8) a) Mori, Y.; Pinnavaia, T. J. *Chem. Mater* **2001**,. *13*, 2173-2178. b) Mercer, L.; Pinnavaia, T. J. *Adv. Mater.* **1997**, *9*, 500. c) Mercer, L.; Pinnavaia, T. J. *Environ. Sci. Technol.*. **1998**, *32*, 2749.

9) Mattigod, S. V.; Fryxell, G. E.; Serne, R. J.; Parker, K. E.; Mann, F. M. *Radiochimica Acta* (in press).

10) a) Fryxell, G. E.; Liu, J.; Gong, M.; Hauser, T. A.; Nie, Z.; Hallen, R. T.; Qian, M.; Ferris, K. F. *Chem. Mat.* **1999**, *11*, 2148-2154. b) Kelly, S.; Kemner, K. M.; Fryxell, G. E.; Liu, J.; Mattigod, S. V.; Ferris, K. F. *J. Phys. Chem.* **2001**, *105*, 6337-6346.

11) Birnbaum, J. C.; Busche, B.; Shaw, W. J.; Fryxell, G. E. *Chem. Comm.* **2002**, 1374-1375.

12) Lin, Y.; Fryxell, G. E.; Wu, H.; Englehard, M. *Env. Sci.& Tech.* **2001**, *35*, 3962-3966.

13) Zemanian, T. S.; Fryxell, G. E.; Liu, J.; Franz, J. A.; Nie, Z. *Langmuir* **2001**, *17*, 8172-8177.

14) a) Wang, R.; Wunder, S. L. *J. Phys. Chem. B.* **2001**, *105,* 173-181. b) Angst, D. L.; Simmons, G. W. *Langmuir*, **1991**, *7*, 2236. c) Brzoska, J. B.; Azouz, I. B.; Rondolez, F. *Langmuir*, **1994**, *10*, 4367.

15) Zemanian, T. S.; Fryxell, G. E.; Liu, J.; Mattigod, S.V.; Shin, Y.; Franz, J. A.; Ustyugov, O.; Nie, Z. Proceedings of the 2001 1st IEEE Conference on Nanotechnology, pp. 288-292: IEEE-NANO 2001 Maui, HI, Oct. 28-30, **2001**.

16) Shin, Y.; Zemanian, T. S.; Fryxell, G. E.; Wang, L. Q.; Liu, J. *Microporous and Mesoporous Materials* **2000**, *37*, 49-56.

Chapter 25

Synthesis of Structured Polymeric Materials Using Compressed Fluid Solvents

A. I. Cooper, R. Butler, C. M. Davies, A. K. Hebb, K. Senoo, and C. D. Wood

Department of Chemistry, Donnan and Robert Robinson Laboratories, Crown Street, Liverpool L69 3BX, United Kingdom

This paper describes the use of supercritical CO_2 and liquid R134a (1,1,1,2-tetrafluoroethane) as alternative solvents for the synthesis of crosslinked polymer materials with fine control over structural features on micro-, meso-, and macroscopic length scales.

Supercritical carbon dioxide ($scCO_2$) has attracted much interest recently as an alternative solvent for materials synthesis and processing (*1-5*). We discuss here four areas where the use of compressed fluid solvents may offer unique advantages:

[1] The synthesis of macroporous polymer beads by suspension polymerization using $scCO_2$ as a 'pressure-adjustable' porogen.

[2] The synthesis of macroporous polymer monoliths using $scCO_2$ as the porogenic solvent.

[3] The synthesis of emulsion-templated polymers using high internal phase CO_2-in-water emulsions (C/W HIPEs)

[4] The synthesis of cross-linked polymer microspheres by dispersion polymerization using liquid R134a as the continuous phase.

Synthesis of Macroporous Polymer Beads by Suspension Polymerization using scCO$_2$ as a 'Pressure-Adjustable' Porogen

Macroporous polymers are important in a wide range of applications such as ion-exchange resins, chromatographic separation media, solid-supported reagents, and supports for combinatorial synthesis (6). Unlike gel-type polymers which swell in the presence of an appropriate solvent, the cross-link density in macroporous polymers is sufficient to form a permanent porous structure which persists in the dry state (7). Macroporous polymers are usually synthesized as beads (typical diameter = 10–1000 μm) by O/W suspension polymerization in the presence of a suitable porogen (i.e., an additive, usually an organic solvent, which induces pore formation in the polymer matrix) (8,9). In general, porogens that are 'good' solvents for the growing polymer network tend to give rise to smaller pores and higher surface areas than porogens that are 'bad' solvents. This is because the degree of solvation imparted by the porogen affects the phase separation process which occurs during polymerization, thus determining the physical structure of the porous channels. To achieve *fine* control over porosity is not always straightforward. In addition, the synthesis of macroporous polymer beads by suspension polymerization is solvent intensive because a large volume of organic solvent (typically 40–60% v/v) is required as the porogen and an even larger volume of a lower boling point solvent is then used to wash out the porogen phase.

O/W Suspension Polymerization using scCO$_2$ as the Porogen

We have developed a method for the synthesis of macroporous polymer beads using no organic solvents whatsoever – just water and CO$_2$ (Figure 1) (10). We have exploited the fact that the solvent strength of scCO$_2$ can be tuned continuously over a significant range by varying the density. As such, scCO$_2$ can be thought of as a 'pressure-adjustable' porogen.

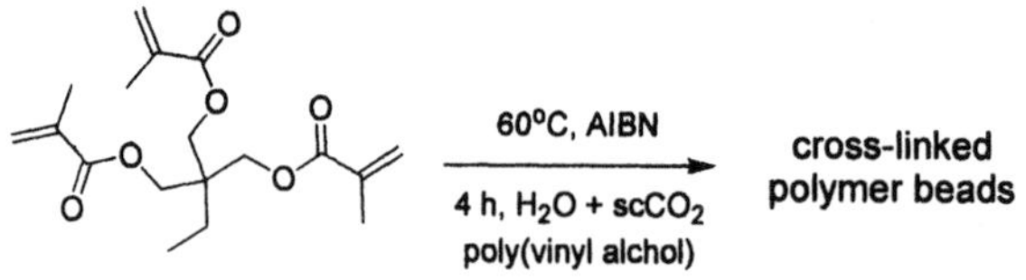

TRIM

Figure 1

In a typical reaction, a mixture of monomers [trimethylolpropane trimethacrylate (TRIM)], initiator [2,2'-azobisisobutyronitrile (AIBN)], and $scCO_2$ was suspended in water with stirring in the presence of a stabilizer [0.5% w/v poly(vinyl alcohol)] to inhibit droplet coalescence. Table 1 summarizes the results of a series of polymerizations carried out under various conditions. In the absence of CO_2, the O/W suspension polymerization of TRIM led to non-porous polymer beads with an average diameter of 180 μm (entry 1). This reaction was repeated in the presence of $scCO_2$ over a range of pressures while keeping all other variables constant (entries 2–5). Uniform spherical macroporous polymer beads were formed when CO_2 was added to the reaction mixture (Figure 2a).

Table 1. Suspension Polymerization of TRIM using $scCO_2$ as the Porogen

	pressure (bar)	mean bead diameter (μm)[a]	intrusion volume (cm^3/g)[b]	median pore diameter (nm)[b]	surface area (m^2/g)[c]	yield (%)
1	1	180	0.00	–	<5	68
2	100	80	0.28	(19)	<5	41
3	200	110	1.05	2206	222	70
4	300	128	1.23	110	253	92
5	400	114	0.69	40	478	90

Reaction conditions: 20% w/v TRIM based on volume of H_2O, 2,2'-azobisisobutyronitrile (AIBN, 2% w/v), 0.5% w/v poly(vinyl alcohol) (88% hydrolyzed, M_w = 88,000 g/mol), 60°C, 6 h. [a] Mean diameter calculated from >100 particles. [b] Measured by mercury intrusion porosimetry over the pore size range 7 nm–20 μm. [c] Measured by N_2 adsorption desorption using the Brunauer–Emmett–Teller method.

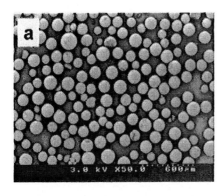
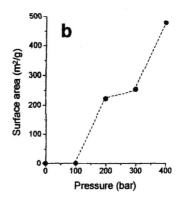

Figure 2. (a) Macroporous polymer beads synthesized using $scCO_2$ as the porogen (entry 4, Table 1); (b) Variation in BET surface area of macroporous beads as a function of CO_2 pressure.

(Reproduced from reference 10. Copyright 2001 American Chemical Society.)

The degree of porosity, the average pore size, and the surface area in the beads could be tuned over a wide range by varying the CO_2 density (Table 1). When the polymerization was carried out at 100 bar, porous beads were generated but the macropore volume was relatively low (0.28 cm^3/g), as was the surface area of the sample (<5 m^2/g). This was explained by observation of the phase behavior, which showed that the monomer phase and the CO_2 phase were not fully miscible at this temperature and pressure. By contrast, at 200 bar the monomer was completely miscible with CO_2, and a homogeneous monomer / CO_2 mixture was dispersed as droplets throughout the aqueous phase. As a result, the beads were found to have increased pore volume (1.05 cm^3/g) and much higher surface area (222 m^2/g). At elevated CO_2 pressures, the products exhibited even greater surface areas, up to a maximum of 478 m^2/g at 400 bar (entry 5). This can be explained by the fact that the pore volume in the materials tended to increase with pressure, whereas the median pore diameter decreased significantly as the pressure was raised (Table 1). The combination of these two trends resulted in a sharp rise in polymer surface area as a function of CO_2 pressure (Figure 2b). We attribute this behavior primarily to variation in the CO_2 solvent strength as a function of density, rather than to changes in the volumetric ratio of monomer to porogen. In all of these experiments, the volume ratio of water to monomer was kept constant, while the CO_2 pressure was varied. Assuming that the water phase and the growing polymer network are both relatively incompressible, the *volume* of compressed CO_2 (though not the density) should remain approximately constant in each experiment, even though the *molar* ratio of monomer to CO_2 changes dramatically. We believe that the variation in pore diameter (and the associated change in surface area) in the beads can be rationalized by the fact that the system is very sensitive to the porogen solvent quality, as found with more conventional porogenic solvents (*11*). The trends observed support this interpretation, with higher CO_2 density (*i.e.*, increased solvent strength) leading to smaller pores and larger surface areas.

Synthesis of Macroporous Polymer Monoliths using scCO_2 as the Porogenic Solvent

Modern HPLC methods frequently involve columns packed with macroporous polymer beads (*12,13*). The flow of the mobile phase between the beads through the large interstitial voids in the column is relatively unimpeded, whereas liquid present in the network of resin pores does not flow and can remain stagnant. The passage of molecules within the pores is controlled by diffusion. Diffusion constants for large molecules, such as proteins or synthetic polymers, are several orders of magnitude lower than for small molecules, causing problems in applications such as chromatography where the separation efficiency is strongly dependent on mass transfer rates. A promising approach to this problem has been the synthesis of continuous, macroporous 'monolithic' polymers which have been developed for a variety of applications (14-20). Typically, a mold is filled with a polymerization mixture containing a cross-linking monomer, functional comonomers, initiator, and a porogenic diluent. This mixture is then polymerized to form a continuous porous monolith which conforms to the shape of the mold. Thus, in applications such as chromatography, all of the solvent is forced to flow through the macropore structure. The porogenic diluent may be either solvating or non-solvating in nature, and carefully chosen ternary solvent mixtures can be used to allow fine control of the porous properties of the monolithic polymers (21-23). A key advantage of this methodology is that the macroporous polymers can be prepared directly within a variety of different containment vessels, including both wide bore chromatography columns and narrow bore capillaries. Disadvantages are that the synthesis is solvent intensive and that it may be difficult to remove solvent residues from the continuous materials after polymerization.

We have synthesised highly cross-linked macroporous polymer monoliths (24,25) using supercritical carbon dioxide (scCO2) as the porogenic solvent. Macroporous cross-linked polymer monoliths were formed in scCO2 by the polymerization of cross-linking monomers such as ethylene glycol dimethacrylate (EGDMA) and trimethylolpropane trimethacrylate (TRIM) (24,25). Continuous polymer monoliths were formed at monomer concentrations in the range 40–60% v/v. These materials conformed to the shape of the reaction vessel (i.e., they were 'molded'). The polymers showed no signs of cracking or breakage when the CO_2 was vented, although these highly porous structures are relatively brittle and fragile when removed from the containment vessel.

Effect of Monomer Concentration on Pore Structure

For polymers formed from TRIM, an increase in monomer concentration led to a marked decrease in the median pore size and a corresponding increase in the specific surface area (24,25). It was found that relatively small changes in the monomer concentration could lead to dramatic changes in the resulting polymer structure (Figure 3). These trends are broadly consistent with studies involving conventional organic solvents as porogens (21-23).

Effect of CO$_2$ Pressure on Pore Structure

Our suspension polymerization studies (*10*) have shown that scCO$_2$ can be used as a 'pressure-adjustable porogen' for the synthesis of macroporous polymer beads (see above). We have now explored this idea in greater detail in order to 'fine-tune' the porous morphology in crosslinked macroporous polymer monoliths. Figure 4 illustrates the effect of CO$_2$ pressure on the porous properties of crosslinked TRIM monoliths synthesized in scCO$_2$ (60°C, 4 h, 2 % w/w AIBN, 5.5 cm^3 TRIM, 4.5 cm^3 CO$_2$). The results show that the effect of CO$_2$ density on the polymer structure is more complex than observed in our initial studies involving macroporous beads (*10*) (see Figure 2b). In particular, physical properties such as total surface area, micropore surface area, macropore diameter, and intrusion volume (macropore volume) all show pronounced maxima or minima in the pressure range 170–190 bar. We can attempt to explain this discontinuous behavior by considering the mechanism of pore formation during gelation.

180–280 bar: Over this pressure range there is a general increase in surface area, an increase in the micropore surface area, a decrease in the average macropore diameter, and a (small) decrease in the total intrusion volume. We attribute this to increased solvation at higher CO$_2$ densities which leads to later phase separation and the formation of a finer pore structure with less in-filling between individual agglomerated microgel particles (*7,10*). This gives rise to a decrease in the average macropore diameter, an increase in the number of micropores, and consequently higher surface areas. The small decrease in intrusion volume as measured by Hg porosimetry can be attributed to the fact that this technique cannot detect pores smaller than about 7 nm and that an increasing percentage of the pore volume falls below this limit.

140–180 bar: Clearly, the arguments outlined above are contradicted by the sharp increase in surface area and micropore surface area that is observed as the CO$_2$ pressure is reduced from 180 bar to 140 bar. We rationalize this by noting that the system is approaching the demixing pressure (*i.e.*, at pressures below about 140 bar / 60°C, TRIM and CO$_2$ are no longer fully miscible). Thus, at pressures approaching 140 bar one might expect to observe pronounced solute-solute clustering (*26*). We believe that (i) clustering may change the effective monomer concentration during microgelation; (ii) poor solvent quality may influence nucleation and size of microgel particles; and; (iii) solvent quality will affect partitioning of monomer from the porogen phase into the swollen polymer phase. We suggest that these factors override any effects relating to phase separation later in the reaction, and that the very poor solvent quality conditions are in essence similar to reaction conditions at a *higher* volumetric monomer concentration, even though the total reaction volume remains constant in these SCF systems (*10*).

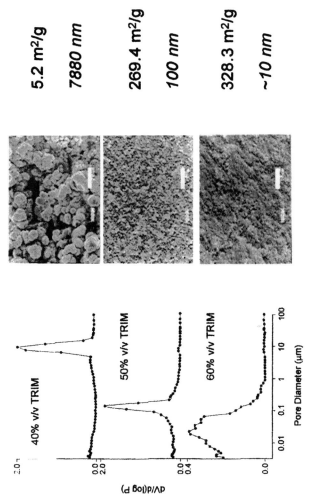

Figure 3. *Effect of monomer concentration on morphology of macroporous crosslinked polymer monoliths. Scale bar in electron micrographs = 10 μm.*

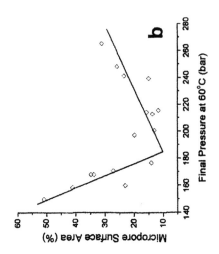

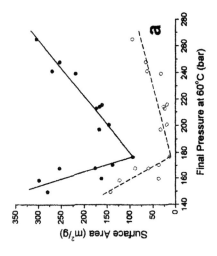

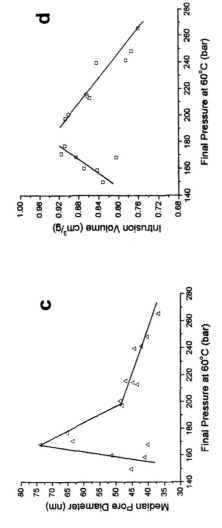

Figure 4. Effect of CO_2 pressure on on morphology of macroporous crosslinked polymer monoliths. (a) BET surface area (continuous line = total surface area, dashed line = micropore surface area); (b) Percent micropore volume; (c) Median pore diameter; (d) Intrusion volume (macropore volume).

Synthesis of Emulsion-Templated Polymers from C/W HIPEs

Emulsion templating is a versatile method for the preparation of well-defined porous polymers and inorganic materials (27,28). The technique involves forming a high internal phase emulsion (HIPE) (>74.05 v/v internal phase) and then locking in the structure *via* polymerization of monomers dissolved in the external phase. Removal of the internal phase (*i.e.*, the emulsion droplets) leaves a skeletal replica of the emulsion.

Porous hydrophilic polymers have applications such as separation media and biological tissue scaffolds. In principle, these types of material can be produced by oil-in-water (O/W) emulsion templating. However, this process is very solvent intensive due to the large volume of the oil phase (usually an organic solvent). Furthermore, removal of the organic phase after reaction may be difficult. The latter is especially important for biological / biomedical applications where solvent residues are undesirable. We have addressed this problem by using $scCO_2$ as the internal droplet phase. Removal of the template phase is simple since the CO_2 reverts to the gaseous phase upon depressurization.

Supercritical CO_2 has been used previously for production of microcellular polymeric foams (29-31) and biodegradable composite materials (32). Both of these supercritical fluid (SCF) techniques involve a foaming mechanism. This limits the range of porous materials that can be produced because many materials cannot be foamed. We have developed an entirely new approach involving the synthesis of porous materials from high internal phase SCF emulsions (Figure 5) (33). This new technique provides a route to materials with well-defined pore structures without the use of any volatile organic solvents – just water and CO_2.

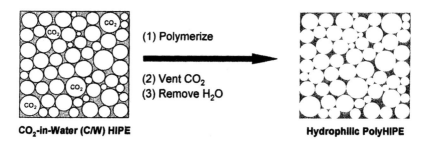

CO₂-in-Water (C/W) HIPE   Hydrophilic PolyHIPE

(1) Polymerize

(2) Vent CO_2
(3) Remove H_2O

Figure 5. Synthesis of porous materials by polymerization of C/W HIPEs.

We chose to use a relatively low molecular weight PFPE ammonium carboxylate surfactant (M_w = 567 g mol⁻¹, Figure 6), since Johnston has demonstrated that surfactants of this type exhibit significant solubility in water and have a propensity to form C/W rather than W/C emulsions (34).

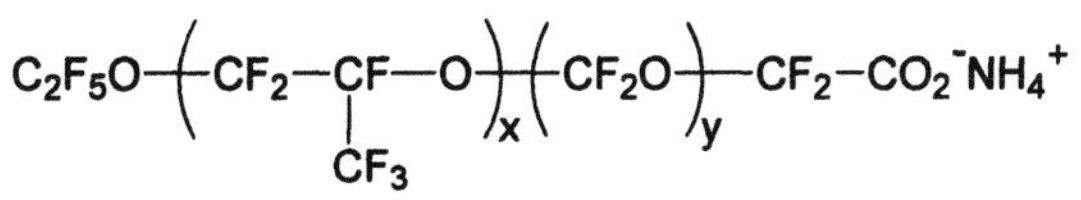

$$C_2F_5O-\left(CF_2-\underset{\underset{CF_3}{|}}{CF}-O\right)_x\left(CF_2O\right)_y-CF_2-CO_2^-NH_4^+$$

MW = 550 g/mol

Figure 6

First of all, we studied the emulsification of pure water and CO_2 (temperature = 25°C, CO_2 pressure = 100–300 bar, volume fraction of CO_2 = 70%, surfactant concentration = 1–10% w/v based on H_2O). Milky-white C/W emulsions were formed which filled the entire reaction vessel and were stable for some hours. When pure water was replaced with a 40% w/v solution of acrylamide (AM) and *N,N*-methylene bisacrylamide (MBAM) (AM:MBAM = 8:2 w/w), it was found that the corresponding emulsions were much less stable. In the absence of stirring, rapid separation occurred over a period of a few minutes leading to two distinct phases: a transparent upper phase (CO_2) and a white milky lower phase (H_2O with a small amount of emulsified CO_2). This instability became even more pronounced as the mixture was heated to 60°C.

Clearly, the presence of the organic monomers had caused significant destabilization of the C/W emulsion. This may be because the monomers act as de-emulsifiers by adsorbing at the water–CO_2 interface, thus reducing interfacial tension gradients and hence droplet stability (*34*). It was found that this destabilization could be counteracted by the addition of poly(vinyl alcohol) (PVA) to the aqueous phase prior to polymerization (10% w/v relative to H_2O, see Table 2), probably because this increases the viscosity of the aqueous phase. Again, the kinetic stability of the C/W emulsions tended to decrease somewhat with increasing temperature, but in the presence of PVA the systems were sufficiently stable for templating to occur and for open-cell porous materials to be produced (Figure 6). The materials conformed closely to the interior of the reaction vessel and no significant shrinkage was observed upon venting the CO_2, although some shrinkage always occurred when the polymers were dried.

At a CO_2 phase volume fraction of 70%, open-cell porous polymers were formed with pore volumes in the range 1.8–2.6 cm^3 g^{-1} and median pore diameters in the range 1.5–5.4 µm (Table 2, samples 1–6). The median pore size as measured by mercury intrusion porosimetry agreed qualitatively with the size of the holes observed in the cell walls between the templated CO_2 emulsion droplets. In general, higher surfactant concentrations led to more open, interconnected structures with an increased number of interconnecting pores in the cell walls. Substitution of acrylamide with 2-hydroxyethyl acrylate (sample 4) also led to porous, open-cell materials, suggesting that this technique might be applied to a wide range of hydrophilic polymers and hydrogel materials. Moreover, we have recently shown that these structures can be produced by using inexpensive (and even biocompatible) hydrocarbon surfactants, instead of the non-degradable perfluoropolyethers used in our preliminary studies.

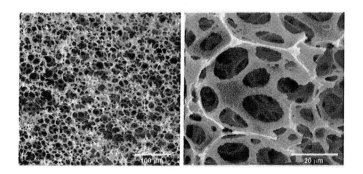

*Figure 6. Porous AM / MBAM polymer synthesized from C/W HIPEs;
left: scale bar = 100 µm; right: scale bar = 20 µm*

Table 2. Synthesis of Emulsion-Templated Polymers from C/W HIPEs

	volume of CO_2 (% v/v)^a	PFPE (% w/v)^a	V_{pore} $(cm^3/g)^b$	pore diameter $(µm)^b$
1	70	0.25	2.0	3.8
2	70	0.5	2.6	3.8
3	70	1	2.1	5.4
4^c	70	1	1.8	1.5
5	70	5	2.0	1.8
6^d	70	5	2.4	3.1
7	75	1	1.2	2.3
8	80	1	5.9	55.0
9	80	2	3.9	3.9
10	80	3	3.8	4.0

Reaction conditions: Acrylamide (AM) + N,N-methylene bisacrylamide (MBAM) (40% w/v in H_2O, AM:MBAM = 8:2), $K_2S_2O_8$ (2% w/v), poly(vinyl alcohol) (10% w/v based on H_2O, M_w = 9–10 kg mol^{-1}), 60°C, 250–290 bar, 12 h. [a] Relative to H_2O. [b] Measured by mercury intrusion porosimetry over the pore size range 7 nm–100 µm. [c] 2-hydroxyethyl acrylate used in place of AM. [d] AM:MBAM = 9:1 w/w.

Polymer Synthesis using Hydrofluorocarbon Solvents: Synthesis of Cross-Linked Polymer Microspheres by Dispersion Polymerization in R134a

Supercritical CO_2 has been researched quite extensively as a solvent for dispersion polymerization since it is non-toxic, non-flammable, and is widely available from a number of inexpensive sources (2,35-40). Nonetheless, it is important to consider all potential economic and environmental impacts – not just the cost of the solvent. In this respect, the relatively high pressures associated with the use of scCO$_2$ may present a problem. In the short term, high pressures translate into increased capital equipment costs. In the longer term, large pressure differentials contribute to operating costs and to overall energy consumption.

R134a (1,1,1,2-tetrafluoroethane) is a hydrofluorocarbon (HFC) that, like CO_2, is non-flammable, non-toxic, and has zero ozone depletion potential (41,42). R134a has found widespread use as a CFC-replacement in refrigeration and in auto air conditioning systems. In addition, the low toxicity of R134a has led to approval as a propellant in metered dose inhalers (43). We have found that liquid R134a (T_c = 101.1 °C, P_c = 40.6 bar) can be used as a solvent for dispersion polymerization at much lower pressures than are possible with scCO$_2$ (44). The ease of solvent separation is retained because the boiling point of R134a (–26.5 °C) is well below ambient temperature. Lower pressures could equate to lower operating costs for certain processes. On the other hand, HFCs are much more expensive than CO_2, and efficient recycling of these fluids would certainly be a prerequisite for industrial-scale use.

Figure 7 shows electron micrographs for cross-linked MMA / TRIM copolymer microparticles prepared by dispersion polymerization in R134a. Both MMA and TRIM (and mixtures thereof) are freely soluble in liquid R134a at room temperature and above, at least up to concentrations of 50% v/v. Using AIBN (5% w/v) as the initiator, free-radical polymerization produced cross-linked powders in good yields by precipitation polymerization in the absence of any stabilizer (entry 1). The unstabilized precipitation polymers contain microparticles, some of which are spherical, but these exist in a highly agglomerated state (Fig. 7a). The reaction was repeated in the presence of a monofunctional perfluoropolyether (PFPE) carboxylic acid stabilizer (M_n = 550 g/mol) of the type developed by Howdle for the dispersion polymerization of MMA in scCO$_2$ (40). This stabilizer is soluble in R134a, at least at the low concentrations (0.5% w/v) used for dispersion polymerization. A uniform white latex was observed in the presence of the PFPE stabilizer, and discrete polymer microspheres were produced in good yield (Fig. 7b). The latex was observed to be stable although some particle precipitation was observed on the sapphire view

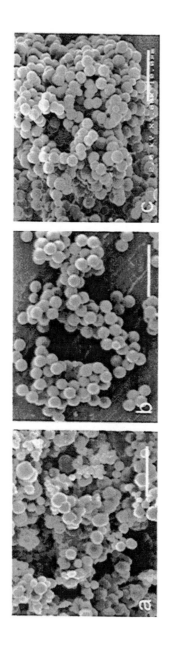

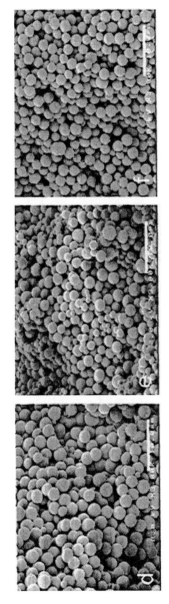

Figure 7. Electron micrographs of cross-linked polymers synthesized by dispersion polymerization in R134a (scale bar = 10 μm). Typical yields = 80–95%. (a) 60°C, 14 bar, no stabilizer. (b) 60°C, 19 bar, perfluoropolyether carboxylic acid stabilizer. (c) 60°C, 14 bar, perfluoroundecanoic acid stabilizer. (d) 90°C, 36 bar, perfluoropolyether carboxylic acid stabilizer. (e) 90°C, 200 bar, pentadecafluorooctanoic acid stabilizer. (f) 90°C, 200 bar, poly(butyl methacrylate) stabilizer.
(Reproduced from reference 44. Copyright 2002 American Chemical Society.)

window towards the end of the reaction. These results are consistent with a dispersion polymerization mechanism and the phase behavior is similar to that observed in comparable systems involving $scCO_2$ (*37*).

From a practical perspective, the most important feature of these polymerizations is the relatively low reaction pressure (typically in the range 10–20 bar at 60 °C). Comparable dispersion polymerizations carried out using $scCO_2$ have involved much higher pressures (170–350 bar) in order to achieve reasonable solvent densities at similar temperatures. At even higher reaction temperatures (>80°C), the pressures required for the use of $scCO_2$ as a solvent may sometimes be prohibitive. Figure 8d shows the product morphology for a cross-linked polymer synthesized in R134a at 90 °C using the PFPE stabilizer and BPO as the initiator. This sample was produced by dispersion polymerization at a pressure of 36 bar. The critical temperature for R134a is 101.1°C: thus, it is possible to carry out reactions in the liquid state at quite moderate pressures (5–50 bar) at temperatures between ambient and 100 °C. Moreover, the solvent density under these conditions (for pure R134a) ranges between 0.9 and 1.3 g/cm^3. This is significant because many polymerization reactions proceed at temperatures that are well above ambient. Liquid CO_2 can be used as a solvent at relatively low pressures (<60 bar), but only for reactions occurring at temperatures below 31.1°C. In this case, that would require the use of low-temperature initiators that are difficult to transport and handle.

Acknowledgement

We thank the Royal Society (for provision of a University Research Fellowship to AIC) and Bradford Particle Design, Avecia, and EPSRC (GR/R15597 + GR/23653) for financial support. Acknowledgement is made to the Donors of The Petroleum Research Fund, administered by the American Chemical Society, for partial travel support to AIC.

References

1. Jessop, P. G.; Leitner, W. *Chemical Synthesis using Supercritical Fluids*; Wiley, VCH: Weinheim, 1999.
2. Kendall, J. L.; Canelas, D. A.; Young, J. L.; DeSimone, J. M. *Chem. Rev.* **1999**, *99*, 543-563.
3. Cooper, A. I. *Adv. Mater.* **2001**, *13*, 1111-1114.
4. Cooper, A. I. *J. Mater. Chem.* **2000**, *10*, 207-234.
5. Desimone, J. M. *Science* **2002**, *297*, 799-803.
6. Hodge, P.; Sherrington, D. C. *Syntheses and Separations using Functional Polymers*; Wiley: New York, 1989.
7. Sherrington, D. C. *Chem. Commun.* **1998**, 2275-2286.
8. Yuan, H. G.; Kalfas, G.; Ray, W. H. *J. Macromol. Sci., Rev. Macromol. Chem. Phys.* **1991**, *C31*, 215-299.

9. Vivaldo-Lima, E.; Wood, P. E.; Hamielec, A. E.; Penlidis, A. *Ind. Eng. Chem. Res.* **1997**, *36*, 939-969.
10. Wood, C. D.; Cooper, A. I. *Macromolecules* **2001**, *34*, 5-8.
11. Rohr, T.; Knaus, S.; Gruber, H.; Sherrington, D. C. *Macromolecules* **2002**, *35*, 97-105.
12. Lewandowski, K.; Murer, P.; Svec, F.; Fréchet, J. M. J. *Chem. Commun.* **1998**, 2237-2238.
13. Xu, M. C.; Brahmachary, E.; Janco, M.; Ling, F. H.; Svec, F.; Fréchet, J. M. J. *J. Chromatogr. A* **2001**, *928*, 25-40.
14. Svec, F.; Fréchet, J. M. J. *Science* **1996**, *273*, 205-211.
15. Peters, E. C.; Svec, F.; Fréchet, J. M. J. *Adv. Mater.* **1999**, *11*, 1169-1181.
16. Petro, M.; Svec, F.; Fréchet, J. M. J. *Biotech. Bioeng.* **1996**, *49*, 355-363.
17. Tennikov, M. B.; Gazdina, N. V.; Tennikova, T. B.; Svec, F. *J. Chromatogr. A* **1998**, *798*, 55-64.
18. Rohr, T.; Yu, C.; Davey, M. H.; Svec, F.; Fréchet, J. M. J. *Electrophoresis* **2001**, *22*, 3959-3967.
19. Yu, C.; Xu, M. C.; Svec, F.; Fréchet, J. M. J. *J. Polym. Sci. Polym. Chem.* **2002**, *40*, 755-769.
20. Whitcombe, M. J.; Vulfson, E. N. *Adv. Mater.* **2001**, *13*, 467.
21. Peters, E. C.; Petro, M.; Svec, F.; Fréchet, J. M. J. *Anal. Chem.* **1998**, *70*, 2296-2302.
22. Peters, E. C.; Petro, M.; Svec, F.; Fréchet, J. M. J. *Anal. Chem.* **1998**, *70*, 2288-2295.
23. Peters, E. C.; Petro, M.; Svec, F.; Fréchet, J. M. J. *Anal. Chem.* **1997**, *69*, 3646-3649.
24. Cooper, A. I.; Wood, C. D.; Holmes, A. B. *Ind. Eng. Chem. Res.* **2000**, *39*, 4741-4744.
25. Cooper, A. I.; Holmes, A. B. *Adv. Mater.* **1999**, *11*, 1270-1274.
26. Brennecke, J. F.; Chateauneuf, J. E. *Chem. Rev.* **1999**, *99*, 433-452.
27. Imhof, A.; Pine, D. J. *Nature* **1997**, *389*, 948-951.
28. Cameron, N. R.; Sherrington, D. C. *Adv. Polym. Sci.* **1996**, *126*, 163-214.
29. Parks, K. L.; Beckman, E. J. *Polym. Eng. Sci.* **1996**, *36*, 2404-2416.
30. Parks, K. L.; Beckman, E. J. *Polym. Eng. Sci.* **1996**, *36*, 2417-2431.
31. Shi, C.; Huang, Z.; Kilic, S.; Xu, J.; Enick, R. M.; Beckman, E. J.; Carr, A. J.; Melendez, R. E.; Hamilton, A. D. *Science* **1999**, *286*, 1540-1543.
32. Howdle, S. M.; Watson, M. S.; Whitaker, M. J.; Popov, V. K.; Davies, M. C.; Mandel, F. S.; Wang, J. D.; Shakesheff, K. M. *Chem. Commun.* **2001**, 109-110.
33. Butler, R.; Davies, C. M.; Cooper, A. I. *Adv. Mater.* **2001**, *13*, 1459-1463.
34. Lee, C. T.; Psathas, P. A.; Johnston, K. P.; deGrazia, J.; Randolph, T. W. *Langmuir* **1999**, *15*, 6781-6791.
35. DeSimone, J. M.; Maury, E. E.; Menceloglu, Y. Z.; McClain, J. B.; Romack, T. J.; Combes, J. R. *Science* **1994**, *265*, 356-359.
36. Canelas, D. A.; Betts, D. E.; DeSimone, J. M. *Macromolecules* **1996**, *29*, 2818-2821.
37. Cooper, A. I.; Hems, W. P.; Holmes, A. B. *Macromolecules* **1999**, *32*, 2156-2166.
38. Giles, M. R.; Hay, J. N.; Howdle, S. M. *Macromol. Rapid Commun.* **2000**, *21*, 1019-1023.

39. Christian, P.; Howdle, S. M.; Irvine, D. J. *Macromolecules* **2000**, *33*, 237-239.
40. Christian, P.; Giles, M. R.; Griffiths, R. M. T.; Irvine, D. J.; Major, R. C.; Howdle, S. M. *Macromolecules* **2000**, *33*, 9222-9227.
41. McCulloch, A. *J. Fluorine Chem.* **1999**, *100*, 163-173.
42. Corr, S. *J. Fluorine Chem.* **2002**, 118, 55-67.
43. Blondino, F. E.; Byron, P. R. *Drug Dev. Ind. Pharm.* **1998**, *24*, 935-945.
44. Wood, C. D.; Senoo, K.; Martin, C.; Cuellar, J.; Cooper, A. I. *Macromolecules*, **2002**, *35*, 6743-6746.

Reactions in Supercritical Carbon Dioxide

Chapter 26

Catalytic Hydrogenation of Olefins in Supercritical Carbon Dioxide Using Rhodium Catalysts Supported on Fluoroacrylate Copolymers

Roberto Flores[1], Zulema K. Lopez-Castillo[1], Aydin Akgerman[1,*], Ibrahim Kani[2], and John P. Fackler, Jr.[2]

[1]Chemical Engineering Department, Texas A&M University, College Station, TX 77843
[2]Department of Chemistry, Texas A&M University, College Station, TX 77842

A novel catalyst soluble in $scCO_2$ was synthesized by grafting rhodium ligands to a fluoroacrylate copolymer. Different compositions of copolymer, Rh/copolymer molar ratios, and substrate/catalyst molar ratios were studied for the hydrogenation of 1-octene in $scCO_2$ at 70°C and 172 bar. Also, the effect of temperature at 172 bar, and the effect of pressure at 70°C were analyzed for the same reaction. The products for the hydrogenation of 1-octene were mostly n-octane together with isomerization products. Finally, a comparative study of the hydrogenation of 1-octene and cyclohexene was carried out at 70°C and 172 bar.

Organometallic molecular catalysts have a complex tunable molecular structure that offer high activity related to metal content, high selectivity, low sensitivity toward catalyst poisons, mild reaction conditions, absence of internal diffusion problems, variability of steric and electronic properties, and the possibility of understanding the reaction mechanism [1]. However, their main constraints are the expensive catalyst recovery and purification processes after reaction, and the employment of toxic solvents as reaction media. Efforts in the recovery and reuse of organometallic catalysts have lead to the development of reactions in aqueous and fluorous systems [2–5]. However, the use of volatile fluorocompounds are prohibited due to global warming problems, and the use of water as a solvent would generate contaminated aqueous waste, which could not be released into the natural cycle without expensive cleanup procedures. Supercritical carbon dioxide (scCO$_2$) has shown significant potential for replacing conventional solvents as reaction media because of its unique properties, which near the critical region are easily tunable by small variation in pressure and/or temperature allowing the possibility of engineering the reaction environment to improve the reactivity and selectivity [6-8]. Additionally, it is easily separated from the other components in the process stream, nonflammable, non-toxic, and capable of performing the duties of a non-polar solvent while allowing manipulation of its solvent strength through a wide range of polarities. Organic compounds and gases are mostly soluble in scCO$_2$, but conventional catalysts are only scarcely soluble in it. Therefore, the development of new catalytic materials soluble in scCO$_2$ is required. A novel catalyst soluble in scCO$_2$ has been synthesized by attaching rhodium ligands to a fluoroacrylate copolymer. The catalytic activity of the new material is tested in hydrogenation reactions by analyzing the effect of pressure, temperature, hydrogen concentration, and substrate to catalyst molar ratio.

Experimental Section

Materials

Organic substances were purchased from Sigma-Aldrich and used as received. Carbon dioxide and hydrogen were bought from Brazos Welding Supply and used as received. Catalysts were prepared in the Department of Chemistry at Texas A&M University and used as received.

Catalyst

The preparation and characterization of the catalyst has already been explained [9]. In summary, a copolymer with an easily exchanged succinimide group was prepared by reacting a fluoroacrylate monomer (zonyl TAN, from Dupont) with N-acryloxysuccinimide (NASI) in benzotrifluoride at 100°C during 48 h using AIBN as initiator (reaction 1 in Figure 1). After washing and purifying, the succinimide group was exchanged with phosphines by reacting the copolymer with diphenyl-phosphino-propylamine (DPPA) in 1,1,2-trichlorotrifluoroethane (FC113) at room temperature during 3 h (reaction 2 in Figure 1). Finally, the catalyst was made by reacting the catalyst precursor with [RhCl(COD)]₂ in FC113 at room temperature during 3 h (reaction 3 in Figure 1). By this technique, a number of catalysts with various copolymer composition and rhodium to copolymer molar ratio were prepared.

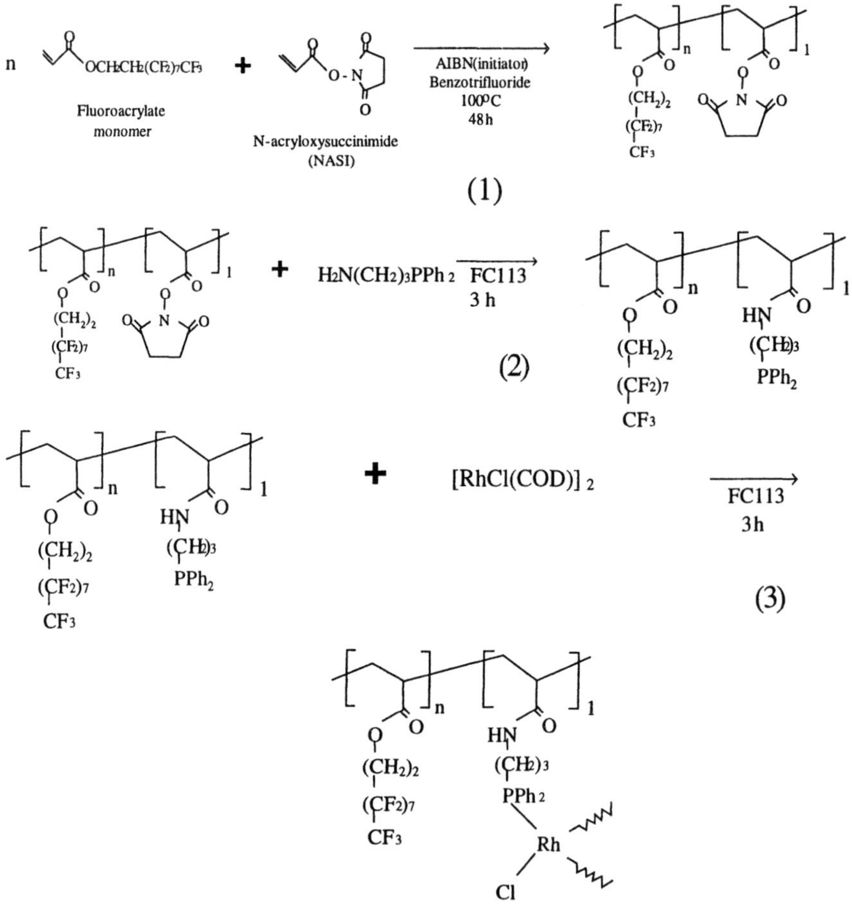

Figure 1. Synthesis route of the Rh catalyst grafted on fluoroacrylate copolymer.

Apparatus and Procedures

The experimental equipment and procedure used for the experimentation have been described [10]. A small visual cell reactor, 8 mL, was employed to visually check the solubility of the catalyst and the miscibility of reactants and products in $scCO_2$ at reaction conditions (Figure 2). Also, a 100 mL batch reactor with sampling on-line was installed to take samples at different times in order to determine the conversion and selectivity profiles against time (Figure 3). After each experiment the reactor was carefully washed and a blank test (without catalyst) was performed to ensure no catalyst was left in the reactor.

The samples taken from the reactors were analyzed by gas chromatography (HP-FID 5890) using a 5% phenyl polyxilosane capillary column. The chromatograph was calibrated using pure samples of the products and reactants. Calculations were done using the response factor method [11].

Experimental Results and Discussion

Effect of Rh to copolymer molar ratio

To study the effect of rhodium to copolymer molar ratio, three different catalysts were prepared. The copolymer composition for all the catalysts were 20:1 (zonyl TAN:NASI), $RhCl(TAN_{20}DPPA)_n$. The copolymer to rhodium molar ratios (n) studied were 3, 10, and 45. Experiments were carried out in the visual cell to also check the miscibility of reactants and products in the

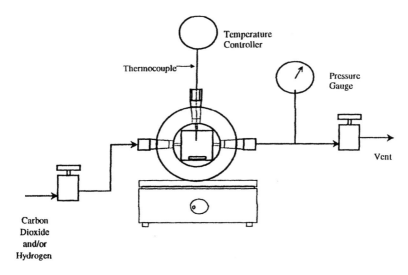

Figure 2. Schematic representation of the visual cell reactor.

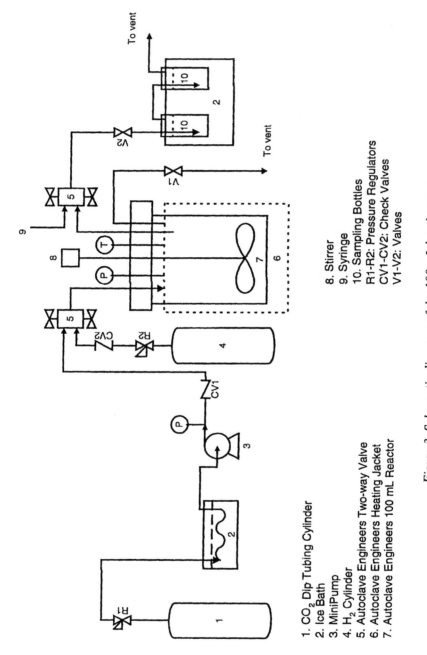

1. CO$_2$ Dip Tubing Cylinder
2. Ice Bath
3. MiniPump
4. H$_2$ Cylinder
5. Autoclave Engineers Two-way Valve
6. Autoclave Engineers Heating Jacket
7. Autoclave Engineers 100 mL Reactor

8. Stirrer
9. Syringe
10. Sampling Bottles
R1-R2: Pressure Regulators
CV1-CV2: Check Valves
V1-V2: Valves

Figure 3. Schematic diagram of the 100 mL batch reactor.
(Reproduced from reference 10. Copyright 2002 American Chemical Society.)

supercritical mixture and the solubility of the catalyst. The experiments were performed at 70°C and 172 bar for 12 h. Hydrogen/1-octene molar ratio was 3, and the amount of catalyst was 5.0 mg. in all the experiments. The catalyst, reactants and products were fully dissolved at experimental conditions. Reactivity results are presented in Figure 4. The catalyst with copolymer/rhodium molar ratio of 3 produced around 72% of total conversion (TON = 3524), with 50% yield to n-octane and 22% yield to isomer products, (E)2-octene and (Z)2-octene; on the other hand, the activity for the other catalysts was very low or nil (TON = 816 for n = 10, and TON = 0 for n = 45).

These results could be attributed to a cross linked phenomena between the copolymer and the Rh, when the last is not present at least in stoichiometric amounts with respect to available phosphine sites contained in the copolymer structure [9], reducing the number of active sites for the hydrogenation reaction. Ideally, the Rh-Cl complex should interact and bond with a DPPA group coming from different copolymer chain (Figure 5a). Nevertheless, cross linked may occur, and the Rh-Cl complex would bind to two or even three DPPA groups coming from the same copolymer chain (Figure 5b and 5c). If the last occur, the electronic and steric properties of the catalytic complex would be affected and it could inhibit the reactivity. Great excess of copolymer to rhodium molar ratio seems to favor the formation of structures 5b and/or 5c. This could be explained considering the Rh-Cl preferentially interacts with the closest DPPA group, which would be attached to the same copolymer chain.

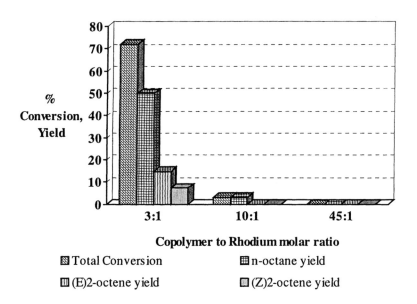

Figure 4. Effect of copolymer to rhodium molar ratio for the hydrogenation of 1-octene at 70°C and 172 bar.

412

On the other hand, Halpern studies about hydrogenation of olefin using the $RhCl(PPh_3)_3$ catalyst showed that the active species was $RhClH_2(PPh_3)_2$ [12, 13]. This complex is formed by oxidative addition of hydrogen to the organometallic compound, and dissociation of one phosphine group according to the following reaction:

$$RhCl(PPh_3)_3 + H_2 \rightleftharpoons RhClH_2(PPh_3)_2 + PPh_3$$

As a result, and based on L'Chatelier principle, a large excess of phosphine group may retard the formation of the active species inhibiting in this way the overall hydrogenation mechanism. This phenomena could occur with the catalysts $RhCl(TAN_{20}DPPA)_{10}$ and $RhCl(TAN_{20}DPPA)_{45}$ resulting in the low activity of the catalysts.

Effect of substrate to catalyst molar ratio

The studies on the effect of substrate to Rh molar ratio were carried out using a catalyst $RhCl(TAN_{20}DPPA)_3$. The experiments were performed in the 100 mL batch reactor and samples were taken at different times. The experimental conditions were 70°C, 172 bar, and the mole fractions of 1-octene and hydrogen were 0.001 and 0.033, respectively. The results are presented in

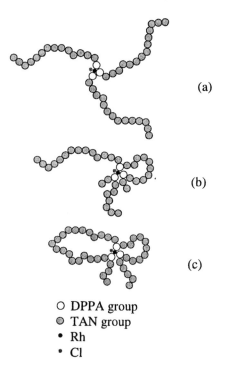

(a)

(b)

(c)

O DPPA group
◉ TAN group
• Rh
• Cl

Figure 5. Possible structures of the Rh-catalytic complex.

Figure 6. As expected, decreasing the substrate to catalyst molar ratio, the reaction rate increases, even though there is no difference when this molar ratio is between 600 and 300. Also, hydrogenation is preferred over isomerization reactions when this molar ratio diminishes. At 120 this selectivity is over 70% and slightly enhances with time suggesting the isomers are also hydrogenated. At 1300 the selectivity to n-octane is low and declines with time. This may mean that the hydrogenation-promoter species are inhibited or transformed to isomerization-promoter complexes during the catalytic cycle.

Effect of Copolymer Composition

To analyze the effect of copolymer composition in the hydrogenation of 1-octene, three catalysts with different zonyl TAN:DPPA molar ratio were prepared: 20:1 [RhCl(TAN$_{20}$DPPA)$_3$], 15:1 [RhCl(TAN$_{15}$DPPA)$_3$], and 7:1 [RhCl(TAN$_7$DPPA)$_3$]. The experiments were carried out in the 100 mL reactor at 70°C, 172 bar, substrate to catalyst molar ratio of 400, and the initial mole fraction of 1-octene and hydrogen were 0.001 and 0.033, respectively. The catalyst RhCl(TAN$_7$DPPA)$_3$ gave the lowest activity because it was not fully dissolved in the supercritical mixture. Experiments in the visual cell reactor confirmed this catalyst was not fully soluble at reaction conditions. For the other two catalysts, there was no difference in the total conversion in the first 20 h. After this time, the activity of the catalyst RhCl(TAN$_{20}$DPPA)$_3$ increased faster compared to the catalyst RhCl(TAN$_{15}$DPPA)$_3$. On the other hand, the selectivity

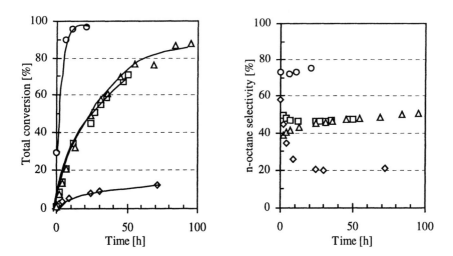

Figure 6. Effect of the substrate to catalyst molar ratio on the hydrogenation of 1-octene at 70°C and172 bar. (○) 120; (□) 300; (Δ) 600; (◊) 1300.

of the catalyst RhCl(TAN$_{20}$DPPA)$_3$ toward n-octane was the smallest, and it increased as the zonyl TAN:DPPA molar ratio decreased.

Effect of Temperature

Experiments were performed at 45, 70 and 95°C, and 172 bar in the 100 mL reactor to determine the effect of temperature at constant pressure on reactivity

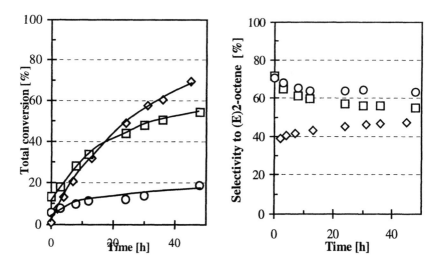

Figure 7. Effect of copolymer composition in the hydrogenation of 1-octene at 70°C and 172 bar. (◊) RhCl(TAN$_{20}$DPPA); (□) RhCl(TAN$_{15}$DPPA); (○) RhCl(TAN$_7$DPPA)

(Reproduced from reference 10. Copyright 2002 American Chemical Society.)

and selectivity for the hydrogenation of 1-octene. With the exception of temperature, all the experimental conditions remained constant. The catalyst was RhCl(TAN$_{20}$DPPA)$_3$. The molar ratio between substrate and catalyst was 600, and the mole fractions of 1-octene and hydrogen were 0.001 and 0.033, respectively. The reactivity of 1-octene was approximately the same at 70 and 95°C and lower at 45°C as can be seen in Figure 8. The lack of temperature effect between 70 and 95°C could be attributed to the attenuation of solvent strength at 95°C. Experiments in the visual cell reactor confirmed a single phase at 70°C, but no experiments were achieved at 95°C due to equipment limitations. On the other hand, the density of the scCO$_2$ decreases from 0.574 g/mL at 70°C to 0.399 g/mL at 95°C. This reduction could provoke that the catalyst is not fully miscible at 95°C because of the retrograde behavior of some solids [7, 14]. This may mean that the catalyst is only active when it is soluble, and the lack of activity of the catalyst due to solubility constraints was compensated by the increase in temperature.

Effect of pressure

The effect of pressure in the hydrogenation of 1-octene was studied at 70°C, and between 172 - 207 bar in the 100 mL reactor. The catalyst employed for this study was $RhCl(TAN_{15}DPPA)_3$ with rhodium/polymer ratio of 3. The initial mole fractions of 1-octene and hydrogen were 0.001 and 0.033, respectively. These mole fractions were kept constant in all the experiments. The substrate to catalyst mole ratio was 400. Figure 9 shows the experimental results. Experiments were run twice to determine reproducibility, the points represent the average, and the error bars the standard deviation. Derivations from the transition-state theory [15] show that the thermodynamic effect of pressure on reactivity and selectivity for reactions at supercritical conditions are related to the following equations, respectively:

$$\left(\frac{\partial \ln k_x}{\partial P}\right)_{T,x} = -\frac{\Delta V^{\ddagger}}{RT}$$

$$\left(\frac{\partial \ln(k_2 / k_1)}{\partial P}\right)_{T,x} = \frac{\overline{v}_{M^*,1} - \overline{v}_{M^*,2}}{RT}$$

where $\Delta V^{\ddagger}$ is the activation volume defined as the difference between the partial molar volume of the activated complex and the sum of partial molar volume of reactants, $\overline{v}_{M^*} - \overline{v}_A - \overline{v}_B$. $\overline{v}_{M^*,1}$ and $\overline{v}_{M^*,2}$ are the partial molar volumes of two transition states for product 1 and 2, respectively. Therefore, the partial molar volume difference of the two transition states will determine how the selectivity would vary with pressure. Since it is not possible to calculate the partial molar volume of a transition state complex, the difference $\overline{v}_{M^*,1} - \overline{v}_{M^*,2}$

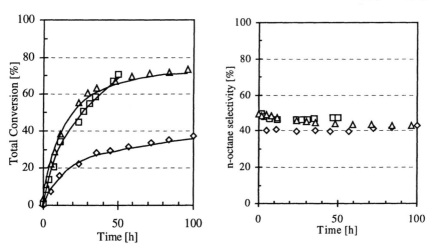

Figure 8. Effect of temperature in the hydrogenation of 1-octene at 172 bar. (◊) 45°C; (□) 70°C; (△)95°C.

has been related to the difference of the partial molar volume of the final products, $\bar{v}_1 - \bar{v}_2$. In our case, we can conclude that the activation volume for the hydrogenation of 1-octene is positive since the reactivity retarded as pressure was increased. In addition, the absence of pressure effect on selectivity could be attributed to very similar partial molar volumes of the n-octane and the 2-octene isomers in our system.

Comparison between hydrogenation of 1-octene and cyclohexene

Hydrogenations of 1-octene and cyclohexene were studied at 70°C and 172 bar in the 100 mL reactor using the same experimental conditions. The catalyst employed was $RhCl(TAN_{15}DPPA)_3$. The mole fractions of the olefins and hydrogen were 0.001 and 0.033, respectively. The substrate to catalyst molar ratio was 400. The total conversion of cyclohexene was slower compared to 1-octene, but because no isomers were produced; the yield profile toward the saturated hydrocarbon (paraffin) is practically the same as is observed in Figure 10. Therefore, the catalytic activity was not affected by the degree of substitution in the double bond in these two olefins at 70°C and 172 bar.

Concluding Remarks
A new $scCO_2$ soluble Rh catalyst supported on fluoroacrylate copolymers was synthesized. The catalysts were active for hydrogenation of 1-octene and cyclohexene. The optimal Rh to copolymer molar ratio was 3. Using

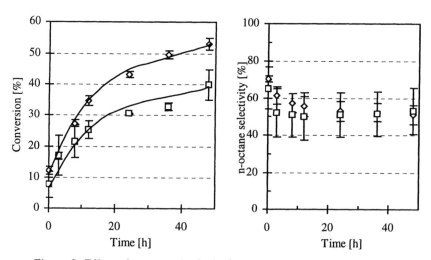

Figure 9. Effect of pressure in the hydrogenation of 1-octene at 70°C.
(◊) 172 bar; (□) 207 bar.

the catalyst $RhCl(TAN_{20}DPPA)_3$ at 70°C, 172 bar and low 1-octene to catalyst molar ratio, the hydrogenation of the 1-octene was favored over its isomerization; however, increasing this ratio the selectivity to hydrogenation decreased. Studies of catalysts with different copolymer composition showed the catalyst $RhCl(TAN_7DPPA)_3$ was partially soluble to reaction conditions yielding to low activity. On the other hand, catalysts $RhCl(TAN_{15}DPPA)_3$ and $RhCl(TAN_{20}DPPA)_3$ were fully soluble giving the same activity at the beginning of the reaction, and after 20 h some inhibition of catalyst $RhCl(TAN_{15}DPPA)_3$ was observed. Significant temperature effect with respect to activity for the 1-octene hydrogenation was observed between 45 and 70°C; however, the selectivity did not change significantly. Increasing the total pressure decreased the activity of the catalyst without affecting the selectivity for the hydrogenation of 1-octene. Hydrogenation rate for 1-octene and cyclohexene was similar, but isomerization of 1-octene occurred increasing the reactivity of 1-octene.

Acknowledgements

This work was supported by The Environmental Protection Agency (EPA), Gulf Coast Hazardous Substance Research Center, Consejo Nacional de Ciencia y Tecnologia (CONACYT) of Mexico, and The Scientific and Technical Research Council of Turkey (TUBITAK-NATO). The authors are grateful for partial support from each organization.

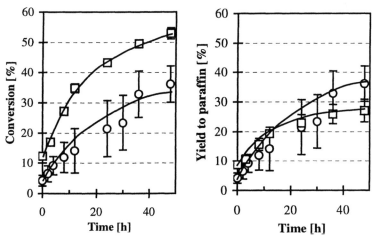

Figure 10. Comparison between the hydrogenation of 1-octene and cyclohexene at 70°C and 172 bar. (□) 1-octene, (○) cyclohexene.

418

References

1. Cornils, B.; Hermann, W. A. in *Applied Homogeneous Catalysis with Organometallic Compounds*; Cornils, B.; Hermann, W. A., Eds.; Wiley-VCH, Verlag GmbH, Weinheim, Germany, 1996, pp 1-25.
2. *Aqueous Phase Organometallic Catalysis*; Cornils, B.; Hermann, W. A., Eds.; Wiley-VCH, Verlag GmbH, Weinheim, Germany, 1998.
3. Lubineau, A.; Augé, J. In *Modern Solvents in Organic Synthesis*; Knochel, P. Ed.; Topics in Current Chemistry; Springler-Verlag Berlin Heidelberg, Germany, 1999, Vol. 206, pp 1-39.
4. Betzemeier, B.; Knochel, P. In *Modern Solvents in Organic Synthesis*; Knochel, P. Ed.; Topics in Current Chemistry; Springler-Verlag Berlin Heidelberg, Germany, 1999, Vol. 206, pp 61-78.
5. Bergbreiter, D. E.; Franchina, J. G.; Case, B. L. Fluoroacrylate-Bound Fluorous-Phase Soluble Hydrogenation Catalysis. *Org. Lett.*, **2000**, 2, 393-395.
6. Leitner, W. In *Modern Solvents in Organic Synthesis*; Knochel, P. Ed.; Topics in Current Chemistry; Springler-Verlag Berlin Heidelberg, Germany, 1999, Vol. 206, pp 107-132.
7. Jessop, P. G.; Ikariya, T.; Noyori, R. Homogeneous Catalysis in Supercritical Fluids. *Chem. Rev.*, **1999**, 99, 475-493.
8. Baiker, A. Supercritical Fluids in Heterogeneous Catalysis, *Chem. Rev.*, **1999**, 99, 453-473.
9. Kani, I.; Omary, M. A.; Rawashdeh-Omary, M. A.; Lopez-Castillo, Z. K.; Flores, R; Akgerman, A.; Fackler, Jr., J. P. Homogeneous catalysis in supercritical carbon dioxide with rhodium catalysts tethering fluoroacrylate polymer ligands. *Tetrahedron*, **2002**, 58, 3923-3928.
10. Lopez-Castillo, Z. K.; Flores, R.; Kani, I.; Fackler, Jr.,J. P.; Akgerman, A. Fluoroacrylate Copolymer Supported Rhodium Catalysts for Hydrogenation Reactions in Supercritical Carbon Dioxide. *Ind. Eng. Chem. Res.*, **2002**, 41, 3075-3080
11. Schomburg, G. *Gas Chromatography. A Practical Course*. New York: VCH Publishers, Inc., 1990
12. Halpern, J., and Wong, C. S. Hydrogenation of Tris(triphenyl-phosphine)chlororhodium(I). *J. C. S. Chem. Comm.*, **1973**, 629-631
13. Halpern, J. Mechanistic Aspects of Homogeneous Catalytic Hydrogenation and Related Processes. *J. Inorg. Chim. Acta*, **1981**, 50, 11-19
14. Darr, J. A.; Poliakoff, M. New Directions in Inorganic and Metal-Organic Coordination Chemistry in Supercritical Fluids. *Chem. Rev.*, **1999**, 99, 495-541.
15. Lin, B.; Akgerman, A. Styrene Hydroformylation in Supercritical Carbon Dioxide: Rate and Selectivity Control. *Ind. Eng. Chem. Res.*, **2001**, 40, 1113-1118

Chapter 27

Hydrogenation Reactions in Supercritical CO_2 Catalyzed by Metal Nanoparticles in a Water-in-Carbon Dioxide Microemulsion

Mariko Ohde, Hiroyuki Ohde, and Chien M. Wai[*]

Department of Chemistry, University of Idaho, Moscow, ID 83844–2343

Water-in-CO_2 microemulsions with diameters in the order of several nanometers are prepared by a mixture of AOT and a PFPE-PO$_4$ co-surfactant. The CO_2 microemulsions allow metal species to be dispersed in the nonpolar supercritical CO_2 phase. By chemical reduction, metal ions dissolved in the water core of the microemulsion can be reduced to the elemental state forming nanoparticles with narrow size distribution. The palladium and rhodium nanoparticles produced by hydrogen reduction of Pd^{2+} and Rh^{3+} ions dissolved in the water core are very effective catalysts for hydrogenation of olefins and arenes in supercritical CO_2.

In the past two decades, supercritical fluid extraction technology has attracted considerable attention from chemists and engineers for its potential applications as a green solvent for chemical processing (*1-3*). One difficulty of using this green solvent for metal dissolution is that metal ions are not soluble in supercritical CO_2 because of the charge neutralization requirement and the weak interactions between CO_2 and metal ions. Converting metal ions to CO_2-soluble metal chelates utilizing organic chelates is one method of extracting them into supercritical CO_2 (*4, 5*). This *in situ* chelation/supercritical fluid extraction method developed in the early 1990s by Wai and coworkers has been widely used today by many researchers for extracting metal species into supercritical CO_2 (*6-12*).

A new development in dissolution of metal species in supercritical CO_2 is based on the observation that metal ions can be stabilized in a water-in-CO_2 microemulsion (*13-19*). The water-in-CO_2 microemulsions with diameters typically in the order of several nanometers (nm) allow metal species and high polarity compounds to be dispersed in nonpolar supercritical CO_2. This type of microemulsion may be regarded as a new solvent for dissolution and transport of metal species in supercritical fluids. It has also been shown that the water-in-CO_2 microemulsion can be used as a reactor for chemical synthesis of nanometer-sized materials in supercritical fluids. The first report on the synthesis of nanometer-sized metal particles using a water-in-CO_2 microemulsion appeared in 1999 (*14*). In this report, Wai and co-workers showed that silver ions in the water core of a water-in-supercritical CO_2 microemulsion could be reduced to nanosized metallic silver particles by a reducing agent dissolved in the fluid phase. The silver nanoparticles can be stabilized and dispersed uniformly in the supercritical fluid phase by the microemulsion for an extended period of time (*15*). Further reports from our research group indicate that by mixing two microemulsions containing different ions in the water cores, exchange of ions can take place leading to chemical reactions (*17, 18, 20*). These reports suggest the possibility of utilizing the water-in-CO_2 microemulsions as nanoreactors for synthesizing a variety of nanoparticles in supercritical CO_2. More recently, using metal nanoparticles synthesized in the CO_2 microemulsion as catalysts for hydrogenation reactions was reported (*21, 22*). This new development opens the door for a wide range of catalysis utilizing microemulsion dispersed nanoparticles in supercritical fluids. This article describes our recent results of metal nanoparticle synthesis using the water-in-CO_2 microemulsion as a template and their potential applications as catalysts for chemical reactions in supercritical CO_2.

Synthesizing Metal Nanoparticles In CO_2 Microemulsion

There has been much interest in recent years to exploit the properties of microemulsion phases in supercritical fluids (*23-33*). A reverse micelle or microemulsion system of particular interest is one based on CO_2 because of its minimum environmental impact in chemical applications. Since water and CO_2

are the two most abundant, inexpensive, and environmentally compatible solvents, the application of such a system could have tremendous implications for the chemical industries °of the 21st century. Reverse micelles and microemulsions formed in supercritical CO_2 allow highly polar or polarizable compounds to be dispersed in this nonpolar fluid. However, most ionic surfactants with long hydrocarbon tails, such as sodium bis(2-ethylhexyl) sulfosuccinate (AOT), are insoluble in supercritical CO_2. Nonionic surfactants have a greater affinity for CO_2. However, they have little tendency to aggregate and take up water due to weak electrostatic interactions of the head groups. The use of ionic surfactants with fluorinated tails provides a layer of a weakly attractive compound covering the highly attractive water droplet cores, thus preventing their short-range interactions that would destabilize the system. However, our experiments indicate that these CO_2-philic ionic surfactants do not form very stable water-in-CO_2 microemulsions especially when the water cores contain high concentrations of electrolytes. A recent communication shows that using a conventional AOT and a fluorinated cosurfactant, a very stable water-in-CO_2 microemulsion containing a relatively high concentration of silver nitrate can be formed (*14-18*).

Using a mixture of AOT and a perfluoropolyether (PFPE) surfactant the resulting water-in-CO_2 microemulsion is stable with W (ratio of water/AOT) values >40 at 40 °C and 200 atm according to the experiments conducted recently in our laboratory. The perfluorinated surfactant obtained from Ausimont has a general structure of $CF_3O[OCF(CF_3)CF_2]_n(OCF_2)_m$ $OCF_2CH_2OCH_2CH_2O-PO(OH)_2$ and an average molecular weight of 870. The initial experiments for synthesizing silver nanoparticles used a mixture of AOT (12.8 mM) and the perfluoropolyether-phosphate (PFPE-PO_4, 25.6 mM) at W = 12. The microemulsion containing $AgNO_3$ (0.33 mM) was optically transparent and stable in supercritical CO_2 for hours. To make metallic silver nanoparticles, a reducing agent such as $NaBH(OAc)_3$ or $NaBH_3CN$ was injected into the superciritcal fluid microemulsion system to reduce Ag^+ in the water core to Ag (*14-16, 18*). The formation of Ag nanoparticls in the microemulsion system was observed within a minute after the introduction of the reducing agent. The formation and stability of the Ag nanopartices was monitored in situ by UV-Vis spectroscopy utilizing the 400 nm band originating from the surface plasmon resonance of nano-sized Ag crystals. The supercritical fluid solution remained optically clear with a yellow color due to the absorption of the Ag nanoparticles. Further spectroscopic studies indicate that the microemulsions are dynamic in nature. Ionic species in the water core of the microemulsion obviously can interact effectively with molecular species dissolved in the fluid phase. Using the same approach copper nanoparticles can be synthesized (*15, 18*).

In a recent communication, Ohde et al. showed the synthesis of palladium nanoparticles by hydrogen reduction of Pd^{2+} ions dissolved in the water core of a CO_2 microemulsion (*18, 21*). The Pd nanoparticles so produced are uniformly dispersed in the supercritical fluid phase and are stable over an extended period of time long enough for catalysis experiments. Reduction of a metal ion to its elemental state in supercritical CO_2 using hydrogen gas is a simple, clean, and effective method for producing nanometer-sized metal particles in the microemulsion. This method is particularly attractive for studying hydrogenation reactions in supercritical CO_2 because H_2 gas is miscible with CO_2 and can serve both as a reducing agent for metal nanoparticles formation as well as the starting material for subsequent hydrogenation. The advantages of performing hydrogenation reactions in supercritical CO_2 compared with conventional solvent systems are known in the literature (*3, 4, 33*). High solubility of hydrogen gas and enhanced diffusion in supercritical CO_2 relative to conventional solvent systems often result in faster and more efficient processes in the supercritical fluid phase. In addition, tunable solvation strength of supercritical CO_2, easy separation of solvent from products, and minimization of waste generation are other attractive features of conducting chemical synthesis in supercritical CO_2.

Metal Nanoparticle Catalyzed Hydrogenation Reactions

The hydrogenation of olefins in supercritical carbon dioxide catalyzed by palladium nanoparticles synthesized in a water-in-CO_2 microemulsion was reported by Ohde et al (*21*). The Pd nanoparticles were prepared by hydrogen reduction of Pd^{2+} ions (a $PdCl_2$ solution) dissolved in the water core of the microemulsion. Effective hydrogenation of both CO_2-soluble olefins (4-methoxycinnamic acid and trans-stilbene) and a water-soluble olefin (maleic acid) catalyzed by the palladium nanoparticles in the microemulsion was demonstrated.

The hydrogenation of 4-methoxy cinnamic acid to 4-methoxy hydrocinnamic acid catalyzed by the Pd nanoparticles was performed first in liquid CO_2 at room temperature (20 °C) (Equation 1). The spectra shown in Figure 1a were taken at 20-second intervals after the injection of hydrogen and 4-methoxy cinnamic acid into the water-in-CO_2 microemulsion with $PdCl_2$ in the water core (W = 20). The first spectrum obtained immediately after the injection (spectrum 1) was identical to that of 4-methoxy cinnamic acid dissolved in CO_2. The broad absorption peak centered around 300 nm decreased gradually and a new absorption peak centered around 270 nm appeared. After about 2 minutes, the absorbance at 300 nm dropped to the baseline level. The absorption peak (270 nm) in spectrum 2 (Figure 1a) was consistent with that of 4-methoxy hydrocinnamic acid. In the absence of $PdCl_2$ in the microemulsion, the absorption peak of 4-methoxy cinnamic acid did not show a measurable decrease after the injection of the olefin and hydrogen into the fiber-optic reactor. Also, in the absence of hydrogen, injection of 4-methoxy cinnamic acid into the reactor with $PdCl_2$ in the water core of the microemulsion did not show any change of absorption at 300 nm either. Figure 1b shows the decrease in the

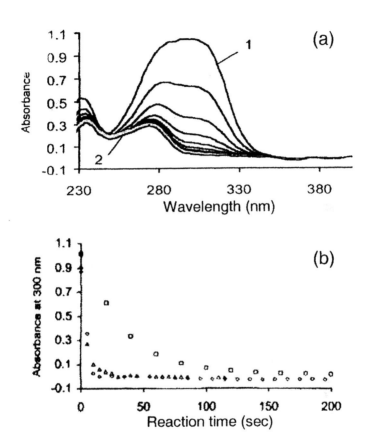

Figure 1. (a) Variation of UV-Vis spectra of 4-methoxy cinnamic acid with time during the hydrogenation process in CO_2 at 20 °C and 200 atm. Each spectrum was taken at 20-second intervals starting from zero time, spectrum 1. (b) Variation of the 4-methoxy cinnamic acid absoption at 300 nm with time at 20 °C and 200 atm (□),35 °C and 200 atm (▲) and at 50 °C and 200 atm (○). (Reproduced with permission from reference 21. Copyright 2002 American Chemical Society.)

absorbance at 300 nm with time for the hydrogenation of 4-methoxy cinnamic acid in liquid CO_2 at 20 °C and in supercritical CO_2 at 35 °C and 50 °C. The speed of the hydrogenation process is much faster in the supercritical CO_2 phase (35 °C and 50 °C) compared with that in the liquid CO_2 phase (20 °C). The hydrogenation process at 50 °C under the specific conditions was virtually completed in 20 seconds. The absorbance in logarithmic scale for both the liquid and supercritical CO_2 experiments decreases linearly with time suggesting the hydrogenation process follows first order kinetics. The apparent rate constants obtained from the slopes are about 1.1×10^{-2} sec^{-1}, 6.9×10^{-2} sec^{-1} and 9.4×10^{-2} sec^{-1} at 20 °C, 35 °C and 50 °C, respectively.

$$\text{(1)}$$

The water-in-CO_2 microemulsion can dissolve hydrophilic organic compounds in the water core. This property can be used to perform hydrogenation of water-soluble olefins in CO_2. It was demonstrated that maleic acid can be converted to succinic acid by catalytic hydrogenation of Pd nanoparticles formed in the CO_2 microemulsion. The experiments were conducted by dissolving 0.04 M $PdCl_2$ (total amount 2.2×10^{-3} mmol in the system) and 1.8 M maleic acid (9.6×10^{-2} mmol in the system) in the water core of a CO_2 microemulsion with W = 20 at 20 °C and 80 atm CO_2. After that, 10 atm of H_2 in 200 atm of CO_2 at 20 °C was injected into the microemulsion system. Two minutes after the injection of hydrogen, the supercritical fluid phase was expanded into an acetone solution for analysis. The acetone solution was evaporated to dryness and D_2O was added for NMR analysis. The NMR spectra showed that only succinic acid was present in the D_2O solution. Washing of the stainless steel reactor with acetone showed no detectable amount of maleic acid by NMR. Based on the detection limit of the maleic acid by NMR, it was estimated that the conversion of maleic acid to succinic acid was greater than 90 %.

It was also reported that the Pd nanoparticles formed in the CO_2 microemulsion could catalyze other hydrogenation reactions such as the conversion of the nitro group (NO_2) to amine (NH_2). One reported example is the hydrogenation of nitrobenzene to aniline catalyzed by the Pd nanoparticles in a CO_2 microemulsion. The conversion was > 99 % within 30 minutes in supercritical CO_2 at 50 °C and 200 atm.

The Pd nanoparticles synthesized in the CO_2 microemulsion are effective for hydrogenation of CO_2-soluble and water-soluble olefins but are not effective for hydrogenation of aromatic compounds. Hydrogenation of arenes is conventionally carried out with heterogeneous catalysts. Bonilla et al. recently reported a Rh catalyzed hydrogenation of arenes in a water/supercritical ethane biphasic system (*35*). Hydrogenation occurred well in this biphasic system with excellent results obtained for a number of arenes after 62 hours of reaction times. However, this approach did not work for a water/supercritical CO_2 biphasic system according to the authors. We have recently explored the possibility of making rhodium nanoparticles in a water-in-CO_2 microemulsion using the hydrogen gas reduction method. To our surprise, the Rh nanoparticles in the microemulsion are capable of catalyzing hydrogenation of arenes in supercritical CO_2 with good efficiencies (22).

Naphthalene was selected as a CO_2-soluble arene for the hydrogenation study because it absorbs in the UV region that could be monitored in situ by the fiber optic cell. Figure 2 shows the variation of the UV-Vis spectra of naphthalene with time after the injection of hydrogen and naphthalene into the fiber optic cell containing the Rh^{3+} ions dissolved in the water core of the CO_2 microemulsion at 50 ^oC and 240 atm. In this experiment, the amount of naphthalene injected was 1.4×10^{-1} mmol which is in large excess relative to the amount of Rh in the system. About 5 minutes after the injection, the absorbance of naphthalene in the reaction cell was reduced to about one half of the initial value. The absorbance was decreased to near background level after 20 minutes. After one hour, the system was depressurized and the materials in the CO_2 phase were collected in $CDCl_3$ for NMR measurements. The NMR results indicated that nearly all of the original naphthalene (> 96 %) injected into the reactor were reacted to form tetralin (Equation 2).

$$\text{(2)}$$

Hydrogenation of a water soluble arene (phenol) was also studied using the Rh nanoparticles formed in a water-in-CO_2 microemulsion as a catalyst. In this case, phenol (3.3×10^{-1} mmol) was dissolved in the water core of a microemulsion together with Rh^{3+} (1.7×10^{-2} mmol). The microemulsion was made of 18.2 mM AOT, 36.4 mM PFPE-PO_4 and W = 30. The hydrogenation experiments were carried out at 50 ^oC and 240 atm with 10 atm of H_2 gas injected into the microemlsion system. Five minutes after the injection of hydrogen, the fluid phase was expanded into a $CDCl_3$ solution for NMR analysis. The ^1H-NMR spectra showed that phenol was not detectable in the solution (Figure 3). The major product was cyclohexanone according to the NMR spectra with minor amounts of cyclohexane also presented in the spectrum (Equation 3). Rhodium catalyzed hydrogenation depends on the formation of the catalyst and the solvation environment. A possible reaction route for the hydrogenation of phenol using Rh nanoparticles in the CO_2 microemulsion is probably by the addition of two moles of hydrogen to form cyclohexene-1-ol, which undergoes tautomerization to form cyclohexanone. Hydrogenation of cyclohexanone to cyclohexanol does not proceed in this reaction system according to the report by Ohde et al. Formation of cyclohexane as a minor product was also observed by hydrogenation of phenol with Rh (III) in 1, 2-dichloroethane (35). More studies are needed in order to understand the

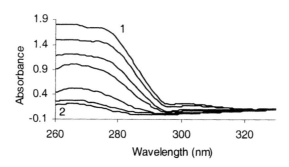

Figure 2. Variation of UV-Vis spectra of naphthalene with time during hydrogenation process in CO_2 at 50 °C and 240 atm. Each spectrum was taken at zero time (spectrum 1), 10 s, 1 min, 5min,10 min, 20 min and 60 min (spectrum 2) from top spectrum to bottom spectrum, respectively. (Reproduced with permission from reference 22. Copyright 2002 Royal Society of Chemistry.)

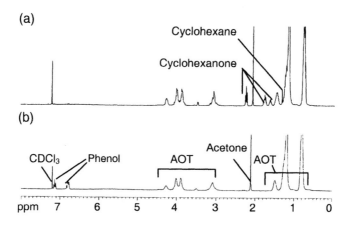

Figure 3. NMR spectra of (a) the products collected from the phenol hydrogenation experiment in $CDCl_3$ (b) control experiment (phenol, AOT, PFPE-PO$_4$ and H$_2$ gas) (Reproduced with permission from reference 22. Copyright 2002 Royal Society of Chemistry.)

mechanisms involved in hydrogenation of phenol and other arenes catalyzed by Rh nanoparticles formed in the CO_2 microemulsion.

$$(3)$$

Conclusion

Our initial studies have demonstrated that the Pd and Rh nanoparticles formed in the CO_2 microemulsions are very effective catalysts for hydrogenation of olefins and arenes in supercritical CO_2. Dispersing metal nanoparticles in supercritical CO_2 utilizing the microemulsion is a new approach for homogenization of heterogeneous catalysis. This approach may have important applications for chemical synthesis in supercritical fluids.

References

1. *Supercritaical Fluid Technology*; Penninger, J. M. L.; Radosz, M.; Mchugh, M. A.; Krukonis, V. J., Eds.; Process Technology Proceedings 3; Elsevier: Amsterdam, The Netherlands, 1985.
2. *Supercritical Fluid Extraction*; McHugh, M. A.; Krukonis, V. J., Eds.; Principles and Practice. 2nd ed; Stoneham: Butterworth-Heineman, 1993.
3. Phelps, C. L.; Smart, N. G.; Wai, C. M. *J. Chem. Edu.* **1996**, *73(12)*, 1163-1168.
4. Wai, C. M.; Laintz, K. E. US patent 5,356,538, 1994.
5. Laintz, K. E.; Wai, C. M.; Yonker, C. R.; Smith, R. D. *J. Supercritical Fluids* . **1991**, 4, 194-198.
6. Lainta, K. E.; Wai, C. M.; Yonker, C. R.; Smith, R. D. *Anal. Chem.* **1992**, 64, 2875-2878.
7. Lin, Y.; Smart, N. G.; Wai, C. M. *Trends in Anal. Chem.* **1995**, 14, 123-133.
8. Darr. J.; Poliakoff, M. *Chem. Rev.* **1999**, 99, 495-541.
9. Wai, C. M. *Metal extraction with supercritical fluids*; Bautista, R. G., Ed.; Emerging Separation Technologies for Metals II; the Minerals, Metals & Materials Society: Warrendale, PA. TMS, 1996, pp 233-248.
10. Wai, C. M.; Wang, S. *J. Chromatogr. A*, **1997**, 785, 369-383.

428

11. Smart, N. G.; Carleson, T. E.; Kast, T.; Clifford, A. A.; Burford, M. D.; Wai, C. M. *Talanta.* **1997**, 44, 137-150.

12. Wai, C. M.; Lin, Y.; Ji, M.; Toews, K. L.; Smart, N. G. *Extraction and separation of uranium and lanthanides with supercritical fluids*; Bond, A. H.; Dietz, M. L.; Rogers, R. D., Eds., ACS Symposium Series 716, Progress in Metal Ion Separation and Preconcentration; Amer. Chem. Soc.: Washington, D.C., 1999, Chapter 21, pp 390-400.

13. Johnston, K. P.; Harrison, K. L.; Clarke, M. J.; Howdle, S. M.; Heitz, M. P.; Bright. F. V.; Carlier. C.; Randolph, T. W. *Science.* **1996**, 271, 624-626.

14. Ji, M.; Chen, X.; Wai, C. W.; Fulton. J. L. *J. Amer. Chem. Soc.* **1999**, 121, 2631-2632.

15. Ohde, H.; Hunt, F.; Wai, C. M. *Chem. Mater.* **2001**, 13, 4130-4135.

16. Hunt, F.; Ohde, H.; Wai, C. M. *Rev. Sci. Instru.* **1999**, 70(12), 4661-4667.

17. Ohde, H.; Rodriguez, J. M.; Ye, X.; Wai, C. M. *Chem. Commun.* **2000**, 2353-2354.

18. Wai, C. M.; Ode, H. *J. Chin. Inst. Chem. Engrn.* **2001**, 32(2), 253-261.

19. Sun, Y. P.; Atorngitjawat, P.; Maziami, M. J. *Langmuir.* **2001**, 17, 5707-5710.

20. Ohde, H.; Ohde, M.; Kim, H.; Wai, C. M. *Nano Lett.* **2002**, 271-274.

21. Ohde, H.; Wai, C. M.; Kim, H.; Kim, J.; Ohde, M. *J. Amer. Chem. Soc.* **2002**, 124, 4540-4541.

22. Ohde, M.; Ohde, H.; Wai, C. M. *Chem. Commun.* **2002**, 2388- 2389.

23. Gale, R.W.; Fulton, J. L.; Smith, R. D. *J. Am. Chem. Soc.* **1987**, 109, 920-921.

24. Fulton, J. L.; Pfund, D. M.; McClain, J. B.; Romack, T. J.; Maury, E. E.; Combes, J. R.; Samulski, E. T.; DeSimone, J. M.; Capel, M. *Langmuir.* **1995**, 11, 4241-4249.

25. Smith, R. D.; Matson, D. M.; Fulton, J. L.; Peterson, R. C. *Ind. Eng. Chem. Res.* **1987**, 26, 2298-2306.

26. Tingey, J. M.; Fulton, J. L.; Smith, R. D. *J. Phys. Chem.* **1990**, 94, 1997-2004.

27. Hoefling, T. A.; Enick, R. M.; Beckman, E. J. *J. Phys. Chem.* **1991**, 95, 7127-7129.

28. Eastoe, J.; Robinson, B. H.; Vinsser, A. J. W. G.; Steytler, D. C. *J. Chem. Soc. Faraday Trans.* **1991**, 87(12), 1899-1903.

29. Harrison, K.; Goveas, J.; Johnston, K. P.; O'Rear, E. A. *Langmuir.* **1994**, 10, 3536-3541.

30. Zielinski, Z. G.; Kline, S. R.; Kaler, E. W.; Rosov, N. *Langmuir.* **1997**, 13, 3934-3937.

31. Fulton, J. L.; Pfund, D. M.; McClain, J. B.; Romack, T. J.; Maury, E. E.; Combes, J. R.; Samulski, E. T.; DeSimone, J. M.; Capel, M. *Langmuir.* **1995**, 11, 4241-4249.
32. Cason, J. P.; Khambaswadkar, K.; Roberts, C. B. *Ind. End. Chem. Res.* **2000**, 39, 4749-4755.
33. Clark, N. Z.; Waters, C.; Johnson, K. A.; Satherley, J.; Shiffrin, D. J. *Langmuir.* **2001**, 17(20), 6048-6050.
34. Jessop, P. G.; Ikariya, T.; Noyori, R. *Chem. Rev.* **1999**, 99, 475-493.
35. Bonilla, R. J.; James, B. R.; Jessop, P. G. *Chem. Commun.* **2000**, 941-942.
36. Blum, J.; Amer, I.; Zoran, A.; Sasson, Y. *Tetrahedron letters.* **1983**, 24, 4139-4142.

Chapter 28

Hydroformylation of Olefins in Water-in-Carbon Dioxide Microemulsions

Xing Dong and Can Erkey[*]

Environmental Engineering Program, Department of Chemical
Engineering, University of Connecticut, Storrs, CT 06269

Olefins were hydroformylated in water-in-carbon dioxide
microemulsions in the presence of organometallic catalysts
formed in situ from $Rh(CO)_2acac$ and 3,3',3''-
Phosphinidynetris(benzenesulfonic acid), trisodium salt
(TPPTS) in the presence of synthesis gas. The microemulsions
were supported by sodium salt of bis(2,2,3,3,4,4,5,5-
octafluoro-1-pentyl)-2-Sulfosuccinate $(H(CF_2)_4CH_2OOCCH_2$
$CH(SO_3Na)COOCH_2(CF_2)_4H$ (di-HCF4). The effects of the
presence of salts and acid on the stability of microemulsions
and activity were also investigated.

Background

There is a growing demand for highly selective and efficient catalytic
process due to today's economical and environmental constraints in the
chemical industries. Organometallic homogeneous catalysts seem ideally suited
to answer this challenge since they offer higher activity and chemo-, region-,
and stereo-selectivity than their heterogeneous counterparts for a wide variety of
reactions. In spite of these favorable properties, the number of homogeneous
catalytic processes operating on an industrial scale is scarce [1]. This has
primarily been due to lack of efficient methods to recover expensive
homogeneous catalysts from reaction mixtures and to recycle them. A wide
variety of methods for recovery and recycle of catalysts has been investigated

430   © 2003 American Chemical Society

over the past few decades. These include membrane separations [2], heterogenizing homogeneous catalysts by anchoring them to polymeric supports [3], supported aqueous and liquid phase catalysis [4], fluorous biphasic [5] and aqueous biphasic catalysis [6]. Among these, aqueous biphasic catalysis is a very promising method and has been practiced on an industrial scale since 1994 for production of 330 million pounds n-butyraldehyde per year by hydroformylation of propylene. However, the low solubilities of most organic compounds in water may be prohibitive in extension of the technology to other reaction systems. Furthermore, a biphasic system is not preferable in reactions that are controlled by mass transfer.

Interfacial mass transfer rates can be increased by increasing the surface area of the dispersed aqueous phase. One way to increase surface area is to create a microemulsion. Microemulsions are optically transparent and thermodynamically stable fluids, which are formed by adding surfactant and/or co-surfactant to a system containing two immiscible liquids such as water and oil. There are three general types of microemulsion structures: oil-in-water (O/W), water-in-oil (W/O) and bicontinuous. In W/O microemulsions, the aqueous phase is dispersed as nanosize droplets (typically 5 to 25 nm in diameter) surrounded by a monolayer of surfactant molecules in the continuous hydrocarbon phase resulting in interfacial areas per unit volume that are on the order of 1×10^8 m^{-1}. Numerous studies have been reported in the literature on utilizing water-in-oil microemulsions to carry out a wide range of chemical transformations such as particle formation reactions and enzyme catalyzed reactions. However, we could find only three reports in the literature on the use of such systems in conjunction with water-soluble organometallic catalysts [7-9]. In such systems, recovery and recycle of organometallic catalysts at the end of the reaction would require breaking down the microemulsion and forming a two-phase mixture; an aqueous phase which contains the surfactant and the water soluble catalyst and an organic phase which contains the product mixture. In our opinion, the inherent difficulties associated with microemulsion breakdown have hindered developments in this area.

Such a problem might be solved by utilizing water-in-carbon dioxide (W/CO_2) microemulsions. The conceptual diagram of this method is given in Figure 1. In STAGE 1, a hydrophilic catalyst dissolved in water and liquid and gaseous reactants are fed to a reactor. Subsequently, the reactor is heated and charged with carbon dioxide to the desired temperature and pressure. Catalysis takes place in STAGE 2 in the W/CO_2 microemulsion and the contents are cooled and CO_2 is vented once the reaction is complete. Reduction of pressure causes the microemulsion to breakdown. Consequently, the system separates into an aqueous phase and an organic phase as shown in STAGE 3. The product is easily removed by phase separation. The aqueous solution containing the

432

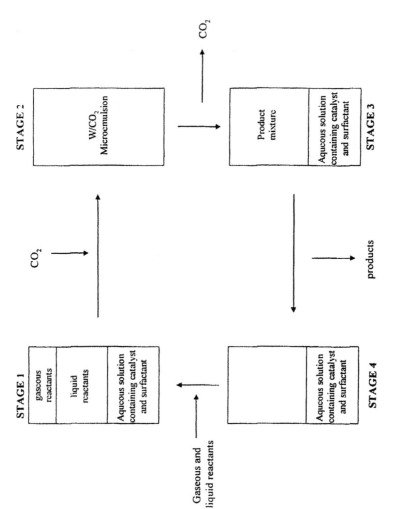

Figure 1. The conceptual diagram for the catalyst recovery and recycle method based on W/CO₂ microemulsions

surfactant and the catalyst is recycled. The reactor is subsequently charged with the gaseous and liquids reactants once again to start the cycle. The breakdown of the microemulsion at the end of STAGE 2 by simple pressure reduction is unique to supercritical fluids which are gases at ambient conditions and is not possible with any other kind of solvent. The possible additional advantages of using $scCO_2$ as the continuous phase over organic solvents for organic synthesis are: (1) The large compressibility of CO_2 in the vicinity of the critical point compared to organic solvents permits one to use pressure as an additional parameter to influence the phase behavior of such systems. (2) High solubility of gases in $scCO_2$ can increase the reaction rate.

W/CO_2 microemulsions were first demonstrated relatively recently by Johnston et al. [10] who showed that an ammonium carboxylate perfluoropolyether [CF$_3$O(CF$_2$CF(CF$_3$)O)$_3$CF$_2$-COO$^-$NH$_4^+$, PFPE] surfactant could support microemulsions in $scCO_2$. The water cores within these microemulsions were probed by elegant techniques including X-band electron paramagnetic resonance (EPR) [11], time-resolved fluorescence depolarization and FTIR [12]. The PFPE-based W/CO_2 microemulsions were subsequently used as nanoreactors for reactions of acidified potassium dichromate with sulfur dioxide, sodium nitroprusside with hydrogen sulfide, cadmium nitrate with sodium sulfide and oxidation of cholesterol by chlosterol oxidase with promising results [13]. Recently, fluorinated ananlogs of AOT were used to form W/CO_2 microemulsions [14]. We reported on synthesis and characterization of Salt of Bis(2,2,3,3,4,4,5,5-Octafluoro-1-pentyl)-2-Sulfosuccinate (H(CF$_2$)$_4$CH$_2$ OOCCH$_2$-CH(SO$_3$Na)COOCH$_2$(CF$_2$)$_4$H, di-HCF4), and Sodium Salt of Bis (2,2,3,4,4,4-Hexafluorobutyl)-2-Sulfosuccinate (CF$_3$CHFCF$_2$CH$_2$OOCCH$_2$-CH (SO$_3$Na)COOCH$_2$CF$_2$CHFCF$_3$, di-HCF3) [15]. We also probed the micelle size and shape for di-HCF4 by SAXS [16]. We also showed that such a microemulsion system is very effective to synthesize nanoparticles [17]. More recently, Wai et al. [18] used palladium nanoparticles formed in W/CO_2 microemulsions to hydrogenate olefins with promising results.

No studies have been reported in the literature on reactions in W/CO_2 microemulsions catalyzed by water soluble organometallic complexes. In our opinion, the difficulties associated with formation of stable microemulsions in the presence of water-soluble organometallic catalysts have prevented developments in this area. Furthermore, a large amount of water needs to be solubilized in CO_2 to achieve sufficiently high catalyst concentrations per unit volume of reactor. Another important issue which should be addressed is the stability of the microemulsions with the addition of additives other than organometallic catalysts. In aqueous phase organometallic chemistry, such additives are used to control the pH of the system and/or aid in formation of active species. For example, it has been reported that pH has a very important effect on the rate and selectivity in hydroformylation of 1-octene [19]. In this article, we report the results of our studies on the stability of W/CO_2 microemulsions supported by di-HCF4 at high water contents and on the stability of these microemulsions in the presence of additives including HCl, NH_4Cl, $CaCl_2$, NaCl and NaOH. We also show that stable W/CO_2 microemulsions can be created in the presence of water soluble organometallic catalysts and in the presence of additives at a wide temperature range. We also provide some data on hydroformylation of some olefins in these W/CO_2 microemulsions.

Experimental Section

The synthesis procedure for this surfactant and the properties of the surfactant are given elsewhere [15]. Experiments were conducted batchwise in the custom-manufactured stainless steel reactor (internal volume of 54 cc) which was manufactured from 316SS and was equipped with two sapphire windows (diameter = 1.25", thickness = 0.5") as shown in Figure 2. The windows were sealed on both sides with PEEK seals. In a typical experiment, a certain amount of di-HCF4, catalyst precursor, ligand, water, additive and a magnetic stir bar were placed in the vessel, which was then sealed. The vessel was then placed on a magnetic stir plate and air was removed from the vessel by flushing it with synthesis gas. The vessel was heated to the desired temperature by a recirculating heater/cooler (Fischer) via a machined internal coil and was charged with synthesis gas and CO_2 from a syringe pump (ISCO, 100D) equipped with a cooling jacket. The temperature was controlled during each experiment with a variation of 0.5 °C. The pressure was measured using a pressure transducer (Omega Engineering Inc, PX01K1-5KGV). During cloud point measurement, the pump was stopped when an optically transparent single-phase solution was obtained. Subsequently, the vessel was slowly depressurized until the cloud point was reached. In the hydroformylation reactions, a certain amount of the olefin was added in the beginning together with other materials. Then the reaction system was heated. The samples were taken periodically by using the sampling system which is shown in Figure 1. The conversion was determined by ^1H NMR (Bruker DRX400).

Results and Discussion

The phase behavior of the W/CO_2 microemulsions at various W/S ratios is given in Figure 3. As expected, increasing the water content increased the cloud point pressures. W/S = 20 translates to a water content of 16 wt % which is the highest water content in the W/CO_2 microemulsions reported so far. Besides high water content, the large temperature range and relatively low cloud point pressures for this microemulsion system are promising.

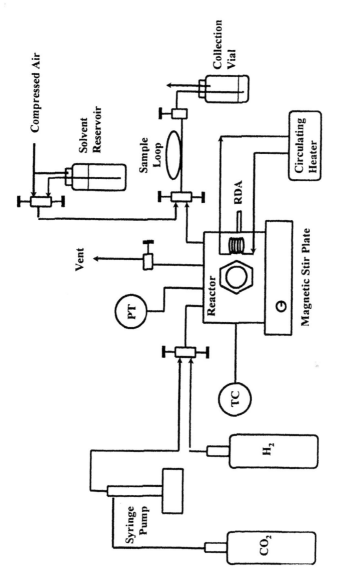

Figure 2. Experimental set-up

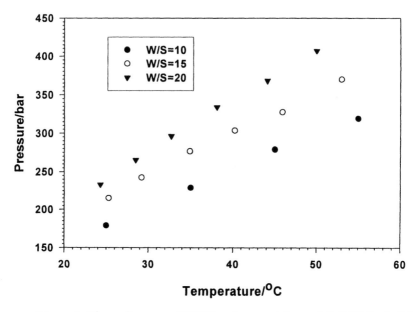

Figure 3. Phase diagram of W/CO$_2$ microemulsion at [di-HCF4]=0.4M

Our system selection for studies on the stability of the W/CO$_2$ microemulsions in the presence of organometallic catalysts is based on the hydroformylation of higher olefins. This reaction involves the formation of branched or linear aldehydes by the addition of H$_2$ and CO to a double bond according to Scheme 1. The linear aldehydes are the preferred products and the selectivity in such reactions is usually expressed in terms of n:iso ratio, which is the ratio of the linear aldehyde to the branched aldehyde. When conducted in the aqueous phase, the reaction is catalyzed with complexes formed in-situ from Rh(CO)$_2$acac and 3,3',3''-Phosphinidynetris (benzenesulfonic acid), trisodium salt (TPPTS) in the presence of synthesis gas.

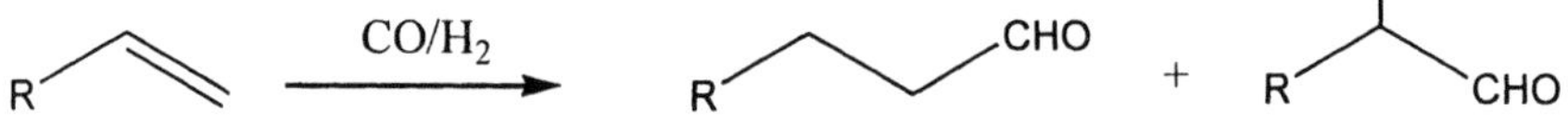

$$R \diagdown \diagup \quad \xrightarrow{CO/H_2} \quad R \diagdown \diagup \diagdown CHO \quad + \quad R \diagdown \diagup^{|} \diagdown CHO$$

Scheme 1. Hydroformylation of olefins

We were able to form stable W/CO$_2$ microemulsions when Rh(CO)$_2$ acac, TPPTS, synthesis gas, water, di-HCF4, olefin and CO$_2$ were combined together at appropriate concentrations. The olefins were also hydroformylated. The detailed compositions of the microemulsion systems which were investigated are listed below Figure 4. The product aldehyde concentrations were determined by sampling the system after a certain period of time before the cloud point measurements. In all runs, the color of the solution became clear and yellow after a period of 10 minutes which indicated the formation of the microemulsion with the catalyst formed in-situ inside the water droplets. The yellow color is perhaps indicative of formation of $Rh(CO)_2(TPPTS)_2$. As shown in Figure 4, the cloud point pressures for this system are very similar to the pressures for the W/CO_2 microemulsion system without any catalyst and reactant. An important consideration in carrying out the hydroformylation reaction in W/CO_2 microemulsion is the drop of the pH of the dispersed aqueous phase to about 3 due to formation of carbonic acid and its derivatives. Studies in the aqueous phase have shown that low pH effects catalytic activity and selectivity adversely. Studies in the literature indicate that pH of the aqueous phase in W/CO_2 microemulsion systems can be increased by the addition of NaOH among other buffers [20]. Therefore, we also carried out stability and activity experiments using 0.25M sodium hydroxide solution instead of pure water. As shown in Figure 4, the cloud point pressures increased slightly at a fixed temperature by the presence of sodium hydroxide.

In our investigations, we found out that that there is a critical concentration of the catalyst beyond which a stable microemulsion can't be formed. This may be due to the change of water droplet properties at high rhodium precursor and ligand concentrations. Furthermore, in the presence of sodium hydroxide, there is also a critical temperature beyond which the microemulsion is broken.

We also investigated the effect of various additives in the presence of the rhodium complex on the phase behavior. As shown in Figure 5, the addition of additives didn't change the phase diagram. The behavior observed with NH_4Cl, NaCl and $CaCl_2$ indicate that Na^+ has no effect on the phase behavior. The addition of HCl excludes that pH has an effect on the phase behavior. The stability of the di-HCF4 supported W/CO_2 microemulsions in the presence of additives is very promising for application of the method to different reaction systems.

Additional data on hydroformylation of 1-pentene, 1-octene and ethyl acrylate are provided in Table 1. In all the runs, the solutions became clear and yellow after a period of 10 minutes, which indicated the formation of the microemulsion with the catalyst formed in situ inside the water droplets. The solutions were clear and homogeneous during the entire run, which definitely excludes reaction via a biphasic pathway. Because of equipment limitations, the highest reaction temperature we investigated was 87.1 °C. The stability of the W/CO_2 microemulsion system at such a high temperature is remarkable. At the conditions employed, conversions ranged from 6 to 75%. The increase of temperature and the addition of NaOH were found to increase the reaction rate. The initial reaction rate for 1-pentene is about two times higher than that of 1-octene. In studies on hydroformylation of different olefins in aqueous biphasic systems, Brady et al. [21] found that there is a marked dependence of the reaction rate on the solubility of the terminal olefins in water. The data shown in

438

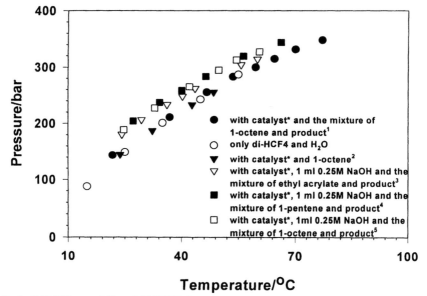

Catalyst* :Rh(CO)₂acac = 0.01mmol, Rh/TPPTS=4
[1] 1.31 mol 1-octene, 0.07 mol nonanal and 2-methyloctanal, 12.8 bar syngas
[2] 1.40 mol 1-octene, 13.6 bar syngas
[3] 0.82 mol ethyl acrylate and 0.26 mol of its aldehyde and enol products, 3.4 bar syngas
[4] 1 mol 1-penten and 0.56 mol of its aldehyde products, 8.7 bar syngas
[5] 1.37 mol 1-octene and 0.03 mol of its aldehyde products, 13.3 bar syngas

Figure 4. Phase diagram for several different microemulsion systems at W/S=10 and [di-HCF4]=0.1 M

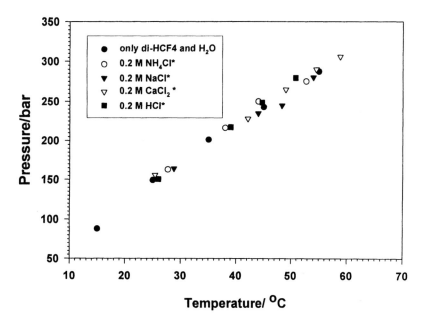

Figure 5. Pressure-temperature phase diagram of W/CO₂ microemulsion stabilized by di-HCF4 with different additives
*:Rh(CO)₂acac = 0.01mmol, Rh/TPPTS = 4, syngas = 13.6 bars, di-HCF4 = 3.6g, 1ml 0.2M aqueous solution

Table 1 indicate that the reaction rates are not dependent on the solubilities of olefins in water. The solubility of 1-pentene is over 50 times higher than that of 1-octene in bulk water, but the initial reaction rate for 1-pentene is only two times higher than that of 1-octene. This indicates that the reactions might be mainly occurring at the interface. Therefore, shifting the reaction medium to a W/CO$_2$ microemulsion may perhaps accelerate many reactions, which are slow in aqueous biphasic systems due to their low solubilities in water. On the other hand, the water solubility of the reactant might still have some effect on reaction rate, which is indicated by the high reaction rate obtained with ethyl acrylate which has a substantially higher water solubility than 1-pentene and 1-octene.

Table 1 Hydroformylation of olefins in W/CO$_2$ microemulsions

substrate		T/$^\circ$C	P/bars	t/hr	conversion%
1-octene	*	87.1	400	43	6
1-pentene	*	85.7	400	43	13
ethyl acrylate	*	85.2	400	22	24
1-pentene	**	66.5	350	40	36
ethyl acrylate	**	65.1	350	22	75

Reaction conditions: Rh(CO)$_2$acac = 0.01mmol, Rh/TPPTS=4, substrate = 0.2 ml, water or 0.25 M NaOH solution=1 ml, water/surfactant=10, synthesis gas (CO/H$_2$=1) 13.6 bars. Reactor volume: 54 ml. Conversion was measured by NMR.
*: water
**: 0.25 M NaOH solution

The data presented in Table 1 also show that pH has an effect on activity in the W/CO$_2$ microemulsion system. When 0.25M sodium hydroxide solution was used instead of pure water, the reaction rates for the hydroformylation of 1-pentene increased even though the temperature was decreased from 87 $^\circ$C to 66 $^\circ$C. The pH of the dispersed phase of a W/CO$_2$ microemulsion system has been determined as 5 at a NaOH concentration of 0.25M and as 2.8 without NaOH [20]. The increase in the rate might be due to the higher pH value aiding the formation of active species, such as the complete dissociation of the precursor to the monomeric species [19].

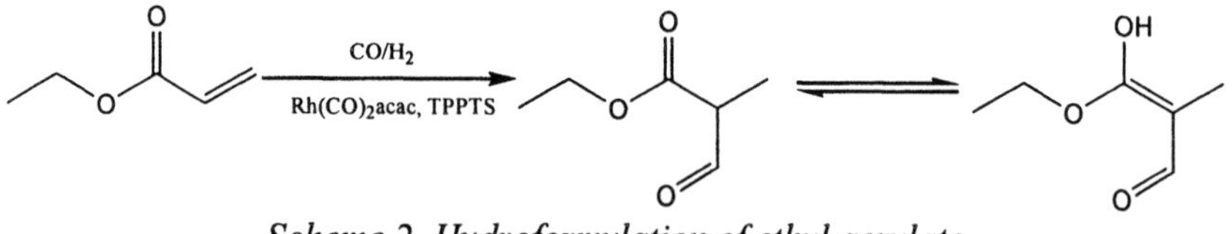

Scheme 2. Hydroformylation of ethyl acrylate

The hydroformylation of ethyl acrylate (Scheme 2) was also followed as a function of time and the data are shown in Figure 4. The ^{1}H NMR indicated that the product existed as a mixture of the aldehyde and enol forms. Almost all of the aldehyde was in branched form with very little or no linear aldehyde. During the reaction, the ratio of aldehydes and enol form remained almost one. The rhodium hydride was also found in the ^{1}H NMR spectrum located around -16.4

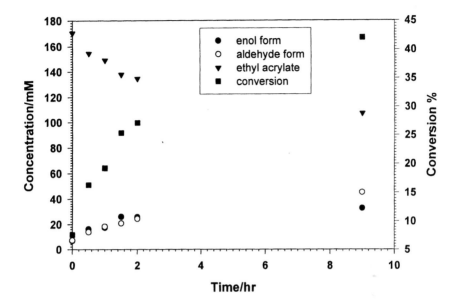

Figure 6. Hydroformylation of ethyl acrylate in W/CO$_2$ microemulsion

Reaction conditions: Rh(CO)$_2$acac = 0.01mmol, Rh/TPPTS=4, di-HCF4 = 3.6g, 1ml 0.2M NaOH and 1 ml ethyl acrylate, CO/H$_2$ = 14.4 bars (at room temperature), reaction temperature = 65.6 °C, total pressure = 348.4 bars (at reaction temperature)

ppm, which suggests that the reaction in the microemulsion system proceeds via hydrido intermediates.

Conclusions and Future Work

In summary, we demonstrated that it is possible to carry out reactions in W/CO$_2$ microemulsions catalyzed by water soluble organometallic complexes. The microemulsions were stable in the presence of several electrolytes and the cloud points were not affected even at high electrolyte concentrations. In the future, we would like to investigate the factors controlling activity and selectivity and determine the feasibility of a catalyst recovery and recycle system based on such a microemulsion system.

442

Acknowledgements

The authors gratefully acknowledge financial support from Connecticut Innovations, Inc., Grant 98CT013.

Literature Cited

[1] Parshall, G. W.; Ittel, S. D. *Homogeneous Catalysis*, 1992.
[2] Kragl, U.; Dreisbach, C.; Wandrey, C. Membrane reactors in homogeneous catalysis. *Appl. Homogeneous Catal. Organomet. Compd.* **1996**, 2, 832-843.
[3] Hartley, F. R. *Supported Metal Complexes. A New Generation of Catalysts.* **1985**, 318 pp
[4] Arhancet, J. P.; Davis, M. E.; Merola, J. S.; Hanson, B. E. Hydroformylation by supported aqueous-phase catalysis: a new class of heterogeneous catalysts. *Nature (London)* **1989**, 339(6224), 454-5.
[5] Horvath, I. T. Fluorous Biphase Chemistry. Accounts of Chemical Research 1998, 31(10), 641-650.
[6] Cornils, B.; Herrmann, W. A.; Editors. *Aqueous-Phase Organometallic Catalysis: Concepts and Applications.* **1998**, 615 pp
[7] Haumann, M.; Koch, H.; Hugo, P.; Schomacker, R. *Applied Catalysis, A: General* **2002**, 225(1-2), 239-249.
[8] Schomacker, R.; Haumann, M.; Koch, H. *PCT Int. Appl.* **1999**.
[9] Tinucci, L.; Platone, E. *Eur. Pat. Appl.* **1990**
[10] Harrison, K. L.; Johnston, K. P.; Sanchez, I. C. Effect of Surfactants on the Interfacial Tension between Supercritical Carbon Dioxide and Polyethylene Glycol. *Langmuir* **1996**, 12(11), 2637-2644.
[11] Heitz, M. P.; Carlier, C.; deGrazia, J.; Harrison, K. L.; Johnston, K. P.; Randolph, T. W.; Bright, F. V. Water Core within Perfluoropolyether-Based Microemulsions Formed in Supercritical Carbon Dioxide. *J. Phys. Chem. B* **1997**, 101(34), 6707
[12] Clarke, M. J.; Harrison, K. L.; Johnston, K. P.; Howdle, S. M. Water in Supercritical Carbon Dioxide Microemulsions: Spectroscopic Investigation of a New Environment for Aqueous Inorganic Chemistry. *J. Am. Chem. Soc.* **1997**, 119(27), 6399
[13] Jacobson, G. B.; Lee, C. T., Jr.; Johnston, K. P. Organic Synthesis in Water/Carbon Dioxide Microemulsions. *Journal of Organic Chemistry* **1999**, 64(4), 1201-1206.
[14] Eastoe, Julian; Cazelles, Beatrice M. H.; Steytler, David C.; Holmes, Justin D.; Pitt, Alan R.; Wear, Trevor J.; Heenan, Richard K. Water-in-CO2 Microemulsions Studied by Small-Angle Neutron Scattering. *Langmuir* **1997**, 13(26), 6980-6984.

[15] Liu, Z.; Erkey, C. Water in Carbon Dioxide Microemulsions with Fluorinated Analogues of AOT. *Langmuir* **2001**, 17(2), 274-277.

[16] Dong, X.; Erkey, C.; Dai, H.; Li, H.; Cochran, H. D.; Lin, J. S. Phase Behavior and Micelle Size of an Aqueous Microdispersion in Supercritical CO_2 with a Novel Surfactant. *Industrial & Engineering Chemistry Research* **2002**, 41(5), 1038-1042.

[17] Dong, X.; Potter, D.; Erkey, C.. Synthesis of CuS Nanoparticles in Water-in-Carbon Dioxide Microemulsions. *Industrial & Engineering Chemistry Research* ACS ASAP.

[18] Ohde, H.; Wai, C. M.; Kim, H.; Kim, J.; Ohde, M.. *J. Am. Chem. Soc.* **2002**, 124(17), 4540-4541.

[19] Deshpande, R. M.; Purwanto; D. H.; Chaudhari, R. V. *Journal of Molecular Catalysis A: Chemical* **1997**, 126(2-3), 133-140.

[20] Holmes, J. D.; Ziegler, K. J.; Audriani, M.; Lee, C. Ted Jr.; Bhargava, P. A.; Steytler, D. C.; Johnston, K. P. *Journal of Physical Chemistry B* **1999**, 103(27), 5703-5711.

[21] Cornils, B.; Herrmann, W. A.; Editors. *Aqueous-Phase Organometallic Catalysis: Concepts and Applications.* **1998**.

Indexes

Author Index

Subject Index